Modeling and Simulation of
Mineral Processing Systems

This book is dedicated to my
wife Helen

Modeling and Simulation of Mineral Processing Systems

R.P. King

Department of Metallurgical Engineering
University of Utah, USA

Boston   Oxford   Auckland   Johannesburg   Melbourne   New Delhi

Butterworth-Heinemann
Linacre House, Jordan Hill, Oxford OX2 8DP
225 Wildwood Avenue, Woburn, MA 01801-2041
A division of Reed Educational and Professional Publishing Ltd

A member of the Reed Elsevier plc group

First published 2001

British Library Cataloguing in Publication Data
King, R.P.
    Modeling and simulation of mineral processing systems
    1. Ore-dressing – Mathematical models   2. Ore-dressing – Computer simulation
    I. Title
    622.7

Library of Congress Cataloging in Publication Data
King, R.P. (Ronald Peter), 1938–
    Modeling and simulation of mineral processing systems/R.P. King.
    p. cm.
    Includes bibliographical references and index
    ISBN 0 7506 4884 8
    1. Ore-dressing–Mathematical models.   2. Ore-dressing–Computer simulation.
    I. Title.
    TN500.K498 2001
    622'.7'015118–dc21   2001037423

ISBN 0 7506 4884 8

For information on all Butterworth-Heinemann
publications visit our website at www.bh.com

Typeset at Replika Press Pvt Ltd, Delhi 110 040, India
Printed and bound in Great Britain.

FOR EVERY TITLE THAT WE PUBLISH, BUTTERWORTH-HEINEMANN
WILL PAY FOR BTCV TO PLANT AND CARE FOR A TREE.

Preface

Quantitative modeling techniques and methods are central to the study and development of process engineering, and mineral processing is no exception. Models in mineral processing have been difficult to develop because of the complexity of the unit operations that are used in virtually all mineral recovery systems. Chief among these difficulties is the fact that the feed material is invariably a particulate solid. Many of the conventional mathematical modeling techniques that are commonly used for process equipment have limited application to particulate systems and models for most unit operations in mineral processing have unique features. Common ground is quite difficult to find. The one obvious exception is the population balance technique and this forms a central thread that runs throughout the modeling techniques that are described in this book. The models that are described are certainly incomplete in many respects, and these will be developed and refined by many researchers during the years ahead. Nevertheless, the models are useful for practical quantitative work and many have been widely used to assist in the design of new equipment and processes. Some of the newer models have not yet been seriously tested in the industrial environment.

The book is written at a reasonably elementary level and should be accessible to senior undergraduate and graduate students. A number of examples are included to describe the application of some of the less commonly used models, and almost all of the models described in the book are included in the MODSIM simulator that is available on the companion compact disk. The reader is encouraged to make use of the simulator to investigate the behavior of the unit operations by simulation.

The main advantage of using quantitative models is that they permit the complex interactions between different unit operations in a circuit to be explored and evaluated. Almost all of the models described are strongly nonlinear and are not usually amenable to straightforward mathematical solutions, nor are they always very convenient for easy computation using calculators or spread sheets. In order to investigate interactions between models, the simulation method is strongly recommended, and the focus throughout this book has been on the development of models that can be used in combination to simulate the behavior of complex mineral dressing flowsheets. Simulation techniques are popular because they allow complex problems to be tackled without the expenditure of large resources. All the models described here can be used within the MODSIM simulator so they are readily accessible to the reader. The models are transparent to the user and the models can be tried in isolation or in combination with other unit operations. MODSIM has proved itself to be an excellent teaching tool both for conventional courses and, in recent years, to support an Internet course delivered from the University of Utah. I hope that distributing MODSIM widely, together with this book, will encourage researchers and engineers to make use of this technique, which brings to every engineer the results of many man-years of research endeavor that have contributed to the development of the models. Simulation does have its limitations and the reader is reminded that a simulator can not be relied upon to provide an exact replica of any specific real plant operation. The reliability of the simulator output is limited by the accumulated reliability of the component models.

The work of many researchers in the field has been inspiration for the models that are described in this book. I have tried to extract from the research literature those modeling techniques that have proved to be useful and applicable to practical situations and I have tried to identify those modeling methods and process features that have been confirmed, at least partially, by careful experimental observation. In this I am indebted to many students and colleagues whose work has provided data, ideas and methods.

The preparation of a book is always a time-consuming task and I wish to thank my wife, Ellen, for her patience and encouragement while it was being written and for typing much of the final version of the manuscript. The manuscript went through many revisions and the early versions were typed by Karen Haynes. Kay Argyle typed the illustrative examples. It is a pleasure to acknowledge their contributions for which I am very grateful.

R.P. King
Salt Lake City

Contents

1
Introduction

This book covers the quantitative modeling of the unit operations of mineral processing. The population balance approach is taken and this provides a unified framework for the description of all of the unit operations of mineral processing. Almost all of the unit operations, both separation and transformation operations, can be included. Many ore dressing operations are sufficiently well understood to enable models to be developed that can be usefully used to describe their operation quantitatively. Experimental data that has been obtained by many investigators over the past couple of decades has provided the basis for quantitative models that can be used for design and simulation of individual units in any flowsheet. Quantitative methods are emphasized throughout the book and many of the old empirical methods that have been in use since the early years of the twentieth century are passed over in favor of procedures that are based on an understanding of the behavior of the particulate solids that are the basic material of mineral processing operations.

The focus is quantitative modeling. All mineral processing equipment exhibits complex operating behavior and building quantitative models for these operations is not a straightforward task. In some cases the basic fundamental principles of a particular type of equipment are ill understood. In most cases the complexity of the operation precludes any complete analysis of the physical and chemical processes that take place in the equipment and control its operation. In spite of these difficulties much progress has been made in the development of useful models for almost all of the more important types of mineral processing equipment.

The style of modeling usually referred to as phenomenological is favored in the book. This means that the physical and, to a limited extent, the chemical phenomena that occur are modeled in a way that reflects the physical realities inherent in the process. This method cannot always be made to yield a complete model and often some degree of empiricism must be used to complete the model. This approach has a number of advantages over those that are based on empirical methods. Phenomenological models are not based entirely on available experimental data and consequently do not need to be substantially modified as new data and observations become available. If the basic principles of a particular operation are properly formulated and incorporated into the models, these can be continually developed as more information and greater understanding of the principles become available. Generally speaking, phenomenological models are not so prone to catastrophic failure as operating conditions move further from known experimental situations. Overloaded and underloaded operating conditions will often emerge as the natural consequences of adding more material to the model and scale-up principles can usually be more confidently applied if these are based on a sound description of the operating principles of the equipment concerned and an understanding of the detailed behavior of solid particles in the equipment. Extrapolation of operation outside of the normal region is less hazardous if the phenomena have been modeled with due care having been taken to ensure that the models have correct asymptotic behavior at more extreme conditions.

A common feature of all of the models that are developed in the book is the role played by solids in the particulate state in the unit operations that are considered. The very essence of mineral processing is the physical separation of minerals and, given the most common occurrence patterns of minerals in naturally occurring exploitable ore bodies, successful separation techniques require the reduction of the solid raw material to the particulate state with particle dimensions commensurate with the scale of the mineralogical texture. Both the size and the composition of the particles play important roles in the models because the behavior of the particles in the equipment is influenced in a significant way by these parameters. Often the effect of particle composition is indirect because the mineralogical content of a particle directly influences its density, surface characteristics, magnetic and electrical properties. These physical attributes in turn influence the behavior in the equipment that is designed to exploit differences in physical and chemical properties among the particles to effect a separation.

Physical separation of minerals requires that the various mineral species must be liberated by comminution and this imposes some strong requirements on the models that are developed. Mineral liberation is quite difficult to model mainly because of the complexity of naturally occurring mineralogical textures and the associated complexity of the fracture processes that occur when an ore is crushed and ground prior to processing to separate the minerals. In spite of the complexity, some effective modeling techniques have been developed during the past two decades and the models that are used here are effective at least for describing the liberation of minerals from ores that have only a single mineral of interest. Modeling of liberation phenomena has been greatly facilitated by the development of computer based image analysis systems. These methods have made it possible to examine the multicomponent mineral bearing particles and to observe the disposition of mineral phases across the particle population at first hand. Modern instruments can reveal the mineral distribution in considerable detail and digital images are the rule rather than the exception now. Computer software for the analysis of digital images is readily available and these range from comparatively simple programs that provide rudimentary image processing and image analysis capabilities to special purpose programs that provide facilities to manipulate and process images and subsequently analyze their content using a wide variety of algorithms that have been developed specifically to measure mineral liberation phenomena in naturally occurring ores.

Mineral liberation is the natural link between comminution operations and mineral recovery operations, and it is not possible to model either type effectively without proper allowance for the liberation phenomenon. The approach that is taken to liberation modeling in this book keeps this firmly in mind and mineral liberation is modeled only to the extent that is necessary to provide the link between comminution and mineral recovery.

The population balance method is used throughout to provide a uniform framework for the models. This method allows the modeler to account for the behavior of each type of particle in the processing equipment and at the same time the statistical properties of the particle populations are correctly described and accounted for. In the unit operations such as grinding machines where particles are transformed in terms of size and composition, the population balance models are particularly useful since they provide a framework within which the different fracture mechanisms such as crushing and attrition can be modeled separately but the effects of these separate subprocesses can be accounted for in a single piece of equipment. Population balance methodology is well developed and the models based on it can be coded conveniently for computation.

Models for the performance of mineral processing equipment are useful for many purposes – plant and process design and development, performance evaluation and assessment, equipment and process scale-up but most importantly for simulation. Many of the models that are described in this book were developed primarily to provide the building blocks for the simulation of the operation of complete mineral processing plants. As a result the models all have a common structure so that they can fit together seamlessly inside a plant simulator. The population balance method facilitates this and it is comparatively simple for the products of any one unit model to become the feed material for another.

All of the models developed are particularly suited to computer calculations. Simulation of a wide range of engineering systems is now accepted as the only viable procedure for their analysis. Purely theoretical methods that are aimed at complete and precise analytical solutions to the operating equations that describe most mineral processing systems are not general enough to provide useful working solutions in most cases. Very fast personal computers are now available to all engineers who are required to make design and operating decisions concerning the operation of mineral processing plants. A variety of software packages are available to analyze data within the framework of the quantitative models that are described in this book.

All of the models described here are included in the MODSIM mineral processing plant simulator. This is a low-cost high-performance simulation system that is supplied on the compact disc that is included with this book. It can be used to simulate the steady-state operation of any ore dressing plant. It has been tested in many ore dressing plants and has been shown to be reliable. It is fully documented in a user manual, which is included in digital form on the compact disc. The theoretical and conceptual base for the simulation method that is embodied in MODSIM is described in detail in this book and the models used for simulation are fully described. The simulator should be regarded as a resource that can be used by the reader to explore the implications of the models that are developed in the book. With very few exceptions, all of the models that are discussed in the book are included in MODSIM. It is a simple matter to simulate single unit operations and the reader is thus able to observe the effect of changing parameter values on the models and their predictions. In many cases multiple interchangeable models are provided for individual unit operations. This makes it easy for the reader to compare predictions made by different models under comparable conditions and that can be useful when deciding on the choice of unit models to include in a full plant simulation. A number of interesting plant simulations are included on the compact disc and the reader is encouraged to run these and explore the various possibilities that the simulator offers.

Many people have contributed to the development of models for mineral processing operations. No comprehensive attempt has been made in this book to attribute models to individual researchers and formal referencing of specific papers in the text has been kept to a minimum to avoid this distraction for the reader. A brief bibliography is provided at the end of each chapter and this will provide the user with the main primary references for information on the models that are discussed.

Bibliography

This book is not meant to be a primary textbook for courses in mineral processing technology and it should be used together with other books that provide descriptions of the mineral processing operations and how they are used in industrial practice. Wills' *Mineral Processing Technology* (1997) is specially recommended in this respect. The two-volume *SME Mineral Processing Handbook* (Weiss, 1985) is an invaluable source of information on operating mineral processing plants and should be consulted wherever the mechanical details of a particular unit operation need to be clarified. Theoretical modeling principles are discussed by Kelly and Spottiswood (1982) and Tarjan (1981, 1986). Woolacott and Eric (1994) provide basic quantitative descriptions for several of the mineral processing operations. Gaudin's superlative 1939 text, although dated, is still an excellent source of fundamental scientific information on the unit operation of mineral processing.

References

Gaudin, A.M. (1939) *Principles of Mineral Dressing*. McGraw-Hill, New York.
Kelly, E.G. and Spottiswood, D.J. (1982) *Introduction to Mineral Processing*. Wiley, New York.
Tarjan, G. (1981) *Mineral Processing*. Vol 1. Akademai Kiado, Budapest.
Tarjan, G. (1986) *Mineral Processing*. Vol 2. Akademai Kiado, Budapest.
Weiss, N.L. (ed.) (1985) *SME Mineral Processing Handbook*. Vols 1 and 2. SME, Lyttleton, CO.
Wills, B.A. (1997) *Mineral Processing Technology*. Butterworth-Heinemann, Oxford.
Woolacott, L.C. and Eric, R.H. (1994) *Mineral and Metal Extraction. An Overview*. S. Afr. Inst. Min. Metall. Johannesburg

2
Particle populations and distribution functions

2.1 Introduction

The behavior of ore dressing equipment depends on the nature of the individual particles that are processed. The number of particles involved is very large indeed and it would be quite impossible to base computational procedures on any method that required a detailed description of the behavior of each particle. The complexity of such procedures would mean that any useful or meaningful models would be entirely out of the question. But the characteristics of individual particles do have to be taken into account and useful models cannot be developed if these are to be based entirely on average properties of all the particles in the population.

Individual particles differ from each other in many respects. The differences that are of interest in ore dressing operations are those physical properties that influence the behavior of a particle when subject to treatment in any ore dressing equipment. The two most important fundamental properties are the size of the particle and its mineralogical composition. Other properties such as shape, specific gravity, fracture energy, surface area, contact angle and so on are also important and, in some ore dressing operations, can be of overriding significance. The operations of comminution and classification are primarily dependent on the size of the particles treated but the composition, specific gravity, brittleness and other properties can also influence the behavior of the particles to a greater or lesser extent during treatment. Gravity concentration operations exploit primarily the differences in specific gravity between particles and thus different mineral species can be separated from each other.

The various physical properties are not necessarily independent of each other. For example the specific gravity of a single particle is uniquely fixed once the mineralogical composition is specified. Likewise the surface properties of a particle will be specified by the mineral components that are exposed on the surface of the particle.

Some definite scheme for the description of the properties of the particles in the particle population is required that will allow enough detail to permit the models to be sufficiently sensitive to individual particle properties but at the same time sufficiently comprehensive to allow the economy of not having to define the properties of each individual particle. Such a scheme is provided by a description using distribution functions.

2.2 Distribution functions

The distribution function for a particular property defines quantitatively how the values of that property are distributed among the particles in the entire population. Perhaps the best known and most widely used distribution function is the particle size distribution function $P(d_p)$ defined by $P(d_p)$ = mass fraction of that portion of the population that consists of particles with size less than or equal to d_p. The symbol d_p is used throughout this book to represent the size of a particle.

The function $P(d_p)$ has several important general properties:

(a) $P(0) = 0$
(b) $P(\infty) = 1$
(c) $P(d_p)$ increases monotonically from 0 to 1 as d_p increases from 0 to ∞.

Properties (a) and (b) are obvious because no particle in the population can have a size less than or equal to 0 and all the particles have a size less than infinity. Property (c) reflects the fact that the fraction of the population having size less than or equal to d_{p1} must contain at least all those particles of size d_{p2} or smaller, if $d_{p2} \leq d_{p1}$.

Of course the concept of particle size is ambiguous. Particles that are of interest in mineral processing do not have regular definable shapes such as spheres and cubes. The size of a spherical particle can be unambiguously defined as the diameter. Likewise the size of a cube can be defined unambiguously as the length of a side but another dimension could be equally well used such as the longest diagonal. Particle size clearly does not have a unique meaning even for particles with regular shapes. In mineral processing technology an indirect measure of size is used. The size of a particle is defined as the smallest hole opening in a square-mesh screen through which the particle will fall. Sometimes it is necessary to work with particles that are too small to measure size conveniently by means of screening. Then other appropriate indirect measures are used such as the terminal falling velocity in a fluid of specified viscosity and density.

In practical applications it is convenient and often essential to make use of a discrete partioning of the length scale so that the particle population is divided conceptually into groups each identified by the smallest and largest size in the group.

The value of P can be measured experimentally at a number of fixed sizes that correspond to the mesh sizes of the set of sieves that are available in the laboratory. This data is usually presented in tabular form showing mesh size against the fraction smaller than that mesh. Graphical representations are useful and are often preferred because it is generally easier to assess and compare particle size distributions when the entire distribution function is immediately visible. A variety of different graphical coordinate systems have become popular with a view to making the distribution function plot as a straight line or close to a straight line. The particle size axis is usually plotted on a logarithmic coordinate scale. A variety of calibrations for the ordinate scale is used. Specially ruled graph papers are available for this purpose and these can be easily drawn by computer.

The mesh sizes in the standard sieve series vary in geometric progression because experience has shown that such a classification will leave approximately equal amounts of solids on each of the test sieves in a screen analysis. Thus each mesh size is a constant factor larger than the previous one. The constant factor is usually $2^{1/4}$ or $\sqrt{2}$. The mesh sizes in such a series will plot as equidistant points on a logarithmic scale.

Although the distribution function $P(d_p)$ is perfectly well defined and is amenable to direct measurement in the laboratory, it is not directly useful for modeling ore dressing unit operations. For this purpose the derived density function is used. The discrete particle size density function $p_i(d_p)$ is defined as follows:

$$p_i(d_p) = \int_{D_i}^{D_{i-1}} dP(d_p) = P(D_{i-1}) - P(D_i) = \Delta P_i \tag{2.1}$$

= mass fraction of that portion of the particle population that consists of particles that have size between D_i and D_{i-1}

$p_i(d_p)$ is called the fractional discrete density function and the argument d_p is often dropped if there is no risk of confusion.

$\Delta d_p = D_{i-1} - D_i$ is the so-called size class width and is usually not constant but varies from size to size. The finite width of the size class defined by Δd_p is important in the development of the modeling techniques that are used. The idea of a particular size class is central to the development of our modeling procedure. The size class is considered conceptually to include all particles in the entire population that have size falling within the class boundaries D_i and $D_i + \Delta d_p$. It is customary to designate the class boundary by means of a subscript and in order to distinguish the class boundaries clearly they will always be denoted by the symbol D_i which indicates the lower boundary of size class i. Thus the entire particle population is conceptually classified into classes each one of which is defined by its upper and lower boundary. It is conventional to run the number of the classes from larger sizes to smaller. Thus $D_i \geq D_{i+1}$. The top size class has only one boundary D_1 and it includes all particles which have size greater than D_1.

The concept of the particle classes effectively allows us to formulate models for mineral processing systems by describing the behavior of classes of particles rather than the behavior of individual particles. A representative size is associated with each particle size class and it is assumed that all particles in the class will behave in the processing systems as if it had a size equal to the representative size. Clearly this will be a viable working assumption only if the size class is sufficiently narrow. It is not possible to define the concept 'sufficiently narrow' precisely but it is generally assumed that a $\sqrt{2}$ series for the class boundaries is the largest geometric ratio that can be safely used. The key to the success of this approach to the modeling of particulate systems is the use of narrow size intervals. This in turn implies that a large number of particle classes must be considered. From a practical point of view this increases the amount of computation that is required if realistically precise descriptive models are to be developed for particulate processes. Consequently this approach requires efficient computer code if it is to be implemented as a viable practical tool.

2.2.1 Empirical distribution functions

Several empirical distribution functions have been found to represent the size distribution of particle populations quite accurately in practice and these are useful in a number of situations. The most common distributions are:

Rosin-Rammler distribution function defined by:

$$P(D) = 1 - \exp[-(D/D_{63.2})^\alpha] \tag{2.2}$$

$D_{63.2}$ is the size at which the distribution function has the value 0.632.

Log-normal distribution defined by:

$$P(D) = G\left(\frac{\ln(D/D_{50})}{\sigma}\right) \tag{2.3}$$

where $G(x)$ is the function

$$G(x) = \frac{1}{\sqrt{2\pi}} \int_{-\infty}^{x} e^{-t^2/2} \, dt$$

$$= \frac{1}{2}\left[1 + erf\left(\frac{x}{\sqrt{2}}\right)\right] \tag{2.4}$$

which is called the Gaussian or normal distribution function. It is tabulated in many mathematical and statistical reference books and it is easy to obtain values for this function. In this distribution D_{50} is the particle size at which $P(D_{50}) = 0.5$. It is called the median size. σ is given by

$$\sigma = \tfrac{1}{2}(\ln D_{84} - \ln D_{16}) \tag{2.5}$$

The log-normal distribution has a particularly important theoretical significance. In 1941, the famous mathematician A.N. Kolmogorov[1] proved that if a particle and its progeny are broken successively and if each breakage event produces a random number of fragments having random sizes, then, if there is no preferential selection of sizes for breakage, the distribution of particle sizes

---

[1]Kolmogorov, A.N. (1941) *Uber das logarithmisch normale Verteilungsgestz der Dimesionen der Teilchen bei Zerstuckelung. Comtes Rendus (Doklady) de l'Academie des Sciences de l'URSS*, Vol. 31, No. 2 pp. 99–101. Available in English translation as 'The logarithmically normal law of distribution of dimensions of particles when broken into small parts'. NASA Technical Translations NASA TTF 12, 287.

will tend to the log-normal distribution after many successive fracture events. Although this theoretical analysis makes assumptions that are violated in practical comminution operations, the result indicates that particle populations that occur in practice will have size distributions that are close to log-normal. This is often found to be the case.

Logistic distribution defined by:

$$P(D) = \frac{1}{1 + \left(\dfrac{D}{D_{50}}\right)^{-\lambda}} \tag{2.6}$$

These three distributions are two-parameter functions and they can be fitted fairly closely to measured size distributions by curve fitting techniques.

The Rosin-Rammler, log-normal and logistic distribution functions have interesting geometrical properties which can be conveniently exploited in practical work.

The Rosin-Rammler distribution can be transformed to:

$$\ln \ln \left(\frac{1}{1 - P(D)}\right) = \alpha \ln(D) - \alpha \ln(D_{63.2}) \tag{2.7}$$

which shows that a plot of the log log reciprocal of $1 - P(D)$ against the log of D will produce data points that lie on a straight line whenever the data follow the Rosin-Rammler distribution. This defines the Rosin-Rammler coordinate system.

The log-normal distribution can be transformed using the inverse function $H(G)$ of the function G. This inverse function is defined in such a way that if

$$G(x) = g \tag{2.8}$$

then

$$x = H(g) \tag{2.9}$$

From Equation 2.3

$$H(P(D)) = \frac{\ln(D/D_{50})}{\sigma} \tag{2.10}$$

and a plot of $H(P(D))$ against log D will be linear whenever the data follow the log-normal distribution. This is the log-normal coordinate system.

The logistic distribution can be transformed to:

$$-\log \left(\frac{1}{P(D)} - 1\right) = \lambda \log D - \lambda \log D_{50} \tag{2.11}$$

which shows that the data will plot on a straight line in the logistic coordinate system whenever the data follow the logistic distribution. Plotting the data in these coordinate systems is a convenient method to establish which distribution function most closely describes the data.

2.2.2 Truncated size distributions

Sometimes a particle population will have every particle smaller than a definite top size. Populations of this kind occur for example when a parent particle of size D' is broken. Clearly no progeny particle can have a size larger than the parent so that the size distribution of the progeny particle population is truncated at the parent size D'. Thus

$$P(D') = 1.0 \qquad (2.12)$$

The most common truncated distribution is the logarithmic distribution.

Logarithmic distribution function is defined by the function:

$$P(D) = \left(\frac{D}{D'}\right)^{\alpha} \quad \text{for} \quad D \leq D' \qquad (2.13)$$

which clearly satisfies equation 2.12.

D' is the largest particle in the population and α is a measure of the spread in particle sizes.

Other truncated distributions are the Gaudin-Meloy and Harris distributions.

Gaudin-Meloy distribution is defined by:

$$P(D) = 1 - (1 - D/D')^n \quad \text{for} \quad D \leq D' \qquad (2.14)$$

Harris distribution is defined by:

$$P(D) = 1 - (1 - (D')^s)^n \quad \text{for} \quad D \leq D' \qquad (2.15)$$

Truncated versions of the Rosin-Rammler, log-normal and logistic distributions can be generated by using a transformed size scale. The size is first normalized to the truncation size:

$$\xi = D/D' \qquad (2.16)$$

and the transformed size is defined by:

$$\eta = \frac{\xi}{1 - \xi} \qquad (2.17)$$

The **truncated Rosin-Rammler** distribution is:

$$P(D) = 1 - \exp\left(-\left(\frac{\eta}{\eta_{63.2}}\right)^{\alpha}\right) \quad \text{for} \quad D \leq D' \qquad (2.18)$$

The **truncated log-normal** distribution is:

$$P(D) = G\left(\frac{\ln(\eta/\eta_{50})}{\sigma}\right) \qquad (2.19)$$

with

$$\sigma = \tfrac{1}{2}(\ln(\eta_{84}) - \ln(\eta_{16})) \qquad (2.20)$$

The **truncated logistic** distribution is:

$$P(D) = \frac{1}{1 + \left(\dfrac{\eta}{\eta_{50}}\right)^{-\lambda}} \tag{2.21}$$

Straight line plots can be generated for truncated data using the appropriate coordinate systems as was done in Section 2.2.

The logarithmic distribution can be transformed to

$$\log[P(D)] = \alpha \log(D) - \alpha \log(D') \tag{2.22}$$

which shows that a plot of $P(D)$ against D on log-log coordinates will produce data points that lie on a straight line whenever the data follow the logarithmic distribution.

The Gaudin-Meloy distribution can be transformed to

$$\log(1 - P(D)) = n \log(D' - D) - n \log D' \tag{2.23}$$

Data will give a linear plot in the log-log coordinate system if plotted as $1 - P(D)$ against $D' - D$. In order to make such a plot it is necessary to know the value of D' and this is a disadvantage.

The truncated Rosin-Rammler, log-normal and logistic distributions can be linearized by using the appropriate coordinate systems as described in the previous section but using the variable η in place of D. In every case, these straight line plots can be constructed only after the truncation size D' is known.

A typical set of data measured in the laboratory is shown in Table 2.1.

Table 2.1 A typical set of data that defines the particle size distribution of a population of particles

Mesh size (mm)	Mass % passing
6.80	99.5
4.75	97.5
3.40	93.3
2.36	86.4
1.70	76.8
1.18	65.8
0.850	55.0
0.600	45.1
0.425	36.7
0.300	29.6
0.212	23.5
0.150	18.3
0.106	13.9
0.075	10.0
0.053	7.1
0.038	5.0

The data of Table 2.1 is plotted in six different coordinate systems in Figure 2.1. It is useful when plotting by hand to use graph paper that is already scaled for the coordinate that is chosen. Most of the popular rulings are

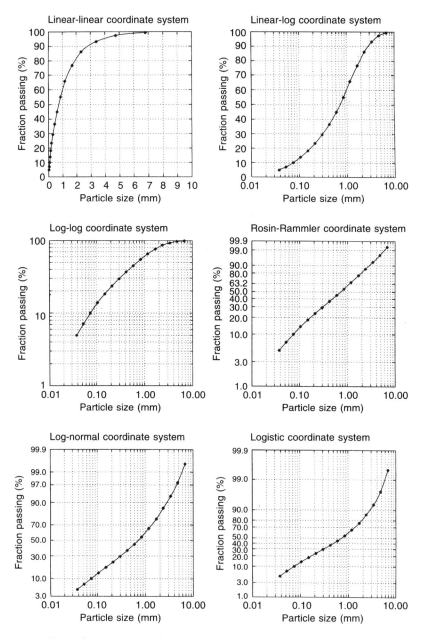

Figure 2.1 Typical particle size distribution plots showing the data of Table 2.1 plotted in six different coordinate systems

readily available commercially. It is even more convenient to use a computer graph plotting package such as the PSD program on the companion CD that accompanies this text and which offers all of the coordinate systems shown as standard options.

2.3 The distribution density function

In much of the theoretical modeling work it will be convenient to work with a function that is derived from the distribution function by differentiation. Let x represent any particle characteristic of interest. Then $P(x)$ is the mass fraction of the particle population that consists of particles having the value of the characteristic less than or equal to x. The distribution *density* function $p(x)$ is defined by

$$p(x) = \frac{dP(x)}{dx} \tag{2.24}$$

The discrete density function defined in Equation 2.1 is related to the density function by

$$p_i = \int_{D_i}^{D_{i-1}} p(x)dx$$

$$= P(D_{i-1}) - P(D_i) \tag{2.25}$$

A usual but imprecise interpretation of the distribution density function is that $p(x)dx$ can be regarded as the mass fraction of the particle population that consists of particles having the value of the characteristic in the narrow range $(x, x + dx)$.

An important integral relationship is

$$\int_0^\infty p(x)dx = P(\infty) - P(0) = 1 \tag{2.26}$$

which reflects that the sum of all fractions is unity.

2.4 The distribution by number, the representative size and population averages

Because all particle populations contain a finite number of particles it is also possible to describe the variation of particle characteristics through the number fraction. The number distribution function for any characteristic (having values represented by the variable x) is defined as the function $\Psi(x)$ which is the fraction by number of particles in the population having size equal to x or less. The associated number density function is defined by

$$\psi = \frac{d\Psi(x)}{dx} \tag{2.27}$$

The discrete number density is given by

$$\psi_i = \Psi(X_{i-1}) - \Psi(X_i) = \Delta\Psi_i \qquad (2.28)$$

where the upper case letters represent the class boundaries.

Often it is useful to have average values for any characteristic with the average taken over all members of the population. The average value of any characteristic property is given by

$$\bar{x}_N = \frac{1}{N_T} \sum_{j=1}^{N_T} x^{(j)} \qquad (2.29)$$

where $x^{(j)}$ is the value of the characteristic property for particle j and N_T is the total number of particles in the population. Equation 2.29 is unwieldy because the summation must be taken over a very large number, N_T, of particles. The number of terms in the summation is greatly reduced by collecting particles that have equal values of x into distinct groups. If the number of particles in group i is represented by $n^{(i)}$ and the value of x for particles in this group is represented by x_i, then the average value of the property x in the whole population is given by

$$\bar{x}_N = \frac{1}{N_T} \sum_{i=1}^{N} n^{(i)} x_i \qquad (2.30)$$

where N represents the total number of groups that are formed. The ratio $n^{(i)}/N_T$ is the fraction by number of the particle population having size x_i. This allows an alternative and even more convenient way of evaluating the average

$$\bar{x}_N = \sum_{i=1}^{N} x_i \psi_i \qquad (2.31)$$

Other averages are sometimes used. For example the average could be weighted by particle mass rather than by number

$$\bar{x} = \frac{1}{M_T} \sum_{i=1}^{N} m^{(i)} x_i \qquad (2.32)$$

In Equation 2.32 M_T represents the total mass of material in the population and $m^{(i)}$ the mass of particles in the group i having representative value x_i. The ratio $m^{(i)}/M_T$ is the fraction by mass of particles in the group i and this is related to the distribution function

$$\frac{m^{(i)}}{M_T} = P(x_{i+1}) - P(x_i) = \Delta P_i = p_i \qquad (2.33)$$

$$\bar{x} = \sum_{i=1}^{N} x_i \Delta P_i \qquad (2.34)$$

$$= \sum_{i=1}^{N} x_i p_i(x) \qquad (2.35)$$

In the limit as the group widths decrease to zero, Equation 2.34 becomes

$$\bar{x} = \int_0^1 x \, dP(x) \tag{2.36}$$

$$= \int_0^\infty x p(x) \, dx \tag{2.37}$$

In a similar way the variance of the distribution can be obtained

$$\sigma^2 = \int_0^\infty (x - \bar{x})^2 \, p(x) \, dx \tag{2.38}$$

The distribution density function is useful for the evaluation of the average of any function of the particle property x.

$$\overline{f(x)} = \int_0^\infty f(x) \, p(x) \, dx \tag{2.39}$$

In the same way, the average value of the property, x, weighted by number is obtained from

$$\overline{x_N} = \int_0^\infty x \psi(x) \, dx \tag{2.40}$$

or more generally

$$\overline{f(x)_N} = \int_0^\infty f(x) \psi(x) \, dx \tag{2.41}$$

For example, if all particles in the population are spherical, the average particle volume is the average value of $\pi d_p^3 / 6$. Thus

$$\text{Average particle volume} = \int_0^\infty \frac{\pi d_p^3}{6} \psi(d_p) \, dd_p \tag{2.42}$$

In order that it is possible to describe the behavior of the particles adequately, the concept of a representative size for each size class is introduced. A representative size for size class i is defined through the expression

$$d_{pi}^3 = \frac{1}{\psi_i(d_p)} \int_{D_i}^{D_{i-1}} d_p^3 \psi(d_p) \, dd_p \tag{2.43}$$

where $\psi(d_p)$ is the number distribution density function and $\psi_i(d_p)$ is the number fraction of the population in size class i. Other definitions of the representative size can be used and the precise definition will depend on the context in which the representative size will be used. It is important that the representative size be such that a single particle having the representative size would behave in a way that will adequately represent all particles in the class.

It is also possible to estimate the representative size from

$$d_{pi} = \frac{1}{p_i} \int_{D_i}^{D_{i-1}} d_p p(d_p) \, dd_p$$

$$= \frac{1}{p_i} \int_{D_i}^{D_{i-1}} d_p dP(d_p) \tag{2.44}$$

which weights the individual particles in the class by mass.

These two definitions of the representative size require the size distribution function to be known before the representative size can be established. In many circumstances this will be unsatisfactory because it would be more convenient to have the size classes together with their representative sizes defined independently of the size distribution. A common method is to use the geometric mean of the upper and lower boundaries for the representative size.

$$d_{pi} = (D_i D_{i-1})^{1/2} \tag{2.45}$$

Since $D_N = 0$ and D_0 is undefined, Equation 2.45 cannot be used to calculate the representative sizes in the two extreme size classes. These sizes are calculated using

$$d_{p1} = \frac{d_{p2}^2}{d_{p3}}$$

$$d_{pN} = \frac{d_{pN-1}^2}{d_{pN-2}} \tag{2.46}$$

These formulas project the sequence d_{pi} as a geometric progression into the two extreme size classes.

The arrangements of mesh and representative sizes is shown in Figure 2.2.

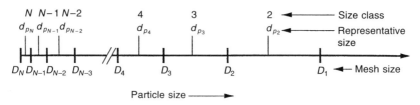

Figure 2.2 Arrangement of class sizes, representative sizes and mesh sizes along the particle size axis

2.5 Distributions based on particle composition

The mineralogical composition of the particles that are processed in ore dressing operations varies from particle to particle. This is of fundamental importance in any physical separation process for particulate material. The primary objective of ore dressing processes is the separation of materials on the basis of mineralogical composition to produce concentrates having a relative abundance of the desired mineral. The objective of comminution operations is the physical separation of minerals by fragmentation. Unfortunately, except in favorable cases, the minerals do not separate completely and many particles, no matter how finely ground, will contain a mixture of two or more mineral species. Some particles will, however, always exist that are composed of a single mineral. These are said to be perfectly liberated. The amount of mineral that is liberated is a complex function of the crystalline structure and mineralogical texture of the ore and the interaction between these and the comminution fracture pattern.

The mineralogical composition of a particle can be unambiguously defined by the fractional composition of the particle in terms of the individual mineral components that are of interest. Generally, more than one mineral species must be accounted for so that the mineralogical composition is described by a vector g of mineral fractions. Each element of the vector g represents the mass fraction of a corresponding mineral in the particle. The number of elements in the vector is equal to the number of minerals including gangue minerals. Thus in a particle that is made up of 25% by mass of chalcopyrite, 35% of sphalerite and 40% of gangue would be described by a mineral fraction vector $g = (0.25\ 0.35\ 0.40)$. A number of discrete classes of mineral fractions can be defined and the range of each element fraction, i.e. the range of each component of the vector g, must be specified for each class of particles. The fractional discrete distribution function can be defined as was done for particle size.

A special class exists for mineral fractions at the extreme ends of the composition range. In ore dressing operations it is usual to work with particle populations that have some portion of the mineral completely liberated. Thus a definite non-zero fraction of the particle population can have a mineral fraction exactly equal to zero or unity and a separate class is assigned to each of these groups of particles. These classes have class widths of zero. If only a single valuable mineral is considered to be important, g is a scalar and the distribution function $P(g)$ will have the form shown in Figure 2.3.

The concentration of particles in the two extreme classes representing completely liberated gangue and mineral respectively is represented by the step discontinuities in the distribution functions. When more than one mineral is significantly important, the simple graphical representation used in Figure 2.3 is no longer available and a multidimensional description is required.

2.6 Joint distribution functions

It often happens that more than one property of the particle is significant in influencing its performance in an ore dressing operation. In that case it is essential to use a description of the particle population that takes all relevant properties into account. The appropriate description is provided by the joint

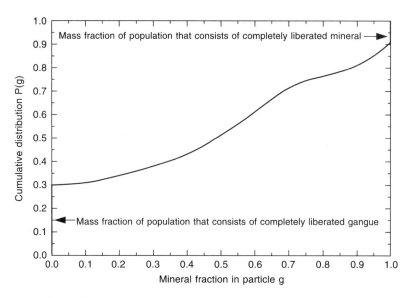

Figure 2.3 Typical cumulative grade distribution function for a particle population containing completely liberated and unliberated particles

distribution function. If the two properties concerned are the size and the mineralogical composition the joint distribution function would be defined as

$P(D, g)$ = mass fraction of the total particle population that has size $\leq D$ *and* mineral fraction $\leq g$.

Just as with single properties, the population can be divided into a finite number of discrete groups – each identified by a unique pair of the variables g and d_p. Thus the space of variable g and d_p can be sectioned on a rectangular grid and the discrete fractional distribution density function is defined by

$p_{ij}(d_p, g)$ = Fraction of material with size in the range (D_i, D_{i-1}), *and* composition in the range (G_{j-1}, G_j)

      = mass fraction of material in area (b) in Figure 2.4.

The relationship between the discrete fractional distribution density function and the two-dimensional cumulative distribution function is illustrated by reference to Figure 2.4.

$$P(D_{i-1}, G_j) = \text{fraction of material in areas a + b + c + d}$$
$$P(D_i, G_{j-1}) = \text{fraction of material in area d}$$
$$P(D_i, G_j) = \text{fraction of material in area a + d}$$
$$P(D_{i-1}, G_{j-1}) = \text{fraction of material in area c + d}$$

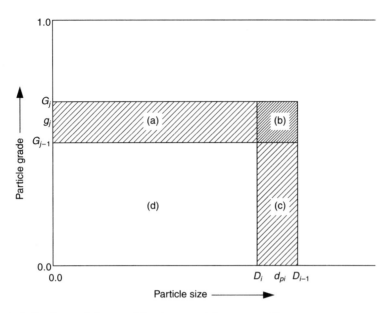

Figure 2.4 Regions of the particle size, particle composition space

Therefore

$$p_{ij}(d_p, g) = P(D_{i-1}, G_j) - P(D_i, G_j) - P(D_{i-1}, G_{j-1}) + P(D_i, G_{j-1})$$

and this is the mass fraction of particle population that has

$$D_i < d_p \leq D_{i-1} \text{ and } G_{j-1} < g \leq G_j$$

The ideas developed above for two-dimensional distributions are extended in the obvious way to higher dimensions. In particular, such an extension would be required for a multidimensional composition vector g if more than one mineral species is present.

It is clear from its definition that the value of $p_{ij}(d_p, g)$ does not change if the order of its arguments d_p and g, and therefore the indices i and j, is changed. Thus

$$p_{ij}(d_p, g) = p_{ji}(g, d_p) \tag{2.47}$$

2.7 Conditional distribution functions

When dealing with a collection of particles it is often convenient to separate them into groups according to a single property in spite of the variability of other important properties among the particles. An obvious example of this is a classification based on size, which is achieved by screening in the laboratory. The separation into size classes occurs in spite of any distribution of mineralogical composition so that particles of widely differing composition will be trapped on the same test sieve. Each batch of material on the different test sieves will have a different distribution of compositions. For example, the batch of particles in the finest size class will be relatively rich in completely liberated material. There is a unique composition distribution function for each of the size classes. The screening is called a conditioning operation and the distribution function for each size class is called a conditional fractional distribution function.

The conditional discrete density function $p_{ji}(g \mid d_p)$ is defined as the mass fraction of the particles in size class i (i.e. have size between D_i and D_{i-1}), that are in composition class j. These conditional distribution functions can be related to the joint distribution functions that have already been defined.

The concept of the conditional distribution is illustrated schematically in Figures 2.5 and 2.6. In Figure 2.5 a representative sample of the particle population is screened and is thereby separated into the required size classes. The material on each screen is then separated on the basis of particle mineral content using, for example, dense-liquid fractionation. The two discrete distribution functions are defined in terms of the masses of material produced by these two sequential operations.

$$p_{35}(g, d_p) = \frac{M_4}{M_T} \tag{2.48}$$

$$p_{35}(g \mid d_p) = \frac{M_4}{M_1} \tag{2.49}$$

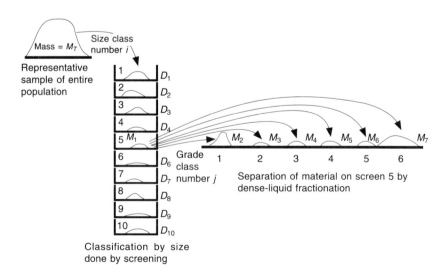

Figure 2.5 Schematic representation of the measurement of the distribution of particle grade conditioned by particle size

and

$$p_5(d_p) = \frac{M_1}{M_T} \qquad (2.50)$$

It is easy to see that

$$p_{35}(g|d_p) = \frac{M_4}{M_1} = \frac{M_4/M_T}{M_1/M_T} = \frac{p_{35}(g, d_p)}{p_5(d_p)} \qquad (2.51)$$

and that $p_{j5}(g | d_p)$ shows how the material on screen 5 is distributed with respect to the particle composition.

In Figure 2.6 the separation is done on the basis of mineral content first, then each particle grade class is classified by screening. This produces the size distribution conditioned by particle grade.

$$p_{53}(d_p | g) = \frac{M_6'}{M_1'} = \frac{M_6'/M_T}{M_1'/M_T} = \frac{p_{53}(d_p, g)}{p_3(g)} \qquad (2.52)$$

It is clear that

$$M_2 + M_3 + M_4 + M_5 + M_6 + M_7 = M_1 \qquad (2.53)$$

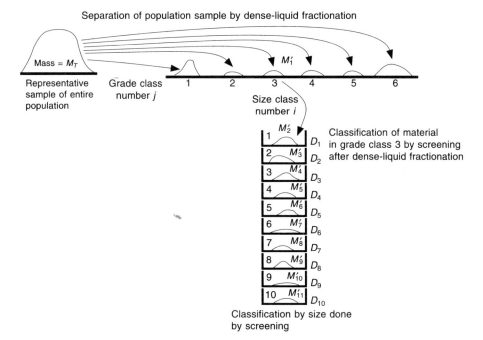

Figure 2.6 Schematic representation of the measurement of the distribution of particle size conditioned by particle grade

so that

$$\sum_{j=1}^{6} p_{ji}(g \mid d_p) = 1 \tag{2.54}$$

and

$$\sum_{j=1}^{6} p_{j5}(g, d_p) = p_5(d_p) \tag{2.55}$$

These ideas can be generalized to develop the following relationships. If M is the mass of the total population, the mass of particles that fall in the two classes j and i simultaneously is just $Mp_{ji}(g, d_p)$. When this is expressed as a fraction of only those particles in the d_p class, namely $Mp_i(d_p)$, the conditional distribution is generated.

Thus

$$\begin{aligned} p_{ji}(g \mid d_p) &= \frac{Mp_{ji}(g, d_p)}{Mp_i(d_p)} \\ &= \frac{p_{ji}(g, d_p)}{p_i(d_p)} \end{aligned} \tag{2.56}$$

Equation 2.56 is important chiefly because it provides a means for the determination of the theoretically important joint discrete distribution function $p_{ji}(g, d_p)$ from the experimentally observable conditional distribution function $p_{ji}(g \mid d_p)$

$$p_{ji}(g, d_p) = p_{ji}(g \mid d_p)p_i(d_p) \tag{2.57}$$

Since

$$p_{ji}(g, d_p) = p_{ij}(d_p, g) \tag{2.58}$$

we note that

$$p_{ji}(g, d_p) = p_{ij}(d_p \mid g)p_j(g) \tag{2.59}$$

and

$$p_{ji}(g \mid d_p)p_i(d_p) = p_{ij}(d_p \mid g)p_j(g) \tag{2.60}$$

Equation 2.57 corresponds to an experimental procedure in which the particle population is first separated on the basis of size by screening followed by a separation of each screened fraction into various composition groups. Equation 2.60 on the other hand corresponds to a separation on the basis of composition (by magnetic, electrostatic, or dense liquid techniques perhaps) followed by a sieve analysis on each composition class. Either way the same joint distribution function is generated but the former experimental procedure is in most instances much less convenient than the latter because of the experimental difficulties associated with separation by composition. It is usually more efficient to combine one composition separation with many size separations (which are comparatively simple to do in the laboratory) than the other way about.

The density functions satisfy the following general relationships which may be verified using the same simple principles that are used above.

$$\sum_i \sum_j p_{ij}(x,y) = 1 \tag{2.61}$$

$$\sum_j p_{ij}(x,y) = \sum_j p_{ij}(x \mid y) p_j(y) = p_i(x) \tag{2.62}$$

$$\sum_i p_{ij}(x,y) = \sum_i p_{ji}(y \mid x) p_i(x) = p_j(y) \tag{2.63}$$

$$\sum_i p_{ij}(x \mid y) = 1 \tag{2.64}$$

The principles developed in this section can be used to define the conditional distribution functions $P(g \mid d_p)$ and $P(g \mid d_p)$ as well as the associated density functions $p(g \mid d_p)$. These are related by

$$p(x \mid y) = \frac{dP(x \mid y)}{dx} \tag{2.65}$$

$$p(x \mid y) = \frac{p(x,y)}{p(y)} \tag{2.66}$$

and satisfy the following relationships analogous to Equations 2.61–64.

$$\iint p(x,y)\, dx\, dy = 1 \tag{2.67}$$

$$\int p(x,y)\, dy = p(x) \tag{2.68}$$

$$\int p(x,y)\, dx = \int p(y \mid x) p(x)\, dx = p(y) \tag{2.69}$$

$$\int p(x \mid y)\, dx = 1 \tag{2.70}$$

2.7.1 Practical representations of conditional grade distributions – the washability curve

Conditional grade distributions have been used for many years in practical mineral processing and a number of standard representational methods have evolved. Of these, the most widely used is the washability distribution and the associated washability curve. This method was developed initially to analyze coal washing operations and it is based on the dense-liquid fractionation laboratory method. This procedure is based on the use of a sequence of organic liquids having different densities usually in the range 1200 kg/m³ to about 3200 kg/m³ although denser liquids can be synthesized and used. The fractionation method relies on careful separation of the floating and sinking fractions when a representative sample of the test particle population is allowed to separate in a quiescent liquid of specified density. The fraction that floats represents the fraction of the particle population that has density less than the density of the test liquid. If the composition of the particles can be related directly to the density of the particle, the measured fraction is equal to the cumulative distribution $P(g)$. It is common practice to undertake dense-liquid fraction testing on specific size fractions in which case the conditional cumulative grade distribution $P(g \mid d_{pi})$ is generated in the experiment. The value of the cumulative distribution function can be measured at different values of g by using a series of liquids whose densities are adjusted to correspond to specific particle grades although it is more common to set up the liquid densities on a regular pattern to suit the material under test. The dense-liquid fractionation test is illustrated in Figure 2.7. Because identical representative samples of the particle population are analyzed in parallel, this method of analysis is called the parallel method.

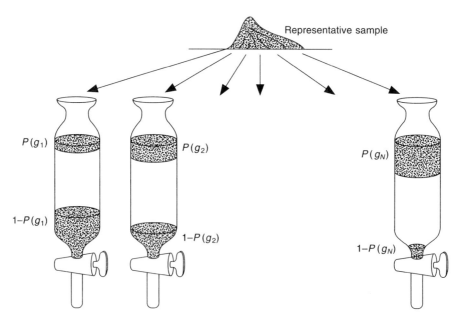

Figure 2.7 Representation of the dense-liquid fractionation experiments. The parallel method generates the cumulative distribution $P(g)$. If a particular fraction size is used, this test generates the conditional distribution $P(g \mid d_p)$

An alternative method for this analysis is often used because it uses smaller quantities of sample and because it yields additional information that is particularly useful and valuable in the analysis and simulation of mineral processing operations. Instead of analyzing N identical samples at N different densities, a single sample is separated sequentially at N different densities.

The sink fraction from the first test is tested at the next higher density after which the sink from the second test passes to the third liquid and so on until separations at all N densities have been completed. This method is illustrated in Figure 2.8.

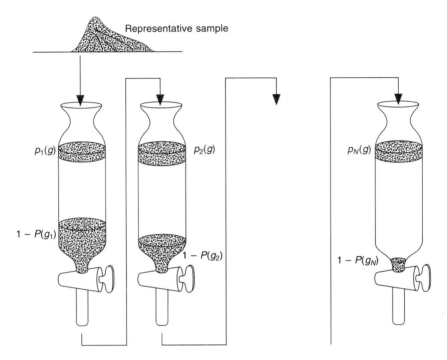

Figure 2.8 Representation of the dense-liquid fractionation experiments. The sequential method generates the discrete distribution $P_j(g)$. If a particular size fraction is used, this test generates the conditional discrete distribution $P_j(g_j \mid d_p)$

Each float fraction is collected and, after removal of any adhering liquid, is available for further analysis. Additional analyses that can be done include average particle density by pycnometry, elemental analysis by X-ray fluorescence analysis or atomic adsorption spectrometry and particle size analysis if a size-composite sample is used.

A typical set of data from a dense liquid test is shown in Table 2.2.

In order to convert the experimental data obtained in the dense-liquid test to the distribution with respect to particle composition it is necessary to relate the density of a particle to its mineralogical composition. If the material is a simple two-component mineral mixture this is

$$\frac{1}{\rho} = \frac{g}{\rho_M} + \frac{1-g}{\rho_G} \tag{2.71}$$

where ρ = density of a particle

Table 2.2 Typical data from a dense-liquid test on magnesite ore

Liquid specific gravity	Total mass yield in this fraction %	% CaO	% SiO$_2$
Float at 2.85	21.6	19.30	18.2
2.85–2.88	5.70	21.76	2.49
2.88–2.91	3.20	10.15	1.52
2.91–2.94	0.90	9.67	2.92
2.94–2.96	7.60	2.95	3.89
2.96–3.03	61.0	0.96	2.55
Sink at 3.03	0.00		

ρ_M = density of the mineral phase
ρ_G = density of the gangue phase
g = mass fraction of mineral in the particle

The inverse of this equation is more useful

$$g = \frac{\rho_G - \rho}{\rho_G - \rho_M} \frac{\rho_M}{\rho} \tag{2.72}$$

which shows that the mineral grade is a linear function of the reciprocal of the particle density.

When the mineralogical texture is more complex than the simple binary mixture of two minerals, additional information is required from the dense-medium test to relate the particle composition to the separating density. This usually requires elemental analysis of the individual fractions that are obtained in the sequential dense-liquid test. Typical data are shown in Table 2.2. From the measured assays, the average mineralogical composition of particles in each fraction can be estimated. In this case the calcite content is estimated from the CaO assay and the magnesite content is estimated by difference assuming that only the three minerals magnesite, calcite and silica are present. The relationship between the particle density and its mineralogical composition is

$$\rho^{-1} = \sum_{m=1}^{M} g_m \rho_m^{-1} \tag{2.73}$$

If the density intervals used in the dense-liquid test are narrow it is reasonable to postulate that the average density of the particles in each density fraction is the midpoint between the end points of the intervals.

The densities calculated from the mineralogical compositions and the known densities of the minerals should correspond quite closely with the midpoint densities as shown in Table 2.3.

In the case of coal, it is usual to measure the ash content and sulfur content of the washability fractions. It is also customary to measure the energy content of the fractions as well because this plays a major role in assessing the utility of the coal for power generation. Even greater detail regarding the make-up

Table 2.3 Data derived from the measured data in Table 2.2

Liquid specific gravity	Magnesite (%)	Calcite (%)	SiO_2 (%)	Calculated density (kg/m³)
Float at 2.85	47.34	34.46	18.2	2828
2.85–2.88	58.65	38.86	2.49	2867
2.88–2.91	80.36	18.13	1.52	2935
2.91–2.94	79.81	17.28	2.92	2933
2.94–2.96	90.85	5.26	3.89	2968
2.96–3.03	95.74	1.71	2.55	2985
Sink at 3.03				

of the coal can be obtained if the full proximate analysis is determined for each washability fraction. It is also possible to distinguish organically bound sulfur from pyritic sulfur in the coal. The greater the detail of the analysis of the washability fractions, the greater the detail in the products that can be calculated by modeling and simulation. These analyses are done on the separated fractions and therefore give values that are conditional on the particle density. This is a useful approach because many coal processing methods separate the particles on the basis of their density. The conditional analyses can then be used to calculate compositions and heating values of the various products of the separation.

2.7.2 *Measurement of grade distributions by image analysis*

In recent years a more direct method of measurement has been developed, namely mineral liberation measurement using automatic image analysis. This technique provides a direct measurement of the distribution of particle grades in a sample from a narrow size fraction. The technique requires the generation of microscopic images of the particles that are mounted in random orientation and sectioned. The images which can be generated by optical or scanning electron microscopy must distinguish each mineral phase that is to be measured. A typical image of a binary mineral system is shown in Figure 2.9. The apparent grade of each particle section in the image can be readily determined when the image is stored in digital form. The apparent areal grade of a particle section is the ratio of mineral phase pixels to total pixels in the section. Figure 2.9 is typical of images obtained from mineral particles that are collected in mineral processing plants and it is easy to see that the lack of liberation will have a profound effect on the unit operations that are used to upgrade the mineral species.

Alternatively the apparent grade of many linear intercepts across a particle section can be measured. It is a simple matter to establish the distribution of apparent linear or areal grades from images containing a sufficiently large number of particle sections. Typical histograms of measured linear grade

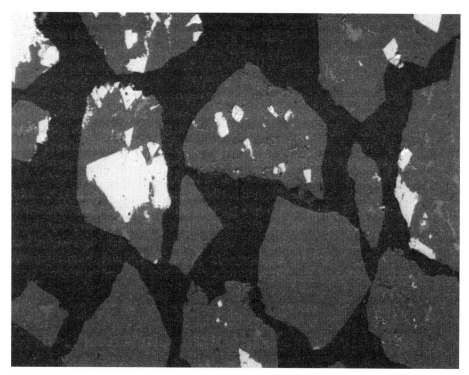

Figure 2.9 Electron microscope image of mineral particle sections showing separate phases

distributions are shown in Figure 2.10. These histograms are based on 12 grade classes, which describe pure gangue in class 1, pure mineral in class 12 and 10 equally spaced grade classes in the grade fraction range from 0.0 to 1.0. Note that 11 separate size fractions were analyzed in this sample.

The measured apparent grade distributions must be stereologically transformed to convert them to the desired grade distribution. The stereological transformation is a typical inverse problem and requires the solution of the integral equation

$$P(g_m \mid d_p) = \int_0^1 P(g_m \mid g, d_p) \, p(g \mid d_p) \, dg \qquad (2.74)$$

where g_m represents the measured apparent grade, either linear or areal, and g represents the true grade of a particle. $P(g_m \mid d_p)$ is the cumulative distribution of apparent grades that is measured in the image. Solution of the integral equation requires care to ensure reliable answers but effective methods are readily available.

2.7.3 *Stereological transformation of measured data*

Equation 2.74 provides the basis of calculating the distribution of volumetric

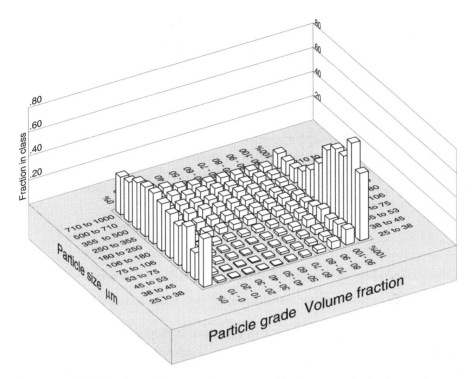

Figure 2.10 Distribution of linear grades measured by image analysis of many images such as that shown in Figure 2.9. Eleven separate size fractions were analyzed

grades $p_v(g_v \mid D)$ in a population of particles when the distribution of linear grades $P_\ell(g_\ell \mid D)$ has been measured or calculated from a theoretical model. It is convenient to use a matrix notation so that Equation 2.74 becomes

$$P_\ell = Kp_v \qquad (2.75)$$

where K is the 12×12 transformation kernel. P_ℓ is the vector of 12 measured cumulative linear grades and p_v is the desired vector of fractional volumetric grades.

The inherently ill-posed nature of the stereological correction problem is reflected in the very high condition number of the transformation matrix K so that the apparently straightforward solution by direct inversion

$$p_v = K^{-1}P_\ell \qquad (2.76)$$

is useless because even very small errors in the measured distribution, which must always be present, are magnified by the inversion and swamp the true solution p_v.

In spite of these difficulties, useful and reliable numerical solutions can be generated by imposing the following appropriate constraints on the solution

$$\sum_{i=1}^{12} p_v^i = 1.0 \tag{2.77}$$

and

$$p_v^i \geq 0 \quad \text{for} \quad i = 1, 2 \dots 12 \tag{2.78}$$

with p_v^i representing the *i*th element of the vector p_v.

The solution is obtained by imposing additional regularization conditions which reflect the relative smoothness that would be expected for the distribution function $p_v(g \mid D)$ in any real sample of ore, and the transformed distribution is obtained as the solution to the constrained minimization problem

$$\underset{p_v^i}{\text{Minimize}} \; \| 100(Kp_v - P_\ell) \| + \lambda \sum_{i=1}^{12} p_v^i \ln p_v^i \tag{2.79}$$

subject to equality constraint 2.77 and inequalities 2.78. The term $-\sum_{i=1}^{12} p_v^i \ln p_v^i$ measures the entropy of the distribution p_v^i and increases as the distribution becomes smoother.

The regularization parameter λ in Equation 2.79 is arbitrary and controls the weight given to maximization of the entropy of the distribution relative to minimization of the residual norm. Large λ favors larger entropy and therefore smoother solutions at the cost of larger residual norms. In practice a value of $\lambda = 1.0$ has been found to be a satisfactory choice in most cases.

In general the kernel matrix should be determined separately for every sample that is analyzed but this is a time-consuming and tedious task. As a result users of this method develop libraries of kernel matrices from which the most appropriate can be selected for a particular application. A useful kernel that will produce good results for many ores is

$P_\ell(g_\ell \mid g_v)$

$$=
\begin{vmatrix}
1.0000 & 0.8190 & 0.5508 & 0.3758 & 0.2633 & 0.1905 & 0.1411 & 0.1041 & 0.0727 & 0.0433 & 0.0144 & 0.0000 \\
1.0000 & 0.9028 & 0.6960 & 0.5101 & 0.3640 & 0.2571 & 0.1809 & 0.1254 & 0.0826 & 0.0470 & 0.0151 & 0.0000 \\
1.0000 & 0.9288 & 0.7645 & 0.5987 & 0.4521 & 0.3318 & 0.2367 & 0.1621 & 0.1033 & 0.0561 & 0.0171 & 0.0000 \\
1.0000 & 0.9460 & 0.8149 & 0.6716 & 0.5329 & 0.4082 & 0.3002 & 0.2087 & 0.1324 & 0.0701 & 0.0205 & 0.0000 \\
1.0000 & 0.9584 & 0.8543 & 0.7334 & 0.6073 & 0.4843 & 0.3688 & 0.2632 & 0.1693 & 0.0894 & 0.0255 & 0.0000 \\
1.0000 & 0.9677 & 0.8857 & 0.7861 & 0.6753 & 0.5590 & 0.4410 & 0.3247 & 0.2139 & 0.1143 & 0.0323 & 0.0000 \\
1.0000 & 0.9745 & 0.9106 & 0.8307 & 0.7368 & 0.6312 & 0.5157 & 0.3927 & 0.2666 & 0.1457 & 0.0416 & 0.0000 \\
1.0000 & 0.9795 & 0.9299 & 0.8676 & 0.7913 & 0.6998 & 0.5918 & 0.4671 & 0.3284 & 0.1851 & 0.0540 & 0.0000 \\
1.0000 & 0.9829 & 0.9439 & 0.8967 & 0.8379 & 0.7633 & 0.6682 & 0.5479 & 0.4013 & 0.2355 & 0.0712 & 0.0000 \\
1.0000 & 0.9849 & 0.9530 & 0.9174 & 0.8746 & 0.8191 & 0.7429 & 0.6360 & 0.4899 & 0.3040 & 0.0972 & 0.0000 \\
1.0000 & 0.9856 & 0.9567 & 0.9273 & 0.8959 & 0.8589 & 0.8095 & 0.7367 & 0.6242 & 0.4492 & 0.1810 & 0.0000 \\
1.0000 & 1.0000 & 1.0000 & 1.0000 & 1.0000 & 1.0000 & 1.0000 & 1.0000 & 1.0000 & 1.0000 & 1.0000 & 1.0000 \\
\end{vmatrix}
$$

$$\tag{2.80}$$

A histogram of the true volumetric distribution of particle grades is shown in Figure 2.11 after stereological correction of the data in Figure 2.10. The importance of transforming the data is immediately apparent from these figures. The untransformed data do not reveal the true nature of the liberation distribution and cannot be used to analyze the behavior of individual particle types in any mineral processing operation. This should be borne in mind when experimental image analysis data is to be used for model calibration purposes.

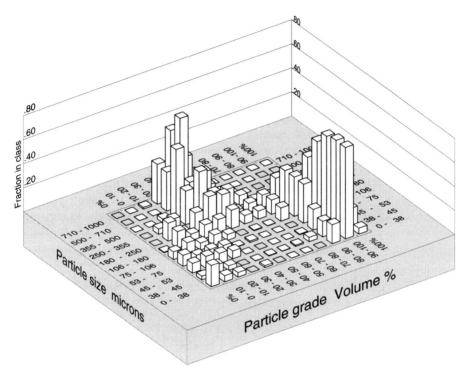

Figure 2.11 Liberation distributions of 11 particle size fractions of a 2-component ore measured by image analysis. These histograms were obtained by stereological transformation of the data shown in Figure 2.10

2.8 Independence

It happens that sometimes two properties can be distributed independently of each other. This idea can be made precise by defining the independence of two properties, say k and d_p, if the following relationship is satisfied.

$$p_{ji}(k \mid d_p) = p_j(k) \tag{2.81}$$

which means that the distribution of k values (these could be flotation rate constants for example) is the same within any size class as it is across the entire population.

This leads to

$$p_{ji}(k, d_p) = p_j(k \mid d_p)p_i(d_p)$$

$$= p_j(k)p_i(d_p) \tag{2.82}$$

which shows that the joint distribution for two properties that are independent can be generated as the product of the two separate distribution functions.

2.9 Distributions by number

In some situations it is useful to use number fractions rather than mass fractions when dealing with particle populations. The relationship between the mass distribution functions and the equivalent number distribution function can be deduced as follows.

The number distribution function $\Psi(d_p)$ is defined to be equal to the number fraction of particles in the entire population with size $\leq d_p$. Number distribution functions and number distribution density functions can be defined in a similar way for each of the distribution types already defined for mass fractions. In particular the discrete number fractional distribution function is defined by

$$\psi_i(d_p) = \frac{\text{number of particles in size class i}}{\text{Total number of particles in the population}}$$

$$= \frac{n_i(d_p)}{N} \tag{2.83}$$

The number distributions can be related to the mass distributions as follows

Let $\quad \psi(m)dm =$ number fraction of particles having mass in $(m, m + dm)$
$\psi(m \mid d_p)dm =$ number fraction of particles of size d_p that have mass in $(m, m + dm)$. That is the distribution density for particle mass conditioned by particle size
$\quad\quad p_i(d_p) =$ mass fraction of particles in size class i
$\quad\quad\quad m =$ mass of a particle of size d_p
$\quad\quad \overline{m}(d_p) =$ average mass of a particle of size d_p
$\quad\quad\quad M =$ total mass of particles in the population
$\quad\quad\quad N =$ total number of particles in the population

$$Mp(d_p) = N \int_0^\infty m\, \psi(m, d_p)dm$$

$$= N \int_0^\infty m\, \psi(m \mid d_p)\psi(d_p)dm \tag{2.84}$$

$$= N\psi(d_p) \int_0^\infty m\, \psi(m \mid d_p)dm$$

$$= N\psi(d_p)\,\overline{m}(d_p)$$

Using Equation 2.25

$$Mp_i(d_p) = M \int_{D_i}^{D_{i-1}} p(d_p)dd_p$$

$$Mp_i(d_p) = N \int_{D_i}^{D_{i-1}} \psi(d_p)\overline{m}(d_p)dd_p$$

(2.85)

If the shape of a particle is statistically independent of size (a good approximation for single particle populations) then

$$\beta = \frac{\overline{m}(d_p)}{d_p^3} = \frac{\overline{m}(d_{pi})}{d_{pi}^3}$$

(2.86)

is constant independent of size.

Equation 2.85 can be simplified by using Equation 2.86 together with Equation 2.43.

$$Mp_i(d_p) = N \frac{\overline{m}(d_{pi})}{d_{pi}^3} \int_{D_i}^{D_{i-1}} d_p^3 \, \psi(d_p)dd_p$$

(2.87)

$$= N\overline{m}(d_{pi}) \, \psi_i(d_p)$$

The relationship between N and M is obtained from

$$M = M \Sigma_i \, p_i(d_p) = N \Sigma_i \, \overline{m}(d_{pi})\psi_i(d_p)$$

$$\frac{M}{N} = \Sigma_i \, \overline{m}(d_{pi}) \, \psi_i(d_p)$$

(2.88)

$$= \beta \, \Sigma_i \, d_{pi}^3 \, \psi_i(d_p)$$

or

$$\frac{M}{N} = \beta \int_0^\infty d_p^3 \, \psi(d_p) \, dd_p$$

(2.89)

where Equation 2.86 has been used.

Substituting Equation 2.89 into Equation 2.87 gives an exact relationship between the discrete distribution by mass $p_i(d_p)$ and the discrete distribution by number $\psi_i(d_p)$

$$p_i(d_p) = \frac{d_{pi}^3 \, \psi_i(d_p)}{\Sigma_i \, d_{pi}^3 \, \psi_i(d_p)}$$

(2.90)

Similarly

$$\psi_i(d_p) = \frac{p_i(d_p)/d_{pi}^3}{\Sigma_i \, p_i(d_p)/d_{pi}^3}$$

(2.91)

2.10 Internal and external particle coordinate and distribution densities

Up to now, the particles have been classified according to two very important properties: the particle size and the mineralogical composition. These are two examples of many properties that can be used to describe the particle. They are certainly the most important descriptive properties so far as ore dressing and hydrometallurgical processes are concerned, but in order to develop effective models for the unit operations it is always necessary to ensure that the particles are described in sufficient detail for the purposes of the modeling study. It is impossible to make a complete description of any one particle – such concepts as particle shapes, surface topography, surface energy etc. cannot be completely described in a quantitative sense using a finite number of variables. Consequently, it is always necessary to choose a finite and comparatively small number of variables that can be used to describe the particle properties in sufficient detail for the purposes of the model.

The properties that describe the nature of the individual particles are called the internal coordinates of the particle phase space. The choice of these coordinates is dictated entirely by the demands of the model. However, it is always necessary to bear in mind that one of the primary objectives of modeling the unit operations is to link these together in a plant flowsheet so that the plant can be simulated. It will often happen that the set of internal coordinates that are required to model one of the unit operations might not be appropriate for the other models. In general the problem is overcome by specifying a set of internal coordinates that will include all internal coordinates required for the separate models as subsets. This means that some of the internal coordinates will be redundant in some of the unit models. This does not cause any problems in general.

In some unit operations, the physical location of the particle in the unit equipment can have a very significant effect on the particle behavior and it is sometimes necessary to track the position of the particles in the equipment in order to describe the operation of the unit as a whole. This is not always necessary but in formulating the general model structure it is convenient to include them.

The distribution functions that were described in Section 2.9 can be related to corresponding distribution density functions and it is convenient to formulate a generalized model in terms of the joint distribution density function for all the particle coordinates external and internal.

The generalized particulate distribution density function $\psi(x)$ is an ordinary function of the vector argument x which comprises all of the external and internal coordinates that are relevant to the problem on hand. All possible vectors x make up the particle phase space. $\psi(x)$ is defined as the number fraction density, i.e. the number fraction of the entire particle population that occupies unit volume of the particle phase space. The finite particle distribution functions defined previously can be constructed from the distribution density as follows.

In the case where the phase space consists only of the particle size (d_p) and the particle grade (g)

$$\psi_{ij}(d_p, g_i) = \int_{D_i}^{D_{i-1}} \int_{G_i}^{G_{i+1}} \psi(d_p, g)\, dd_p\, dg \qquad (2.92)$$

The mathematical description of ψ is sometimes difficult. For example it may not be possible to obtain ψ in terms of any known transcendental functions and ψ may have to include Dirac delta functions. Particular examples are the two extreme ends of the liberation spectrum. Thus we tend to avoid ψ for practical modeling work and use it only formally to develop some necessary model structures.

2.11 Particle properties derived from internal coordinates

The internal coordinates of a particle should be sufficient to describe all significant characteristics of the particle. In many situations it is necessary to obtain some derived quantity. Some examples are: the metal assay from the mineralogical composition, the magnetic susceptibility of the particle from the magnetic susceptibilities of the individual minerals, and the calorific value, volatile content and ash content of a coal particle from the mineral matter content plus the maceral content.

The density of the particle is a particularly important derived property and we note that this is related to mineralogical composition by

$$\frac{1}{\rho} = \sum_{m=1}^{M} \frac{g_m}{\rho_m} \qquad (2.93)$$

where

ρ_m = density of mineral phase
ρ = density of particle
g_m = mass fraction of mineral phase m in the particle (particle grade).

2.12 The population balance modeling method

The characterization of a particle population through the distribution of particles in the external and internal variable spaces provides a powerful formalized modeling procedure. The essential idea is that as the particles move through the processing environment their external and internal coordinates change – the external coordinates because they move physically and the internal coordinates because the particles are altered by the processing.

The collection of external and internal coordinates attached to a particle defines the coordinate location of the particle in the phase space. The external coordinates describe its actual physical location while the internal coordinates describe the characteristics of the particle itself. The internal coordinates must be sufficiently numerous to describe the particle in all the detail necessary for its complete characterization in terms of the processes that must be described by the model.

Some examples of internal coordinates are:

- particle size; this is perhaps the most important of all
- mineralogical composition
- chemical composition
- particle shape
- surface specific energy.

Some of these internal coordinates may themselves be multidimensional. For example the mineralogical composition of the solid material may include more than two minerals in which case the mineralogical composition vector must include one coordinate for each distinct mineral. Sometimes the multi-dimensionality of the internal coordinates must be infinite. This occurs, for example, during leaching processes when the concentration of the species that is being leached varies continuously within the particle and the chemical composition must be known at every point within the particle. In general the concentration profile of the leached component will be low near the surface of the particle and higher further in where lixiviant has not attained high concentrations because of diffusional resistance within the particle. The concentration profile is a functional of the entire processing history of the particle. In practice it is not convenient to use an infinite-dimensional internal coordinate and various finite-dimensional approximations must be made. An important instance when a one-dimensional internal characterization is completely adequate applies to the technically important topochemical processes where the radial position of the topochemical interface inside the particle often carries complete information regarding the concentration profile and its effect on the rate of the chemical process. There is a large class of particulate rate processes that can be modeled topochemically and for which the instantaneous rate of the process over the particle as a whole can be written as a unique function of the position of the topochemical interface. This makes topochemical models very powerful models indeed for the descriptions of particulate rate processes. Furthermore topochemical models permit the calculation of the rate at which the topochemical boundary moves and this rate is important in the formulation of population balance models for the particle population as a whole.

2.13 The fundamental population balance equation

Notation used in this section:

$\psi(x)dx$ is the number fraction of particles per unit volume of phase space, i.e. the number fraction of the particle population that occupies a small volume around the phase coordinate x.

x is the coordinate point in phase space $x \, \varepsilon \, \mathbb{R}^N$

u is a vector of 'velocities' at which particles change their phase coordinates. For example this could be the rate of change of particle size by attrition when particle size is an internal coordinate, the rate of movement of the topochemical interface when the position of the interface is the internal coordinate or the rate of change of particle composition when the particle undergoes chemical change. The corresponding elements for the external coordinate are the physical velocities in three-dimensional space.

$\mathfrak{R}(x)$ is the rate at which particles at coordinate position x are destroyed. This rate is specified as mass per unit volume of phase space per unit time.

W_{in} is the mass rate at which solid material is fed to the system.

W_{out} is the mass rate at which solid material is withdrawn from the system.

The material in the feed and product is distributed in the phase space according to ψ_{in} and ψ_{out} respectively.

$\overline{m}(x)$ is the average mass of a particle at point x in the phase space.

N is the total number of particles in the system.

A, B, D are rates of addition, birth and destruction of particles.

Q is the rate of removal material through product removal streams.

$B(x; x')$ is a distribution function that describes the way material can move suddenly over discrete distances in phase space following physical destruction or breakage. $B(x; x')$ is the mass fraction of solid material that has a phase coordinate less then x after destruction of a particle at phase coordinate x'. The statement that the phase coordinate of a particle is less than x must be interpreted to mean that every phase coordinate of the particle is less than the corresponding element of x. $b(x; x')$ is the corresponding density function.

$a(x; x')$ is the distribution density for particles produced by attrition and other surface wear processes at x'.

In general $a(x; x')$ and $b(x; x')$ are significantly different from each other and in practical applications each needs to be determined separately.

Consider a region R_c of the particle phase space and account for the accumulation of particulate mass in R_c by all processes that change the coordinates of particles. The region R_c is completely enclosed by a surface S_c. These processes are

1. Movement across the boundaries of R_c (convective motion in phase space).
2. Arrivals by finite steps from other regions in the phase space.
3. Destruction of particles in region R_c.
4. Physical additions and withdrawals in feed and product streams.

An accounting of individual particles in the reference region R_c gives:

$$\frac{\partial}{\partial t} \int_{R_c} N\psi(x)dx = - \int_{S_c} N\psi(x)u \cdot n \, d\sigma - D + B - Q + A \qquad (2.94)$$

In this equation, n is the outward pointing normal vector to the surface S_c at point x.

In our mineral processing systems solid mass is conserved and this imposes some important constraints on the formulation of model equations.

The destruction processes can generally be described by a mass rate of destruction per unit volume of phase space

$$D = \int_{R_c} \frac{\Re(\psi(x), x, F[\psi(x)])}{\overline{m}(x)} \, dx \qquad (2.95)$$

and is specified as number of particles broken per unit time in the control volume R_c. The notation $F_1[\psi(x)]$ indicates that the destruction function process D is a function of the entire distribution function $\psi(x)$ and not only of the value of $\psi(x)$ at x. This allows for such effects as the variation of breakage rates in a mill with the amount of fine and coarse particles present.

The birth processes can only result (and must necessarily do so) from the products of the destruction processes since mass is conserved.

$$B = \int_{R_c} \frac{1}{\overline{m}(x)} \int_{R'(x)} \Re(\psi(x'), x', F[\psi(x)]) \, b(x; x')dx' \, dx$$

$$- \int_{R_c} \frac{N}{\overline{m}(x)} \int_{R''(x)} \psi(x')u(x') \cdot \nabla \overline{m}(x') \, a(x; x')dx' \, dx \qquad (2.96)$$

Here $R'(x)$ and $R''(x)$ are the regions of space from which progeny particles can enter the small phase volume dx around the point x by breakage and attrition processes respectively. These are called the feeder regions for point x.

The conservation of mass constraint distinguishes mineral processing systems from other particulate processes such as crystallization and leaching in which mass transfers between the solid phase and the liquid phase and solid phase mass is not conserved.

Mass enters and leaves the processing volume by physical addition and removal through feed and product streams. The addition rate is given by

$$A = W_{in} \int_{R_c} \left(\frac{N}{M} \right)_{in} \psi_{in}(x) \, dx \qquad (2.97)$$

and the removal rate by

$$Q = \sum_j W_{out\,j} \int_{R_c} \left(\frac{N}{M} \right)_j \psi_{out\,j}(x) \, dx \qquad (2.98)$$

In Equation (2.98), j indexes the different physical output streams. The integral over the surface of the reference region is inconvenient and this can be converted to an integral over the volume by application of the divergence theorem.

$$\int_{S_c} N\psi(x)\, u \cdot n\, d\sigma = \int_{R_c} N\nabla \cdot u\, \psi(x)\, dx \qquad (2.99)$$

The final working equation for steady-state operation is

$$N\int_{R_c} \nabla \cdot u\, \psi(x)\, dx + \int_{R_c} \frac{\Re\,(\psi(x),\, x,\, F[\{\psi(x)\}])}{\overline{m}(x)}\, dx$$

$$-\int_{R_c} \frac{1}{\overline{m}(x)} \int_{R'(x)} \Re(\psi(x'),\, x',\, F[\psi(x)])\, b(x; x')dx'\, dx$$

$$(2.100)$$

$$+\int_{R_c} \frac{N}{\overline{m}(x)} \int_{R''(x)} \psi(x')u(x') \cdot \nabla\overline{m}(x')\, a(x; x')dx'\, dx$$

$$= -\sum_j W_{\text{out}\,j}\left(\frac{N}{M}\right)_j \int_{R_c} \psi_{\text{out}\,j}(x)dx + W_{\text{in}}\left(\frac{N}{M}\right)_{\text{in}} \int_{R_c} \psi_{\text{in}}(x)dx$$

This equation can be specialized in a number of ways to suit the requirements of any particular application. It can be written in terms of mass density functions by substituting for $\psi(x)$ in terms of $p(x)$ using equations such as 2.84.

Because the region R_c is arbitrary, Equation 2.100 can be written as a functional integro-differential equation.

$$N\nabla \cdot u\psi(x) + \frac{\Re(\psi(x),\, x,\, F[\psi(x)])}{\overline{m}(x)}$$

$$-\frac{1}{\overline{m}(x)} \int_{R'(x)} \Re(\psi(x'),\, x',\, F[\psi(x')])\, b(x;\, x')dx'$$

$$+\frac{N}{\overline{m}(x)} \int_{R''(x)} \psi(x')u(x') \cdot \nabla\overline{m}(x')\, a(x; x')dx' \qquad (2.101)$$

$$= -\sum_j W_{\text{out}\,j}\left(\frac{N}{M}\right)_j \psi_{\text{out}\,j}(x)dx + W_{\text{in}}\left(\frac{N}{M}\right)_{\text{in}} \psi_{\text{in}}(x)$$

This form of the population balance equation is usually the most convenient for analytical solution.

The population balance equation must be solved subject to the condition

$$\int \psi(x)\, dx = 1 \qquad (2.102)$$

where the integral is taken over the entire phase space.

The two breakage functions $b(x; x')$ and $a(x; x')$ must satisfy two important conditions.

$$\int_{R_{A'}} b(x; x')\, dx = 1$$

$$\int_{R_{A''}} a(x; x')\, dx = 1 \tag{2.103}$$

$R_{A'}$ and $R_{A''}$ are the regions of the phase space that can be reached by progeny particles that are formed by breakage (for b) or attrition (for a) at point x'. These are called the accessible regions of phase space and they are complementary to the feeder regions R' and R''. The determination of the feeder regions R' and R'' and the accessible regions $R_{A'}$ and $R_{A''}$ can be quite difficult and some specific cases are considered later. It is usually easier to define the accessible regions than the feeder regions and this fact can have a significant effect on the choice of computational algorithms that can be used successfully.

In most applications the fundamental population balance equation will be reduced to an appropriate discrete form with the region R_c corresponding to the appropriate particle class.

2.14 The general population balance equation for comminution machines

The population balance equation provides a powerful model for the description of industrial comminution machines. It allows the development of a uniform model that describes the operating behavior of rod, ball, semi-autogenous, and autogenous mills. Because of the great importance of these operations in practice some effort is devoted here to the careful development of the models from the detailed population balance Equation 2.101.

A useful restricted form of the general population balance equation is generated under the following conditions: only one internal coordinate, the particle size, and it is assumed that the breakage and wear processes are not dependent on the position of the particle in the mill so that the external coordinates are irrelevant. The general Equation 2.101 is accordingly written:

$$N\frac{d}{dx}(u(x)\,\psi(x)) + \frac{\Re(\psi(x),\, x,\, F[\psi(x)])}{\beta x^3}$$

$$-\frac{1}{\beta x^3}\int_{R'(x)}\Re(\psi(x'),\, x',\, F[\psi(x')])\,b(x; x')\,dx'$$

$$+\frac{N}{\beta x^3}\int_{R''(x)}\psi(x')u(x')\frac{d\beta x'^3}{dx}\,a(x; x')\,dx' \tag{2.104}$$

$$= -W\left(\frac{N}{M}\right)_{out}\psi_{out}(x) + W\left(\frac{N}{M}\right)_{in}\psi_{in}(x)$$

In Equation 2.104 the scalar x represents the particle size d_p and the average mass of a particle is related to the size by

$$\overline{m}(x) = \beta x^3 \tag{2.105}$$

A commonly-used model for the rate of attrition and other wear processes such as chipping is that this rate is proportional to the surface area of the particle. Using a spherical particle as a model

$$\frac{\pi}{6}\frac{dx^3}{dt} = -\frac{\kappa'\pi x^2}{2} \tag{2.106}$$

which implies that the velocity at which a wearing particle moves in phase space is constant and is given by

$$u(x) = \frac{dx}{dt} = -\kappa' \tag{2.107}$$

A more general model for surface wear processes is

$$\frac{dx}{dt} = -\kappa(x) = -\kappa x^\Delta \tag{2.108}$$

where Δ is a constant between 0 and 1.
    This implies that

$$\frac{dm}{dt} = -\kappa\frac{\pi\rho_s x^{2+\Delta}}{2} \tag{2.109}$$

If $\Delta = 0$ the specific surface wear rate is constant while $\Delta > 0$ means that the specific surface wear rate increases as the size of the particle increases. As shown above $\Delta = 0$ is equivalent to the assumption that surface wear rate is proportional to the surface area of the particle while $\Delta = 1$ means that the surface wear rate is proportional to the mass of the particle.
    It is usual to develop the population balance equation in terms of the distribution by mass and Equation 2.104 is converted using Equation 2.84.

$$\psi(x) = \frac{M}{N}\frac{p(x)}{\beta x^3} \tag{2.110}$$

Equation 2.104 becomes:

$$-\frac{M}{\beta}\frac{d(\kappa(x)\,p(x)/x^3)}{dx} + \frac{\Re(p(x),\,x,\,F[p(x)])}{\beta x^3}$$

$$-\frac{1}{\beta x^3}\int_{R'(x)} \Re(p(x'),\,x',\,F[p(x')])b(x;\,x')dx' \tag{2.111}$$

$$-\frac{M}{\beta x^3}\int_{R''(x)} \frac{p(x')}{x'^3}\,3\kappa(x')x'^2\,a(x;\,x')dx'$$

$$= -W\frac{p_{\text{out}}(x)}{\beta x^3} + W\frac{p_{\text{in}}(x)}{\beta x^3}$$

The rate of destruction is intensive with respect to the average mass density over phase space. Consider two systems having the identical phase space but containing different total masses. In particular $\psi(x)$ is identical for both systems. Then

$$\frac{\Re_1\,(p(x),\,x,\,F[p(x)])}{\Re_2\,(p(x),\,x,\,F[p(x)])} = \frac{M_1}{M_2} \tag{2.112}$$

Therefore

$$\Re(p(x),\,x,\,F[p(x)]) = M\Re'(p(x),\,x,\,F[p(x)]) \tag{2.113}$$

Furthermore a logical assumption relating to the physical breakage process would suggest that

$$\Re(p(x),\,x,\,F[p(x)]) = Mp(x)k(x,\,F[p(x)]) \tag{2.114}$$

$k(x, F[p(x)])$ is the specific rate of breakage of material of size x and it represents the breakage rate when 1 kg of material of size x is in the mill.

The average residence time in the mill is

$$\tau = \frac{M}{W} \tag{2.115}$$

where W is the flowrate through the mill.

Equation 2.111 becomes

$$- \tau\,\frac{d\kappa(x)p(x)}{dx} + 3\tau\kappa(x)\frac{p(x)}{x} + \tau\Re(p(x),\,x,\,F[p(x)])$$

$$- \tau\int_{R'(x)} \Re'(p(x'),\,x'\,F[p(x')])b(x;\,x')dx' - \tau\int_{R''(x)} 3\kappa(x')\frac{p(x')}{x'}a(x;\,x')dx'$$

$$= p_{in}(x) - p_{out}(x) \tag{2.116}$$

This is the fundamental population balance equation for any comminution process and it incorporates both autogeneous and media-induced breakage. It is difficult to solve this integro-differential equation principally because the functions $\Re$, b and a are strongly nonlinear. Under conditions that apply to practically useful situations, only numerical solutions are possible. These are developed for practical situations in Chapter 5.

Bibliography

The application of population balance methods to systems that process particles was described by Hulburt and Katz (1964) which is a good basic reference for population balance methods as they are applied to particulate systems.

Measurement of the distribution of particle composition by image analysis methods is described by Jones (1987) and Barbery (1991). The method used for the stereological transformation used here is based on an idea of Coleman (1988) and is described in detail by King and Schneider (1998).

References

Barbery, G. (1991) *Mineral Liberation. Measurement, Simulation and Practical Use in Mineral Processing*. Editions GB, Quebec.

Coleman, R. (1988) Least squares and its application to mineral liberation. *Arch. Mining Sci.*, Poland. Vol. 33, pp. 270–290.

Hulburt, H.M. and Katz, S. (1964) Some problems in particle technology. A statistical mechanical formulation. *Chemical Engineering Science*, Vol. 19, pp. 555–574.

Jones, M.P. (1987) *Applied Mineralogy: A Quantitative Approach*. Graham and Trotman, London.

King, R.P. and Schneider, C.L. (1998) Stereological correction of linear grade distributions for mineral liberation. *Powder Technology*, Vol. 98, pp. 21–37.

3
Mineral liberation

Recovery of minerals using ore dressing and concentration operations is based on methods that separate particles on the basis of their physical or chemical properties. Individual minerals can be separated completely only if each particle contains only one mineral. Two minerals in the same particle can never be separated using physical separation methods alone. Separating minerals at the particulate level is referred to as liberation since the individual minerals are liberated from each other in a physical way. In practice however, the comminution processes that are used to reduce mineralogical raw materials to the particulate state are, for the most part, unselective and, apart from a few unusual cases, the particles that are formed consist of mixtures of the mineral components that are present in the original ore.

During comminution there is, however, a natural tendency towards liberation and particles that are smaller than the mineral grains that occur in the ore can appear as a single mineral. This happens when the particle is formed entirely within a mineral grain. Obviously this will occur more frequently the smaller the particle size and it is impossible when the particle is substantially larger than the mineral grains in the ore. Typical patterns for the distribution of particles in the size-composition space are shown in Figure 2.11 in Chapter 2. Methods that can be used to model these distributions are presented in this chapter. These methods must necessarily be quite complex because the geometrical structure of any mineralogical material is not uniform and cannot be described by the familiar conventional regular geometrical entities such as spheres and cubes. Mineralogical textures have indeterminate geometries which are to a greater or lesser extent random in size, shape, orientation, and position. Likewise the particles that are generated by comminution operations are irregular in shape and size. Thus the particle population is made up of individuals that have irregular shapes and sizes and which are composed of material which itself has an irregular and complex texture of mineral phases. In spite of this lack of regularity, the distributions of particles with respect to composition do show some regular features particularly with respect to the variation of the distribution with particle size. An empirical but useful distribution is discussed in Section 3.1.

The mineralogical scale of a mineralogical texture is difficult to specify and often the different mineral components are present in the texture at vastly different size scales. In spite of this it is useful to regard each mineral as having a characteristic size that is commensurate with the size of individual grains in the texture. It is not possible to assign a definite value to this size because every mineral grain will be different in size and shape. Except in the most regular crystalline structures, it is not possible to assign a unique size to a grain of irregular shape. In spite of this, the concept of grain size is useful in that it provides some idea of the size to which the material must be reduced by comminution in order to achieve liberation of the phase. The concept of some characteristic grain size has been used in a semi-quantitative sense for many years in mineral processing and this concept is developed here by using a hypothetical liberation size as the main parameter in the development of a quantitative model for mineral liberation during comminution. This concept was introduced in response to the observation that, for many real ore textures, the distribution of particles over the grade range changes from a distribution that is concentrated around the average grade to one that shows considerable dispersion towards the liberated ends over a comparatively small particle size range. This effect is illustrated in Figure 2.11. This rapid change in liberation behavior is evidence that significant liberation occurs at a particle size that is characteristic for the mineral texture.

The models for liberation that are described in this chapter are specific to mineralogical textures that consist of only two minerals – a valuable species and all the other minerals that are present and which are classified as gangue minerals in any one analysis. Although the techniques that are used can be applied to multicomponent ores, the details of a suitable analysis are not yet worked out and they are not included here.

3.1 The beta distribution for mineral liberation

A useful distribution function is developed in this section for the description of the populations of particles that have variable mineral content. This distribution function is based on the beta distribution that is widely used in mathematical statistics.

When describing a population of particles that have a distribution of mineral content, four parameters at least are essential to provide a description of the population that can be usefully used in practice. These are the average grade of the mineral, the dispersion of particle grades about the average value and in addition the fractions of particles that contain only a single mineral – one parameter for each mineral. The latter two quantities are usually referred to as the liberated ends of the distribution. If these latter two quantities are not specified the information is seriously incomplete and the resulting liberation distribution is not very useful because it does not account for the most significant particles of all, namely, those that are liberated.

The following symbols are used to represent the four parameters.

$\bar{g}$ = average grade of mineral in the population (expressed as mass fraction)
σ_g = standard deviation about the mean in the population
L_0 = mass fraction of the population that consists of liberated gangue particles
L_1 = mass fraction of the population that consists of liberated mineral.

The particle population is conceived as consisting of three groups: liberated particles of gangue, liberated particles of mineral and the remainder of the particles which are all composed of mixtures of the two minerals. The distribution of mineral grades over the third group is called the interior grade distribution and the beta distribution function is used as a model. The distribution density for particle grade is given by

$$p(g) = (1 - L_0 - L_1) \frac{g^{\alpha-1}(1 - g)^{\beta-1}}{B(\alpha, \beta)} \quad \text{for } 0 < g < 1 \qquad (3.1)$$

α and β are two parameters that characterize the distribution and $B(\alpha, \beta)$ is the beta function. Two Dirac delta functions must be added at each end to complete this distribution.

Some useful properties of the beta distribution

$$\int_0^1 g^{\alpha-1}(1 - g)^{\beta-1} \, dg = B(\alpha, \beta) \qquad (3.2)$$

The mean of the interior distribution is given by

$$\bar{g}^M = \int_0^1 g \frac{g^{\alpha-1}(1 - g)^{\beta-1}}{B(\alpha, \beta)} \, dg = \frac{B(\alpha + 1, \beta)}{B(\alpha, \beta)}$$

$$= \frac{\alpha}{\alpha + \beta} \qquad (3.3)$$

The variance of the interior distribution is defined as

$$(\sigma^2)^M = \int_0^1 (g - \bar{g}^M)^2 \frac{g^{\alpha-1}(1 - g)^{\beta-1}}{B(\alpha, \beta)} \, dg \qquad (3.4)$$

The parameters α and β are related to the mean and the variance through the expressions

$$\alpha = \bar{g}^M \gamma \qquad (3.5)$$

and

$$\beta = (1 - \bar{g}^M)\gamma \qquad (3.6)$$

where γ is given by

$$\gamma = \frac{\bar{g}^M - (\bar{g}^M)^2 - (\sigma^2)^M}{(\sigma^2)^M} \qquad (3.7)$$

Both α and β must be positive numbers and therefore Equations 3.5 and 3.6 require $\gamma > 0$. This in turn imposes an upper limit on the variance of the distribution by Equation 3.7

$$(\sigma^2)^M < \bar{g}^M (1 - \bar{g}^M) \qquad (3.8)$$

The corresponding cumulative distribution is given by

$$P(g) = L_0 + (1 - L_0 - L_1)I_g(\alpha, \beta) \tag{3.9}$$

where $I_g(\alpha, \beta)$ is the incomplete beta function defined by

$$I_g(\alpha, \beta) = \frac{1}{B(\alpha, \beta)} \int_0^g x^{\alpha-1}(1-x)^{\beta-1}\,dx \tag{3.10}$$

The beta distribution is sufficiently flexible to represent real particle grade distributions in a realistic way. When $\alpha = \beta$ the distribution is symmetrical about the grade $g = 0.5$. In practice particle grade distributions in real materials are generally asymmetric because the average grade of the valuable mineral species is often quite low. The beta distribution function is shown in Figure 3.1 for a number of combinations of the parameters α and β. These distributions

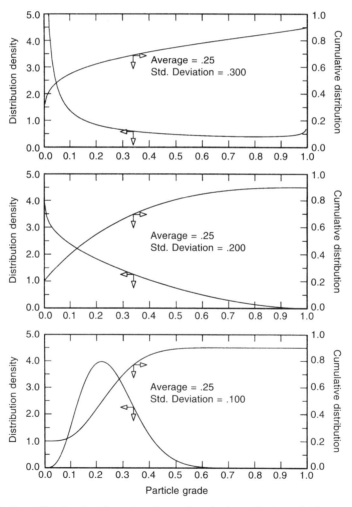

Figure 3.1 Beta distribution function for a description of mineral liberation. Three cases are shown with varying values of $(\sigma^2)^M$. The distribution density changes from bell-shaped to U-shaped as the variance increases

are shown as both distribution densities and cumulative distributions. On the cumulative distribution, the liberated mineral at each end of the distribution represented by the vertical discontinuities at $g = 0$ and $g = 1$. Note how the distribution density changes from bell-shaped to U-shaped as the variance of the inner distribution varies from 0.1 (lower graph) to 0.3 (upper graph). This is an essential requirement for the description of mineral liberation distributions for different particle sizes. When the particle size is distinctly larger than the sizes of the mineral grains within the ore, most particles in the population exhibit a mineral grade close to the mean value for the ore as a whole and the distribution is bell-shaped. On the other hand when the particle size is distinctly smaller than the sizes of the mineral grains, the tendency for liberated and nearly liberated particles to appear is greatly enhanced and the beta distribution function reflects this tendency by exhibiting a strong U-shape. In the limit, as the variance approaches its maximum value $\bar{g}^M(1-\bar{g}^M)$, the beta distribution can also describe a particle population that consists of completely liberated particles only. These properties of the beta distribution are exploited in the following sections to develop models for the liberation process.

In practice it is common to represent the liberated material in terms of the fraction of the available mineral that is liberated:

$$\mathcal{L}_1 = \bar{g}L_1$$

$$\mathcal{L}_0 = (1 - \bar{g})L_0$$

(3.11)

3.2 Graphical representation of the liberation distribution

The distribution density function that is described in Section 3.1 is not particularly useful for practical work. There is no convenient way of representing the liberated material at each end of the distribution density. This deficiency can be overcome by using either the cumulative distribution on which the liberated ends appear as vertical steps at each end of the graph or as a histogram that displays the distribution of particles in a finite number of grade classes. The three distributions in Figure 3.1 are shown as histograms in Figure 3.2.

Although the histogram gives a useful graphical representation of the liberation data, it cannot be used directly to make accurate estimates of the average grade of particles in the population. The usual formula

$$\bar{g} = \sum_{i=1}^{12} g_i p_i(g)$$

(3.12)

will often produce estimates that are significantly in error because of the difficulty in assigning appropriate values to the representative grade g_i of each grade class i. An alternative formula, which avoids this difficulty, is based on the cumulative distribution

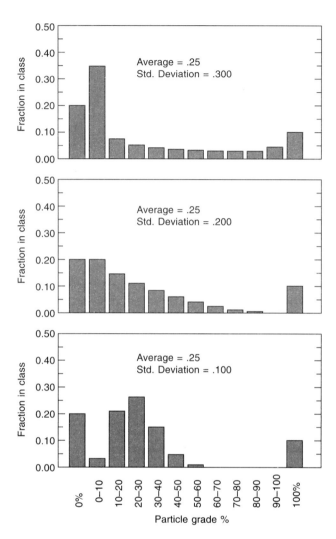

Figure 3.2 The liberation distributions of Figure 3.1 shown as histograms over the grade classes

$$\bar{g} = \int_0^1 gp(g)dg = 1 - \int_0^1 P(g)dg \qquad (3.13)$$

which is obtained using integration by parts. This integral can be evaluated numerically from a knowledge of the cumulation distribution, $P(g)$, at equidistant values of g. This can be generated easily from the histogram data as shown in Example 3.1.

Illustrative example 3.1
The histogram data for the distribution shown in the upper graph in Figure
3.2 is given in Table 3.1. Use Equations 3.12 and 3.13 to make estimates of the
average grade and compare these to the true average.

Table 3.1 Histogram data for the cumulative distribution shown in the top panel of Figure 3.2

G-class	Grade range	$p_i(g)$	$P(g)$	g_i
1	0	0.2000	0.2000	0.00
2	0–10	0.3478	0.5478	0.05
3	10–20	0.0737	0.6215	0.15
4	20–30	0.0512	0.6727	0.25
5	30–40	0.0410	0.7137	0.35
6	40–50	0.0351	0.7488	0.45
7	50–60	0.0315	0.7803	0.55
8	60–70	0.0292	0.8095	0.65
9	70–80	0.0280	0.8375	0.75
10	80–90	0.0282	0.8657	0.85
11	90–100	0.0343	0.9000	0.95
12	100	0.1000	1.0000	1.00

$$\bar{g} = \sum_{i=1}^{12} g_i p_i(g) = 0.285$$

The integral in Equation 3.13 can be conveniently evaluated using any equal-
interval quadrature formula such as Simpson's rule rather than the simple
Euler formula that is used in Equation 3.12. Using Simpson's rule the average
grade is calculated to be 0.281. These values can be compared to the correct
value of $(1 - 0.2 - 0.1)0.25 + 0.1 = 0.275$. The error in calculating the average
has been reduced from 3.8% to 2.2% using the more accurate formula. The
fractional liberation of the two minerals are

$$\mathcal{L}_1 = 0.275 \times 0.1 = 0.0275$$

and

$$\mathcal{L}_0 = (1 - 0.275)0.2 = 0.145$$

3.3 Quantitative prediction of mineral liberation

Minerals are liberated from the host rock of an ore by grinding and in many
mineral processing operations the sole purpose of any comminution operation
is mineral liberation. Grinding operations are usually unselective in the sense

that fractures that are induced in the rock show virtually no correlation with the underlying mineralogical texture of the ore. There are exceptions to this general statement and some of the more important exceptions are examined in some detail later in this chapter. The operations of crushing and grinding can be regarded conceptually as the superimposition of a vast network of fractures through the complex heterogeneous mineralogical texture of the ore. The network of fractures ultimately determines the distribution of particle sizes that are produced in the comminution equipment. The relationship between the network of fractures and the underlying mineralogical texture determines how the mineral phases are distributed among the particles in the population after fracture. This concept can be observed in the particle sections that are shown in Figure 2.9 where the separate mineral phases are displayed at different gray levels so that they can be distinguished easily.

The purpose of any mathematical model of the liberation process is the calculation of the liberation distribution that can be expected when mineral ores are subjected to typical comminution operations. Quantitative models that can be used for this purpose are not easy to develop and some specialized methods must be used. Any model for liberation must begin with some quantitative description of the mineralogical texture of the ore. This is no easy task because the ore texture is always complex in the geometrical sense. The most effective tool for the quantitative characterization of texture is image analysis. Samples of the ore are obtained and these are sectioned and polished to reveal a plane section through the material. The section is examined using optical or electron microscopy and many digital images are collected to represent the ore texture. These digital images are analyzed using specialized image analysis software as described in Section 3.3.1. Typical sections through two ores are shown in Figures 3.3 and 3.4.

3.3.1 Characterization of mineralogical texture by image analysis

Because the texture of ores are so irregular in their geometrical construction it is possible to characterize the texture only in terms of its statistical properties. This requires that many individual observations must be made and the relevant statistical properties of the texture must be estimated from these observations.

The most useful properties to measure are those that will give some measure of the nature of the particles that are formed during grinding. Unfortunately it is not possible to know in advance how any particular piece of ore will fragment during the comminution operations and the fragmentation can only be simulated. The analysis that follows makes use of the concepts of the conditional distributions that are discussed in Section 2.7.

A simulation of the fracture pattern is obtained from a sample of actual particles which have been mounted, sectioned, polished and imaged as shown for example in Figure 3.5. The size and shape of these particles are characterized by the distribution of linear intercepts which is determined by measuring the length of very many intercept lengths across the particles as shown in Figures

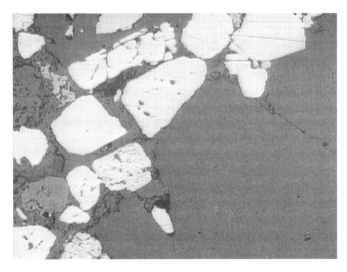

Figure 3.3 Typical image of a section through mineral-bearing ore. Bright phase is pyrite, gray phases are silicates

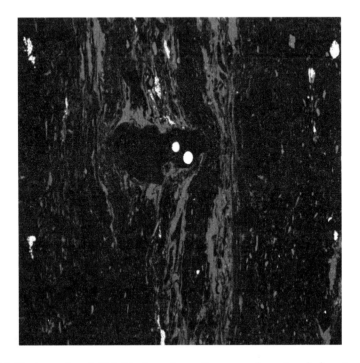

Figure 3.4 Texture of coal. Bright phase is pyrite, gray phase is ash-forming mineral matter and dark phase is the desired carbonaceous material

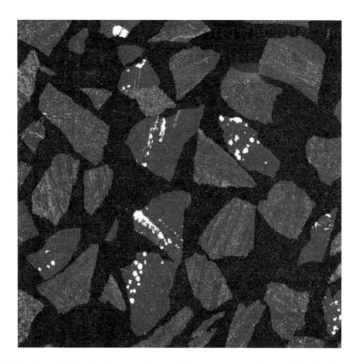

Figure 3.5 Coal particles that were formed by comminution of the texture shown in Figure 3.4

2.9 and 3.5. This linear intercept pattern characterizes the particle population and the linear intercept distribution can be used to generate the particle size distribution by solving the integral equation for the mesh size distribution density $p(D)$:

$$P(\ell) = \bar{\ell} \int_0^\infty P(\ell \mid D) \frac{p(D)}{\bar{\ell}_D} dD \qquad (3.14)$$

where $\bar{\ell}$ is the average intercept over the whole population and $\bar{\ell}_D$ is the conditional average intercept length for particles that have mesh size D. The kernel function $p(\ell \mid D)$ is the conditional distribution of linear intercept lengths for particles that have mesh size D. This kernel function can be measured experimentally but usually only for a sample of particles in a mesh size interval such as the standard $\sqrt{2}$ series. A function that has been found to be representative of particles that are typically found in the products of comminution operations is

$$P(\ell \mid \Delta D_R) = 1 - \left(1 - \frac{\ell}{1.2D_{i-1}}\right) \exp\left(-\frac{R^2\ell}{1.2D_{i-1}}\right) \quad \text{for } \ell \le 1.2D_{i-1}$$

$$= 1.0 \qquad\qquad\qquad\qquad \text{for } \ell > 1.2D_{i-1} \quad (3.15)$$

The conditioning variable ΔD_R indicates that this distribution applies to a sample of particles in a mesh size interval having $D_{i-1}/D_i = R$ rather than at a single size D.

When all particles are exactly of size D, $R = 1$ and

$$P(\ell \mid D) = 1 - \left(1 - \frac{\ell}{1.2D}\right) \exp\left(-\frac{\ell}{1.2D}\right) \tag{3.16}$$

It is usual to use distributions that are weighted by length

$$f(\ell) = \frac{\ell p(\ell)}{\bar{\ell}} \tag{3.17}$$

and

$$f(\ell \mid D) = \frac{\ell p(\ell \mid D)}{\bar{\ell}_D} \tag{3.18}$$

and Equation 3.15 becomes

$$F(\ell) = \int_0^\infty F(\ell \mid D) p(D) dD \tag{3.19}$$

which has the following useful kernel

$$F(\ell \mid D) = 1 - \exp\left[-\left(\frac{\ell}{0.772D}\right)^\lambda\right] \tag{3.20}$$

where the parameter λ is a function of the material and method of comminution.

The distributions given in Equations 3.15 and 3.16 are parameter free and can be used for many ores even when the linear intercept distribution for the particles is not available. When the conditional linear intercept distribution of the particle population is known or can be estimated, the prediction of the expected liberation characteristics of the ore can be calculated using the following straightforward method.

The image of the unbroken ore is sampled by superimposing linear samples drawn from the population defined by Equation 3.14. Each linear sample will cover one or more of the phases and consequently can be characterized by its linear grade g_L which is the fraction of its length that covers the mineral phase. The sample lengths can be sorted into sizes and the distribution of linear grades can be easily estimated for each different linear sample length which generates the conditional linear grade distribution $P(g_L \mid \ell)$ for the ore. This provides the necessary characterization of the ore texture from which its liberation characteristics can be calculated.

The images that are used must be larger than the largest dimension of any texture characteristic of the ore. This can be difficult to achieve if the size scales of the different minerals in the ore are widely different. The pyritic quartzite and coal shown in Figures 3.3 and 3.4 are typical examples. The pyritic quartzite includes large quartz pebbles which are completely barren and the dimensions of the carbonaceous phase in coal are very much larger than the pyrite and ash grains. The images must be collected at sufficiently high resolution to capture all of the essential features of the texture of the finest mineral grains and at the same time they must reveal the full texture of the largest mineral grains. Generally this is not possible in a single image since the field of view of any microscope is limited. This problem can usually be overcome by collecting images that are contiguous and then stitching these together to form a single long image from the sequence of smaller images. The linear sample lines can be sufficiently long to cover the largest feature in the texture while the full resolution of the original image is maintained. An example formed by stitching the pyritic quartzite images is shown in Figure 3.6.

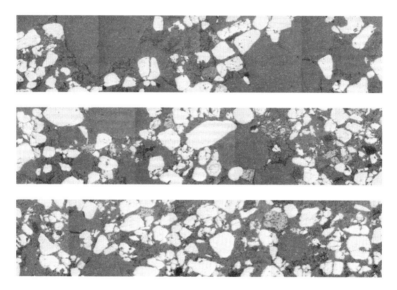

Figure 3.6 Stitched images of unbroken pyritic quartzite ore. Slight gray-level differences between contiguous images have been introduced to reveal where images have been stitched. Images were collected using optical microscopy at 2.5 μm per pixel and each strip shown is 2328 × 480 pixels

The three strips shown in Figure 3.6 actually formed one single strip for analysis and the longest linear sample that covered only one phase was found to be 1042 pixels in this sample. Normally about five strips of stitched images are analyzed to characterize a single polished section and many polished sections are collected to ensure that the ore is adequately sampled. The linear sampling lines are laid down on the image in a head-to-tail pattern to ensure that each portion of the texture is sampled only once since no point in the original ore texture can appear in more than one particle after comminution.

The liberation distribution that can be expected when this ore is comminuted can be calculated using the following two-step procedure. In the first step the measured distribution of linear grades from the linear samples are combined with the linear intercept distribution density $p(\ell \mid D)$ of the particles that are expected to be formed during comminution of the ore. Equations 3.15, 3.16 or 3.20 would be suitable models for the latter distribution. This produces the distribution of linear grades in particles of size D from the equation

$$P(g_L \mid D) = \int_0^\infty P(g_L \mid \ell) p(\ell \mid D) d\ell \qquad (3.21)$$

This equation represents exactly the measurement of the linear grade distribution by sampling particle sections from size D as described in Section 2.7.2. The result of applying this transformation to the distribution of linear grades measured on the texture shown in Figure 3.6 is given in Figure 3.7.

Equation 3.21 implies that $P(g_L \mid \ell, D)$ is independent of the particle size. This is valid provided that the fracture process is random and independent of the texture. The problem is considerably more complex when non-random fracture patterns occur. Non-random fracture is discussed in Section 3.5.

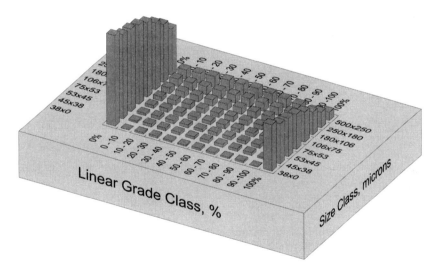

Figure 3.7 Calculated distribution of linear intercepts as a function of particle size for texture shown in Figure 3.6

In the second stage of the calculation, the distribution of linear grades that was calculated using Equation 3.21 is transformed stereologically to generate the distribution of grades in the real three-dimensional particles. This is identical to the problem discussed in Section 2.7.3 for stereologically transforming measured linear grade distributions. The solution method is based on generating solutions to the integral equation

$$P(g_L \mid D) = \int_0^\infty P(g_L \mid g, D) p(g \mid D) dg \qquad (3.22)$$

Unlike Equation 3.21, Equation 3.22 cannot be solved in a straightforward manner because the required solution is the function $P(g \mid D)$ that appears under the integral in the right hand side. However, appropriate solution methods are available and the solution can be generated easily using a number of computer software packages that are readily available as discussed in Section 2.7.3. The result of applying the stereological transformation to the data shown in Figure 3.7 is given in Figure 3.8. The difference in the 'liberation sizes' of the pyrite and silicates is clearly evident. Liberated silica particles can be as large as 250 µm while the largest liberated pyrite particle is no larger than 106 µm.

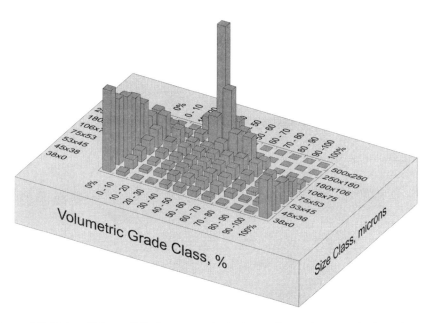

Figure 3.8 Data of Figure 3.7 after stereological transformation to calculate the predicted liberation as a function of particle size after comminution

3.4 Simulating mineral liberation during comminution

3.4.1 *Boundaries of the Andrews-Mika diagram*

The method that was described in Section 3.3 can be used to predict the liberation distribution that will result when an ore is broken in a random fashion and no other processes are involved that will change the make-up of the particle population. Processes such as classification and concentration can have a large effect on the relative distribution of particle types. In particular there must be no preferential selection of particle types during the entire comminution process. These conditions are normally not met in ball mill circuits which usually process feed that includes cyclone underflow and often other concentrates and tailing streams in addition to fresh feed. This complicates the calculation of the mineral liberation during comminution and it is necessary to apply population balance techniques which can describe the fracture of multicomponent particles on a class-by-class basis. In this way the grade classes can be accurately accounted for among the breaking particles and evolution of the liberation distribution can be calculated at all stages in the comminution process.

The effect of the mineral liberation phenomenon is described in the population balance models through the breakage matrix $b(x; x')$ in Equation 2.116. This matrix describes how particles transfer among the internal coordinates during comminution. Two internal coordinates, particle size and particle grade, are required to describe populations of multicomponent particles. The matrix $b(x; x')$ is difficult to model in any particular case because it depends on both the liberation characteristics of the material, which are governed primarily by mineralogical texture, and also by the characteristics of the comminution machine. In spite of the complexities, useful models have been developed and these have proved to be successful in simulations of complex milling and concentration circuits. The approach that is used here is based on a graphical representation of the breakage process called the Andrews-Mika diagram. These diagrams show what type of progeny particles will be generated when a single parent particle breaks in a milling environment. The diagrams are useful because the graphical representation is fairly easy to interpret and the diagrams explicitly include a number of essential constraints that must be applied to the breakage process in the population balance modeling method.

Particle size and particle grade must be included among the internal coordinates to describe mineral liberation and the vector x of internal coordinates is written for a two-component ore as

$$x = (g, d_p) \tag{3.23}$$

Here d_p represents the particle size and g the particle grade. The breakage matrix $b(x; x')$ can be written as

$$b(x; x') = b(g, d_p; g', d_p') \tag{3.24}$$

to reflect the two variables that make up the internal coordinates. This function can be simplified significantly using the rules of conditional distributions discussed in Section 2.7

$$b(g, d_p; g', d_p') = b(g \mid d_p; g', d_p')b(d_p; g', d_p') \tag{3.25}$$

To a large extent this decouples the breakage process from the liberation process and the two breakage functions can be modeled separately and then subsequently combined using Equation 3.25 to generate the transformation function $b(g, d_p; g', d_p')$ which is necessary to apply the population balance method. The function $b(d_p; g', d_p')$ can, in most cases, be assumed to be independent of the grade g' of the parent particle.

$$b(d_p; g', d_p') = b(d_p; d_p') \tag{3.26}$$

This is equivalent to asserting that the breakage function is independent of the grade of the parent particle. This appears to be a realistic assumption in most practical cases. Models for the breakage function $b(d_p; d_p')$ are discussed in Section 5.4. The function $b(g \mid d_p; g', d_p')$ shows how the fracture products of a parent particle of size d_p' and grade g' are distributed with respect to grade for every possible progeny size. A useful practical model for this function is developed in this section.

A geometric approach is taken and the nature of the function is described by reference to the two-dimensional $g - d_p$ plane as shown in Figure 3.9. Any particle is located in this plane according to the values of its mineral grade g and its size d_p. We wish to investigate how the progeny particles of a single parent will dispose themselves in this two-dimensional phase space after breakage when the location of the parent particle is known. Of course this can be determined only in a probabilistic sense because the fracture process is random in character. Thus the liberation distributions for progeny particles from a parent at point g', d_p' is required. To use the simplification that is inherent in Equation 3.25, the distributions that are conditional on the progeny

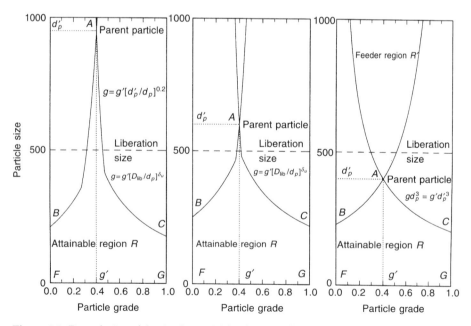

Figure 3.9 Boundaries of the Andrews-Mika diagram for the three cases: parent particle size > liberation size, parent particle size ≈ liberation size, parent particle size < liberation size

size, d_p, are calculated. This gives the function $b(g \mid d_p; g', d'_p)$ which can be combined with the breakage function $b(d_p; d'_p)$ in Equation 3.26 for use in the population balance Equation 2.116. A geometrical model is developed here for $b(g \mid d_p; g', d'_p)$ in terms of a comparatively small number of parameters that can be related to features of the mineralogical texture of the parent particle.

Consider a parent particle at point A in Figure 3.9. When this particle is broken, the progeny particles cannot appear at every point in the phase space. They are restricted by definite physical constraints. Clearly progeny can appear only below point A because every progeny particle must be smaller than the parent. In addition the progeny particles must satisfy the law of conservation of phase volume. This means that no progeny particle can contain a greater volume of mineral than the volume of mineral in the parent particle. Likewise no progeny particle can contain more gangue than was present in the parent particle. These restrictions lead to the two inequalities

$$g v_p \leq g' v'_p \qquad (3.27)$$

and

$$(1 - g) v_p \leq (1 - g') v'_p \qquad (3.28)$$

where v_p is the volume of the progeny particle and v'_p the volume of the parent particle. Since particles that are produced during comminution are more or less geometrically similar, these inequalities can be written approximately as

$$g d_p^3 \leq g' d'^3_p \qquad (3.29)$$

and

$$(1 - g) d_p^3 \leq (1 - g') d'^3_p \qquad (3.30)$$

These inequalities define two regions R and R' in the phase space like those shown in the right hand panel of Figure 3.9. The diagrams shown in Figure 3.9 are called Andrews-Mika diagrams in recognition of the first paper that described the boundaries and explored their significance.

The region R, called the attainable region, shows where progeny particles can appear after fracture of a parent at point A. In particular the liberated edges of the region, represented by BF and CG in the diagram, are included in the attainable region because an unliberated particle can generate completely liberated particles. However, it is physically impossible for a parent at A to generate any liberated particles larger than d_{pC} or liberated gangue particles larger than d_{pB}. The region R' is complementary to R and represents the region of phase space that contains parent particles that could produce a progeny at point A. This region is called the feeder region for progeny at A. Unlike the attainable region, the boundaries of the feeder region neither intersect nor touch the vertical sides of the diagram at $g = 0$ and $g = 1$ because of the obvious fact that an unliberated progeny can never be produced from a liberated parent. The feeder region is important because it appears in the fundamental population balance Equation 2.110 for comminution machines and it is necessary to establish the boundaries of R' to define the region of integration in that equation. The shape of the boundaries of R and R' vary considerably depending on the location of the parent particle in the phase space.

When the parent particle is significantly smaller than the average 'grain size' of the mineral in the ore, the parent will appear under the microscope as shown in image D of Figure 3.10. Under these conditions the parent will typically contain only a single region of mineral and a single region of gangue and inequalities 3.29 and 3.30 are realistic bounds for the regions R and R'.

This is not true when the parent particle is larger than the mineral 'grain size'. Two other cases can be distinguished: the parent size is comparable to the mineral 'grain size' and the situation when the parent size is much larger than the mineral 'grain size'. These situations are illustrated in images B and C respectively in Figure 3.10, which are from the same material but were made from particles of different size and magnification. The texture appears to change from size to size but it is only the apparent texture that changes. The overall texture is of course the same. Textures that display this characteristic are called non-fractal to distinguish them from synthetic textures made using fractal generators and which appear to be qualitatively similar no matter what scale and magnification is used to observe them. If the texture were truly fractal, each apparently uniform mineral grain in Figure 3.10 would appear to be fragmented and made up of much smaller grains.

These geometrical properties of textures affect the boundaries of the feeder and attainable regions. This can be seen most easily by considering point C in the right hand panel of Figure 3.9 which is drawn at the intersection of the boundary of the attainable region and the line $g = 1$. This point reflects the principle that the largest completely liberated progeny particle cannot exceed the total size of the mineral phase in the parent. If the parent particle contains only a single mineral grain as is suggested, for example, by the lower image in Figure 3.11, the largest grain is proportional to $g'd_p'^3$ which is reflected in inequality 3.29. However, when the parent particle contains several separate mineral grains as shown in the upper diagram in Figure 3.11, the intersection of the attainable region boundary with $g = 1$ must reflect the principle that the size of the largest liberated mineral particle in the progeny population cannot exceed the size of the largest single coherent mineral grain in the parent particle. If there is more than one grain in the parent, the size of the largest single grain must be smaller than $g'd_p'^3$ and the upper boundary of the attainable region must fall below $g'd_p'^3$. This is illustrated in the left hand and center panels of Figure 3.9. Similar arguments can be made for the gangue phase, which is reflected in the left hand boundaries of the attainable regions shown in Figure 3.9. Boundaries for the feeder and attainable regions for this situation can be defined by analogy to inequalities 3.29 and 3.30 as

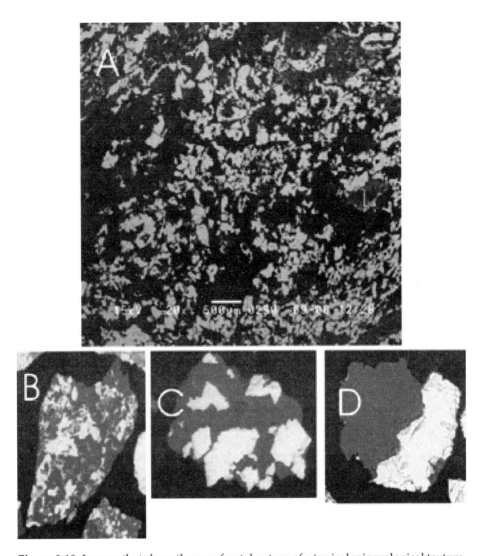

Figure 3.10 Images that show the non-fractal nature of a typical mineralogical texture. A is a section through the unbroken ore. B is a section through a single particle from the 950–1180 µm screen fraction at 20× magnification. C is from the 212–300 µm screen fraction at 70× magnification. D is from the 53–75 µm screen fraction at 300× magnification

$$gd_p^\delta \leq g'd_p'^\delta$$
$$(1 - g)d_p^\delta \leq (1 - g')d_p'^\delta$$

(3.31)

with $\delta \leq 3$. The exponent δ varies with parent size and a model for this variation that has been found to be useful in practice is

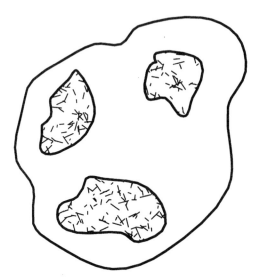

Conservation of total volume

Volume of the largest liberated mineral progeny particle cannot exceed the total volume of all mineral grains in the parent particle.

Conservation of grain volumes

The volume of the largest liberated mineral particle cannot exceed the volume of the largest single mineral grain in the parent particle.

When the parent particle is much larger than the mineral grain size, the boundaries of the Andrews-Mika diagram are controled by the grain volume constraint.

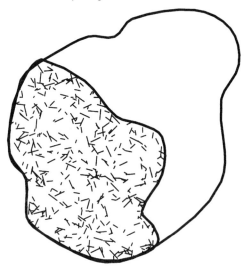

Volume of a single grain of mineral in the parent is the same as the total volume of mineral.

When the parent particle is much smaller than the mineral grain size, both conservation principles give the same limit for the maximum liberated particle size.

Figure 3.11 Images showing the non-fractal nature of mineralogical textures

$$\delta = \min\left[\delta_0\left(\frac{D_{\text{lib}}}{d_p'}\right)^x, 3\right]$$

(3.32)

for $d_p' \leq D_{\text{lib}}$ where D_{lib} is a parameter referred to as the liberation size for the ore. It is roughly equal to the particle size at which the mineral starts to liberate significantly when the ore is comminuted. x is a parameter which is approximately 0.5. The upper limit of $\delta = 3$ represents the law of conservation of phase volume and must never be exceeded.

The symmetry of the texture is also an important factor in fixing the boundaries of the attainable and feeder regions. A symmetrical texture is one in which the different mineral phases cannot be distinguished from geometrical factors alone. The most common asymmetric texture is one in which distinct grains of one mineral are imbedded in a more or less continuous phase of another mineral. To account for this effect the exponent δ can be different for the mineral and the gangue. Thus inequalities 3.29 and 3.30 are written

$$gd_p^{\delta_u} \leq g'd_p'^{\delta_u}$$
$$(1-g)d_p^{\delta_l} \leq (1-g')d_p'^{\delta_l}$$

(3.33)

The asymmetry factor for the texture is the ratio δ_u/δ_l with $\delta_u\delta_l = \delta$.

When the parent size is much greater than the liberation size D_{lib}, and when the progeny particles are also larger than D_{lib}, these will retain the parent composition or at most will be only slightly less or slightly greater than the parent composition. The attainable region is narrow for these progeny and useful approximations to the boundaries are shown in the left hand panel of Figure 3.9 and are given by

$$g_u \leq \max\left[g'\left(\frac{d_p'}{d_p}\right)^{0.2}, g'\left(\frac{D_{\text{lib}}}{d_p}\right)^{\delta_u}\right]$$

(3.34)

$$1 - g_l \leq \max\left[(1-g')\left(\frac{d_p'}{d_p}\right)^{0.2}, (1-g')\left(\frac{D_{\text{lib}}}{d_p}\right)^{\delta_l}\right]$$

(3.35)

In spite of the empirical nature of these relationships, the parameters can be estimated reliably from batch comminution tests on two-component ores.

3.4.2 Internal structure of the Andrews-Mika diagram

The boundaries of the Andrews-Mika diagram are not sufficient to define how progeny particles are distributed when a parent of given size and grade is broken. The internal structure of the diagram contains this information and a useful model for this internal structure is developed here which is based on the beta distribution.

The model is based on the assumption that the progeny particle population at any size will have a beta distribution with respect to particle grade. For progeny having size not much less than the parent, the distribution will be bell-shaped around the parent particle grade and as the progeny size gets further below the parent size, the particles show successively more liberation and the distribution exhibits the characteristic U-shape with more and more particles appearing at or near the liberated ends of the distribution. The beta distribution provides a good model for this type of behavior and the change from bell-shaped to U-shaped distribution is modeled by a steadily increasing variance as the progeny size decreases.

The beta distribution itself does not account for completely liberated material at $g = 0$ and $g = 1$. At any horizontal level in the Andrews-Mika diagram where the boundaries of the attainable region are inside the limits $g = 0$ and $g = 1$, as shown for example by line A–B in Figure 3.12, there are no liberated particles and the total distribution is given by the inner beta distribution alone which extends from A to B. At progeny particle sizes where at least one of the boundaries of the attainable region is a vertical edge at $g = 0$ or $g = 1$, liberated particles are present in the population. It is necessary to calculate what fraction of the particle population appears as liberated particles at grades $g = 0$ and $g = 1$.

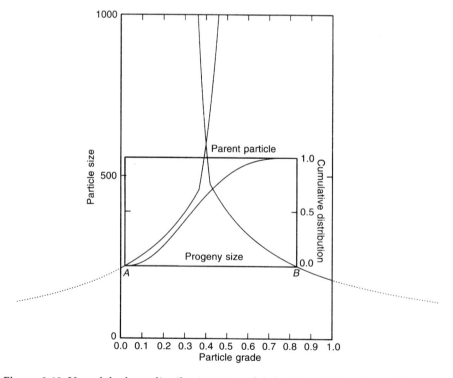

Figure 3.12 Use of the beta distribution to model the internal structure of the Andrews-Mika diagram when neither mineral is liberated

A method to accomplish this is based on the following construction. The curved boundaries of the attainable region are extended out past $g = 0$ and $g = 1$ as shown by the broken lines in Figures 3.13 and 3.14. At any progeny size, the grade distribution is assumed to be beta distributed from one extended boundary to the other but the distribution is truncated at $g = 0$ and $g = 1$ with all the probability mass in the distribution from $g = g_l$ (point A) to $g = 0$ concentrated at $g = 0$ and all the probability mass in the distribution from $g = 1$ to $g = g_u$ (point B) concentrated at $g = 1$. This represents the liberated gangue and liberated mineral respectively. The liberated gangue, L_0, is given by the length of the line CD in Figures 3.13 and 3.14. The liberated mineral is given by the length of the line FE in Figure 3.14. The remainder of the particle population is then distributed internally as a beta distribution as described in Section 3.1 There is an upper limit to the amount of liberated mineral or liberated gangue in the particle population. This is clearly the amount of mineral or gangue respectively in the original parent particle. If the construction described above produces a value larger than either of these it is assumed that the liberation is complete and that the internal distribution is empty.

The internal distribution will not extend over the entire range from $g = 0$ to $g = 1$ if one or both of the boundaries are inside this range at the progeny

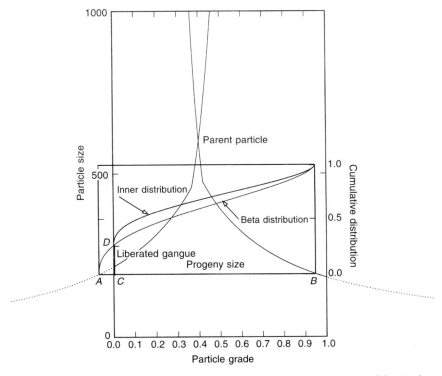

Figure 3.13 Use of the beta distribution to model the internal structure of the Andrews-Mika diagram when the gangue phase is liberated but the mineral is not liberated

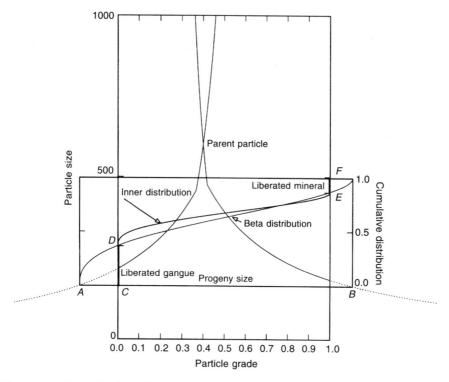

Figure 3.14 Use of the beta distribution to model the internal structure of the Andrews-Mika diagram when both the gangue phase and the mineral phase are liberated

size in question as shown for example in Figures 3.12 and 3.13. This situation is handled by noting that the transformed variable

$$\chi(g) = \frac{g - g_l}{g_u - g_l} \tag{3.36}$$

has a beta distribution over the range $\chi = 0$ to $\chi = 1.0$ with variance σ^2.

The variance of the conditional grade distribution also increases as the progeny size decreases reflecting the greater tendency for liberation as the progeny becomes smaller. The internal distribution changes from bell-shaped to U-shaped as the variance increases.

The variance of the distribution is also influenced by the parent grade. When the parent particle has a grade close to 0.5, the grades of the progeny can spread more easily over the range $g = 0$ to $g = 1$ than when the parent particle is closer to either end of the grade scale. The variation in variance of the progeny distribution with respect to both parent grade and progeny size is modeled by

$$\sigma^2 = g'(1 - g')f(d_p) \tag{3.37}$$

with

$$f(d_p) = \cfrac{1}{1 + \left(\cfrac{d_p}{D_{lib}}\right)^{\lambda}} \qquad (3.38)$$

The function $f(d_p)$ is S-shape in the range 0 to 1 as d_p varies from large sizes to very small sizes. This function has the greatest variation when d_p is close to the mineral liberation size which reflects the tendency of the mineral to liberate over a comparatively narrow range of sizes. This phenomenon has often been observed in typical ore textures.

Substitution of Equation 3.37 into Equation 3.7 gives

$$\gamma = \frac{1 - f(d_p)}{f(d_p)} \qquad (3.39)$$

The parameters of the internal distribution must be chosen so that the average grade of the conditional distribution at every progeny size is equal to the parent grade:

$$\bar{g}^M = \int_{g_l}^{g_u} g p(g \mid d_p; g', d_p') dg = \frac{g' - L_1}{1 - L_1 - L_0} \qquad (3.40)$$

The two parameters of the internal distribution are calculated from $\alpha^M = \bar{g}^M \gamma$ and $\beta^M = (1 - \bar{g}^M)\gamma$. In fact, Equation 3.40 holds only if the fracture process is random and no preferential breakage of one or other of the mineral phases occurs. Preferential breakage is discussed in Section 3.5.

3.5 Non-random fracture

When the fracture pattern that is developed in the ore during comminution is not independent of the mineralogical texture, the fracture process is considered to be non-random. Significant progress has been made on quantifying the effects of non-random fracture when particle fracture is influenced by the mineralogical composition and the texture of the parent particle. Six separate non-random fracture effects have been identified when multi phase particles are broken:

1. *Selective breakage.* This occurs when the different mineral phases have unequal brittleness. The more brittle mineral fractures more easily, and accordingly particles that have a larger content of the more brittle phase will have a larger specific rate of breakage. This phenomenon does not influence the nature of the Andrews-Mika diagram directly and is described entirely through the breakage rate parameters in the population balance equation. The parameter k in Equation 5.94 must be considered to be a function of both the particle size and the grade of the particle. This is discussed further in Section 5.8.3.

2. *Differential breakage*. This occurs when the conventional breakage function depends on the composition of the parent particle. In other words, the size distribution of the progeny from a single breakage event will be influenced by the composition of the parent particle.

3. *Preferential breakage*. This occurs when crack branching occurs more frequently in one of the mineral phases. The most obvious manifestation of preferential breakage is a variation of average composition with particle size in the particle populations. This is often noticed in practice and is usually comparatively easy to detect and measure. Good models are available to incorporate preferential fracture into the Andrews-Mika diagram and comparisons with experimental observations have been satisfactory.

4. *Phase-boundary fracture*. This occurs when cracks have a significant tendency to move along interphase boundaries rather than across the phases. Phase-boundary fracture destroys interphase boundary area in the progeny particles, and this should be measurable by careful image analysis. However, no convincing evidence of significant phase-boundary fracture has been reported in the literature.

5. *Liberation by detachment*. This occurs when mineral grains are comparatively loosely bonded into the ore matrix. Mineral grains become detached from the ore during comminution, which leads to significant and clean liberation of the mineral phases. This is unfortunately a fairly rare phenomenon and is not often encountered with large ore bodies.

6. *Boundary-region fracture*. This occurs when the highly stressed region in the neighborhood of the boundary between two dissimilar minerals is preferentially fractured. This leads to the production of comparatively more smaller particles from the phase-boundary region and therefore to less liberation among finer particles than among coarser particles which originate preferentially from the interior of the mineral phases. Although there is some direct evidence that boundary-region fracture does occur, this phenomenon is usually inferred from indirect observations on the behavior of the particles in concentration plants. For example, preferential fracture in the neighborhood of a minor mineral component will lead to a more than proportionate showing of that mineral on the surfaces of the particles. This can have a profound effect on the response of the particles to flotation. This is a complex phenomenon and no quantitative models have yet been developed to describe the effect convincingly.

Only the first three of these six non-random fracture effects have been modeled successfully and the last three require considerably more research. Selective breakage can be simulated by determining the brittleness of each mineral species and modeling the selection function in terms of the brittleness ratio β as follows

$$k_0(g, d_p) = \frac{2(g + (1 - g)\beta)k_0(d_p)}{1 + \beta} \tag{3.41}$$

In Equation 3.41, $k_0(d_p)$ is the specific rate of breakage as a function only of the size of the parent particle. Models for $k_0(d_p)$ are discussed in Section 5.12. The brittleness ratio can be measured using micro indentation techniques on polished sections of the minerals.

When differential breakage occurs the approximation of Equation 3.26 cannot be used and the size breakage function and a collection of breakage functions for each parent composition must be used. Such breakage functions can be measured in the laboratory using single impact or slow compression testing devices.

If one of the mineral phases breaks preferentially it will end up in greater concentration in the finer sizes after comminution. Although it is not possible to predict this phenomenon from a study of the mineralogy, it can be observed easily in size-by-size assays of a simple laboratory batch mill test. This will reveal the systematic variation of mineral content with size. Preferential breakage has a strong influence on the internal structure of the Andrews-Mika diagram. Equation 3.40 can be relaxed and replaced with the requirement that the average mineral grade over the total particle population must be conserved. Thus

$$\int_0^{d_p'} \int_0^1 gb(g, d_p | g', d_p') dg dd_p = g' \tag{3.42}$$

for every combination of parent grade and size. Using Equation 3.25 this restriction becomes

$$\int_0^{d_p'} b(d_p | g', d_p') \int_0^1 gb(g | d_p; g', d_p') dg dd_p = g' \tag{3.43}$$

The preferential breakage is modeled by finding how the conditional mean

$$\bar{g}(d_p; g', d_p') = \int_0^1 gb(g | d_p; g', d_p') dg \tag{3.44}$$

of the progeny particles varies with their size and parent composition. A model that has been found to satisfy data for several ores is

$$\bar{g}(d_p; g', dp') = g' \pm \varphi g'(1 - g')v\left(\frac{d_p}{d_p'}\right) \tag{3.45}$$

with

$$\int_0^{dp'} b(d_p | g', d_p')v\left(\frac{d_p}{d_p'}\right) dd_p = 0 \tag{3.46}$$

Provided that the function $v(d_p/d_p)$ satisfies Equation 3.46, the conservation constraint will be satisfied. The positive sign is used in Equation 3.45 when the mineral fractures preferentially and the negative sign is used when the gangue fractures preferentially.

A function which satisfies this requirement and which is suggested by typical experimental data is

$$v(u) = u^\alpha \ln(u) \tag{3.47}$$

with

$$\alpha = -\frac{1}{\ln(u^*)} \tag{3.48}$$

$$u = \frac{1 - \dfrac{d_p}{d_p'}}{1 - \Delta_0} \tag{3.49}$$

This model for preferential breakage has two parameters. Values for these parameters must be found from experimental data. Δ_0 is a parameter that is adjusted to satisfy Equation 3.46. u^* can be estimated directly from a plot of mineral grade against particle size in the comminution product. Usually such a plot shows a clear maximum or minimum. u^* is the value of u at the maximum or minimum. The data shown in Figure 3.15 demonstrates this.

3.6 Discretized Andrews-Mika diagram

The models for the boundaries and internal structure of the Andrews-Mika diagram that are developed in the preceding sections are based on the continuous beta distribution. They are not particularly useful for calculations in this form and discrete versions are developed here for use in the discretized population balance models for ball mills that are described in Chapter 5. It is common practice to classify the grade variable into 12 or 22 classes over the range [0, 1]. Twelve classes allow one class at each end of the distribution for liberated mineral and for liberated gangue and 10 intervals in (0, 1). These are often of equal width but this is not essential. Twenty-two classes allow for 20 internal classes which are usually 0.05 grade units in width. The conditional discrete distribution can be obtained from the cumulative conditional distribution using the following equations

$$p_i(g_i|d_{pj}; g_k', d_{pl}') = P(G_{i+1}|d_{pj}; g_k', d_{pl}') - P(G_i|d_{pj}; g_k', d_{pl}')$$

$$\tag{3.50}$$

$$= (1 - L_0 - L_1)\Big(I_{\xi(G_{i+1})}(\alpha^M, \beta^M) - I_{\xi(G_i)}(\alpha^M, \beta^M)\Big)$$

G_i is the grade at the boundary between the $i - 1$ and i grade class.

A shorthand notation is used to describe the discrete versions of these

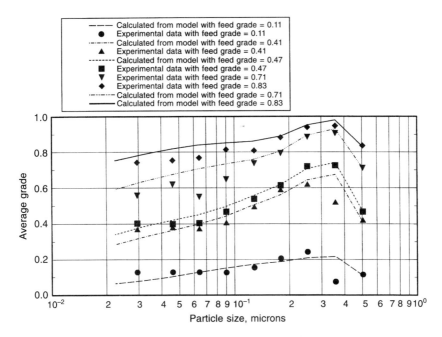

Figure 3.15 Data showing variation of grade with particle size when preferential breakage occurs

distribution functions.

$$a_{i,jkl} = p_i(g_i | d_{pj}; g'_k, d'_{pl})$$ (3.51)

The coefficients $a_{i,jkl}$ make up a discrete version of the attainable region of the Andrews-Mika diagram and examples are shown in Figures 3.16 and 3.17. Each bar in the attainable region shows what fraction of the parent particle transfers to each progeny class on breakage. The bar heights in the diagram are normalized so that the total height of all bars at a particular size is unity. This is consistent with the use of the conditional distributions in the continuous models described in Sections 3.4 and 3.5.

When this model for mineral liberation is inserted into the population balance equations for the ball mill, the feeder region for each point in the phase space rather than the attainable region from each parent location is required. The models developed in Sections 3.4 and 3.5 provide a logical and consistent description of the attainable region and it would be much more difficult to develop models for the feeder region. It is simpler to generate the feeder region from the known structure of the attainable region. Let $b_{i,jkl}$ be the fraction of size l to size j transfers that appear in grade class i when the parent is in grade class k. The relationship between the attainable and feeder regions is

$$b_{i,jkl} = a_{k,jil} \qquad (3.52)$$

Thus to construct a complete feeder region for any progeny particle, a complete set of attainable regions must be available for all possible parents.

Typical examples of the discretized Andrews-Mika diagram are shown in Figures 3.16 and 3.17. It is important to realize that these represent just two of the many discrete Andrews-Mika diagrams that are required to characterize any particular ore. Figures 3.16 and 3.17 show a discretization over 19 size classes and 12 grade classes which requires $19 \times 12 = 228$ separate Andrews-Mika diagrams, one for each possible combination of k and l. In general a theoretical model of the Andrews-Mika diagram is required to generate the appropriate matrices which can be stored before the operation of the comminution equipment can be simulated.

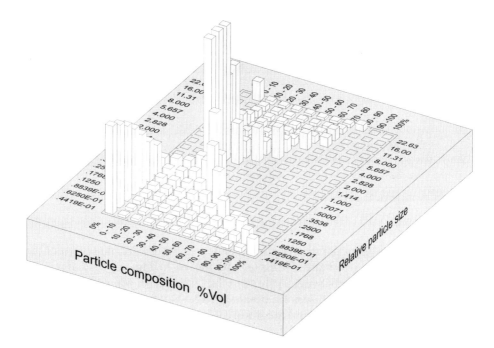

Figure 3.16 Internal structure of a typical Andrews-Mika diagram showing both the feeder and attainable regions. The feeder region is indicated by the shaded bars in the upper half of the diagram and the attainable region by the unshaded bars in the lower half. The height of each bar in the feeder region represents the conditional multi-component breakage function $b_{4,10kl}$ where k and l represent any parent bar in the feeder region. The height of each bar in the attainable region represents the value of $a_{m,n4\ 10}$.

Illustrative example 3.2

Calculate the liberation distribution of progeny particles when parent particles having representative size 3.394 mm and representative grade = 0.504 are

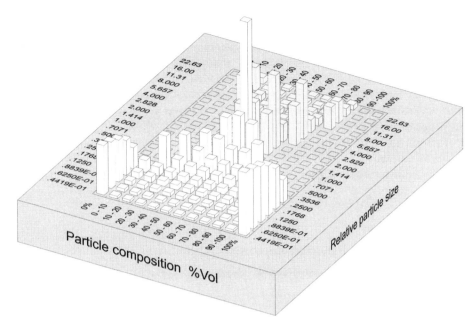

Figure 3.17 Internal structure of a typical Andrews-Mika diagram showing both the feeder and attainable regions. The feeder region is indicated by the shaded bars in the upper half of the diagram and the attainable region by the unshaded bars in the lower half. The height of each bar in the feeder region represents the conditional multi-component breakage function $b_{7,10kl}$ where k and l represent any parent bar in the feeder region. The height of each bar in the attainable region represents the value of $a_{m,n7\ 10}$.

broken by random fracture. Assume that the texture is symmetrical and that the liberation size parameter for the ore is 200 μm. Consider two cases: progeny size is 212 μm which is close to the liberation size, and progeny size = 53 μm which is considerably smaller than the liberation size.

Values for other parameters in the model are $\delta_0 = 2.0$, $x = 0.5$, asymmetry factor = 1.0 and $\lambda = 1.0$.

Case 1: Progeny size = 212 μm

Calculate the boundaries of the Andrews-Mika diagram.

$$\delta = \min\left[2.0\left(\frac{200}{3394}\right)^{0.5}, 3\right] = 0.243$$

Texture is symmetrical, so $\delta_\ell = \delta_u = \delta = 0.243$

The upper boundary is calculated using Equation 3.34

$$g_u = \max\left[0.504\left(\frac{3394}{212}\right)^{0.2}, 0.504\left(\frac{200}{212}\right)^{0.243}\right]$$

$$= \max [0.878, 0.497]$$

$$= 0.878$$

The lower boundary is calculated using Equation 3.35

$$1 - g_\ell = \max \left[0.496 \left(\frac{3394}{212} \right)^{0.2}, 0.496 \left(\frac{200}{212} \right)^{0.243} \right]$$

$$= \max [0.864, 0.489]$$

$$g_\ell = 0.136$$

Both boundaries lie within the range $g = 0$ to $g = 1$, so that the variable $\chi = \dfrac{g - g_\ell}{g_u - g_\ell}$ has a Beta distribution. Thus the cumulative conditional distribution for progeny particles of size 212 μm is

$$P(g \mid 212; 0.504, 3394) = 0.0 \text{ for } g \leq g_\ell$$

$$= I_{x(g)}(\alpha, \beta) \quad \text{for } g_\ell < g < g_u$$

$$= 1.0 \qquad \text{for } g \geq g_u$$

The parameters α and β in this distribution are calculated from

$$\alpha = \chi(g)\gamma$$

$$= \frac{0.504 - 0.136}{0.878 - 0.136} \times \gamma$$

$$= 0.496\,\gamma$$

$$\beta = (1 - \chi(g'))\gamma$$

$$= 0.504\,\gamma$$

The parameter γ determines the variance of the distribution

$$f(d_p) = \frac{1}{1 + \left(\dfrac{212}{200} \right)^{1.0}} = 0.485$$

$$\gamma = \frac{1 - f(d_p)}{f(d_p)} = \frac{1 - 0.485}{0.485} = 1.061$$

$$\alpha = 0.496 \times 1.061 = 0.526$$

$$\beta = 0.504 \times 1.061 = 0.585$$

Since both α and β are less than 1, the distribution is U-shaped, which indicates that the minerals are rapidly approaching liberation as the progeny size passes the mineral grain size.

The cumulative distribution and the corresponding histogram is shown in Figure 3.18

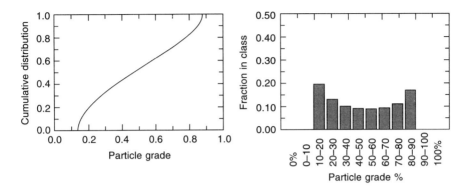

Figure 3.18 Cumulative distribution and histogram of particle grades at progeny size of 212 μm.

Case 2: Progeny size = 53 μm

In this case, the boundaries fall outside the range $g = 0$ to $g = 1$ and completely liberated particles will be found in the progeny population.

$$g_u = \max\left[0.501\left(\frac{3394}{53}\right)^{0.2}, 0.504\left(\frac{200}{53}\right)^{0.243}\right]$$

$$= \max[1.158, 0.696]$$

$$= 1.158$$

$$1 - g_\ell = \max\left[0.496\left(\frac{3394}{53}\right)^{0.2}, 0.496\left(\frac{200}{53}\right)^{0.243}\right]$$

$$= \max[1.140, 0.685]$$

$$= 1.140$$

$$g_\ell = -0.140$$

The liberated factions are calculated from the Beta distribution of the transformed variable,

$$\chi(g) = \frac{g + 0.140}{1.158 + 0.140}$$

which has parameters

$$f(d_p) = \frac{1}{1 + \dfrac{53}{200}} = 0.791$$

$$\gamma = \frac{1 - 0.791}{0.791} = 0.265$$

$$\alpha = g'\gamma = 0.504 \times 0.265 = 0.133$$

$$\beta = (1 - g')\gamma = 0.496 \times 0.265 = 0.131$$

$$L_0 = I_{\chi(0.0)}(0.133, 0.131) = I_{0.108}(0.133, 0.131)$$

$$= 0.382$$

$$L_1 = 1 - I_{\chi(1.0)}(0.133, 0.131) = 1 - I_{0.878}(0.133, 0.131)$$

$$= 1 - 0.603 = 0.397$$

The inner distribution can now be determined.

$$\alpha^M = 0.484 \times \gamma = 0.484 \times 0.265 = 0.128$$

$$\beta^M = (1 - 0.484) \times \gamma = 0.516 \times 0.265 = 0.137$$

$$\bar{g}^M = \frac{g' - L_1}{1 - L_0 - L_1}$$

$$= \frac{0.504 - 0.397}{1 - 0.382 - 0.397}$$

$$= 0.484$$

The cumulative conditional distribution for progeny particles of size 53 μm is

$$P(g \mid 53; 0.504, 3394) = L_0 + (1 - L_0 - L_1)I_g(0.128, 0.137)$$

$$= 0.382 + 0.221 \times I_g(0.128, 0.137)$$

The cumulative distribution and corresponding histogram are shown in Figure 3.19.

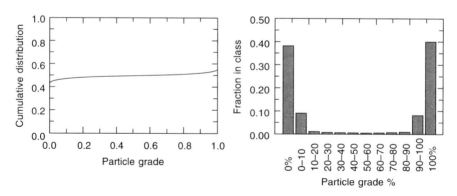

Figure 3.19 Cumulative distribution and histogram of particle grades at progeny size of 53 μm

3.7 Symbols used in this chapter

$B(\alpha, \beta)$	Beta function.
$b(d_p : d_p')$	Breakage function.
$b(g, d_p : g', d_p')$	Breakage function for 2-component material.
d_p	Particle size.
D	Mesh size.
$f(\ell)$	Distribution density (by length) for intercept length.
$f(d_p)$	Variation of variance with progeny size.
g	Mineral grade of particle.
g_L	Linear grade.
I_g	Incomplete beta function.
$\overline{\ell}$	Average intercept length.
$\overline{\ell}_D$	Average intercept length for particles of size D.
L_0	Mass fraction of particle population that consists of liberated gangue.
L_1	Mass fraction of particle population that consists of liberated mineral.
$\mathscr{L}_0$	Fraction of available gangue that is liberated.
$\mathscr{L}_1$	Fraction of available mineral that is liberated.
$p(D)$	Distribution density for particle size.
$p(g)$	Distribution density for particle grade.
$p(\ell)$	Distribution density for intercept length.
$p(\ell \mid D)$	Conditional density for intercept length from particles of size D.
$p(g_L \mid \ell)$	Conditional density for linear grades in intercepts of length ℓ.
ΔD_R	Conditioning variable to indicate particles come from a size interval.

Bibliography

The importance of mineral liberation and its description in terms of the geometry of the mineralogical texture and the geometrical properties of the particles was analyzed by Gaudin (1939) using simple geometrical structures. Barbery (1991) gave an account of most of the more recent research up to 1990. Fander (1985) gives a well illustrated account of many mineral textures as seen under the microscope. Jones (1987) provides an excellent account of quantitative methods that are available for the analysis of mineralogical texture. King (1994) has presented an alternative theoretical method for the prediction of mineral liberation from measurements on sections which can be applied when fracture is not random. Andrews and Mika (1976) analyzed the inter-relationship between comminution and liberation using population balance techniques and this forms the basis of the modeling method discussed in this chapter. King (1990) provided a simple model for the internal structure of the

Andrews-Mika diagram and Schneider (1995) developed the model to the stage where it could be calibrated using image analysis data obtained from samples of ore.

References

Andrews, J.R.G. and Mika, T.S. (1976) Comminution of heterogeneous material. Development of a model for liberation phenomena. *Proc. 11ᵗʰ Int Mineral Processing Congress*, pp. 59–88.

Barbery, G. (1991) *Mineral Liberation. Measurement, Simulation and Practical Use in Mineral Processing*. Editions GB, Quebec.

Fander, H.W. (1985) *Mineralogy for Metallurgists: an Illustrated Guide*. The Institution of Mining and Metallurgy, London.

Gaudin, A.M. (1939) *Principles of Mineral Dressing*. McGraw-Hill, New York.

Jones, M.P. (1987) *Applied Mineralogy: A Quantitative Approach*. Graham and Trotman, London.

King, R.P. (1990) Calculation of the liberation spectrum in products produced in continuous milling circuits. *Proc. 7ᵗʰ European Symposium on Comminution*, Vol. 2, pp. 429–444.

King, R.P. (1994) Linear stochastic models for mineral liberation. *Powder Technology*. Vol. 81, pp. 217–234.

Schneider, C.L. (1995) The Measurement and Calculation of Liberation in Continuous Grinding Circuits. Ph.D Thesis, University of Utah.

4
Size classification

It is always necessary to control the size characteristics of particulate material that is fed to process equipment that separates the mineralogical components. It is not possible within the production environment to exercise precise control over the size of all particles in a population and, in most cases, size classification equipment is designed to split a feed of particulate material into a coarse and a fine product. Occasionally one or two product streams of intermediate size may be produced such as by a double or triple deck screen. Probably the most common application of size classification is its use to prevent oversize material leaving a comminution circuit. Oversize material is recycled to a comminution stage for further size reduction before passing to subsequent stages of processing.

The most significant consideration when evaluating classification equipment is the lack of a clean cut at a particular size. Even the most efficient industrial size classifiers will pass a proportion of oversize material and will retain a portion of undersize material.

Size classification equipment is also subject to capacity limitations and these must be considered when evaluating the performance of new or existing classification equipment. In fact older methods of performance evaluation concentrate entirely on the capacity limitations. However, more modern procedures recognize that the efficiency of separation must also be considered if the performance of a size classifier in a plant circuit is to be accurately evaluated. The procedures that follow take this into account.

There are essentially two types of classifier available for process operation: screens and classifiers that rely on the variation of terminal settling velocities of particles of varying size when these are immersed in a viscous fluid. In general the former are used for the classification of drier material at coarser sizes.

4.1 Classification based on sieving–vibrating screens

The basic method of operation of a screen is simple. The screen presents a barrier to the passage of oversize material while it readily passes undersize material. It is only necessary to ensure that each particle has an opportunity to reach the screen. In practice each particle is given several opportunities to pass through the screen. Screens can be stationary or the screen can vibrate which increases the rate of presentation of each particle and assists in moving oversize material over and away from the screening surface.

4.1.1 Models based on screen capacity

The traditional method of evaluation of screen performance is the use of a capacity measure. This represents the ability of the screen to accept and handle the feed tonnage of material. The most important assumption in this approach is that the capacity of a screen is directly proportional to its surface area so that the basic capacity is specified at tons of feed per hour per square meter of screen. This quantity is represented by I_u. The basic capacity of any screen is determined under standard operating conditions using a predefined standard feed material. As the nature of the feed material changes and as the operating conditions change so the actual capacity of the screen changes, it will increase for conditions less arduous than the standard and decrease for conditions more arduous than the standard. These modifications are represented by capacity factors which multiply the standard unit screen capacity to get the actual screen capacity under conditions that the screen will actually meet in its position in the operating plant

$$\text{Rated screen feed capacity} = I_u K_1 K_2 \ldots$$

$$= I_u \prod_i K_i \quad \text{tons/hr m}^2 \tag{4.1}$$

where the separate K_i are the capacity factors for deviations from the standard conditions with respect to a number of individual conditions.

The basic unit capacity varies primarily with the size of the screen opening, screens with larger openings being able to handle large quantities of feed material. A typical relationship between I_u and mesh size is:

$$I_u = 0.783h + 37 \qquad \text{for } h \geq 25 \text{ mm} \tag{4.2}$$

$$= 20.0h^{0.33} - 1.28 \qquad \text{for } h < 25 \text{ mm} \tag{4.3}$$

where I_u is in tons/hr m^2 and h is the mesh size in mms. Each screen manufacturer has its own basic capacity-size relationships for its range of screens. The above expressions are only meant to define a typical trend.

The individual capacity factors are provided by screen manufacturers in tabular or graphical form although the availability of computers is encouraging the presentation of these factors in algebraic form. The following factors are typical for woven wire mesh screens.

The open area factor K_1

The standard condition is usually 50% open area and the capacity is proportional to the open area available.

$$K_1 = \frac{\% \text{ open area}}{50} \tag{4.4}$$

For material having a bulk density less than 800 kg/m^3 the standard open area is 60% rather than 50% and Equation 4.4 should be modified accordingly.

The half-size factor K_2

Feed that contains a large proportion of material that is considerably smaller than the screen mesh size will be handled more easily by a screen. The standard condition is defined as feed material having 40% smaller than one half of the mesh size. If the feed has more than 40% smaller than one half of the screen mesh size, the half-size factor will exceed unity and vice versa.

$$K_2 = 2P^F(0.5h) + 0.2 \tag{4.5}$$

The oversize factor K_3

A screen can handle a greater tonnage of feed material that contains large quantities of oversize material because this material passes directly over the screen and need not be transmitted through the mesh. This is accounted for by the oversize factor K_3, which has a value of unity for a standard feed containing 25% oversize material. This factor increases very quickly as the fraction of oversize increases and is given by

$$K_3 = 0.914 \exp \exp (4.22\overline{P}^F(h) - 3.50) \tag{4.6}$$

In Equation 4.6, $\overline{P}^F(h)$ is the fraction of material in the feed that has size greater than the screen mesh size h. It is related to the cumulative size distribution function by

$$\overline{P}^F(h) = 1 - P^F(h) \tag{4.7}$$

The bulk density factor K_4

Denser materials will be transmitted more easily than lighter materials. A factor K_4 accounts for this effect when bulk density differs from the standard of 1600 kg/m^3.

$$K_4 = \frac{\rho_B}{1600} \tag{4.8}$$

The deck position factor K_5

Screens that are lower down in the deck receive undersize from the screen above and can handle less material than a screen that takes fresh feed. The capacity decreases with position according to capacity factor K_5.

$$K_5 = 1.1 - 0.1S \tag{4.9}$$

where S represents the deck position; 1 for top deck, 2 for 2nd deck and so on.

The screen angle factor K_6

The standard inclined screen has an angle of inclination of 15°. Lower angles of inclination increase the projected area of the screen aperture in the horizontal plane and the screen can handle a greater load. This is accounted for by capacity factor K_6.

$$K_6 = 1.0 - 0.01(\alpha - 15) \qquad (4.10)$$

where α is the angle of inclination in degrees.

The wet screen factor K_7

Screening at finer mesh sizes can be improved by spraying the screen load with water. The factor K_7 accounts for this effect

$$K_7 = 1.0 + 2.4 \times 10^{-4}(25 - h)^{2.5} \qquad \text{for } h \leq 25 \text{ mm}$$
$$= 1.0 \qquad\qquad\qquad\quad \text{for } h > 25 \text{ mm} \qquad (4.11)$$

The aperture shape factor K_8

The standard screen has square openings and other shapes influence the capacity as shown in Table 4.1.

Table 4.1 Screen capacity factor for different apertures

Shape of screen opening	K_8
Round	0.8
Square	1.0
2 to 1 rectangular slot	1.15
3 to 1 rectangular slot	1.2
4 to 1 rectangular slot	1.25

Particle shape factor K_9

Slabby and elongated particles are more difficult to screen than particles that are essentially isometric. If the feed contains about 15% of slabby or elongated particles K_9 should be set at 0.9. Larger amounts of this type of material would give significant problems and would need to be investigated specially.

The surface moisture factor K_{10}

Surface moisture tends to make the particles adhere and screen capacity is reduced. Factor K_{10} accounts for this and can be evaluated from Table 4.2.

Table 4.2 Surface moisture capacity factor for screens

Condition of feed	K_{10}
Wet, muddy or sticky material	0.75
Wet surface quarried and material from surface stockpiles with up to 15% moisture by volume	0.85
Dry crushed material	1.0
Naturally or artificially dried material	1.25

4.1.2 Screen transmission efficiency

Ideally the screen should be able to transmit all of the undersize material in the feed. In practice, however, not all of the undersize material passes through the screen and the fraction of the feed undersize that does pass through is referred to as the screen efficiency. The efficiency is determined primarily by the actual feed loading on the screen relative to the rated feed capacity as calculated by Equation 4.1. The efficiency of transmission decreases if the screen must handle feed in excess of 80% of the rated tonnage because the access of individual particles to the screen surface is hindered to a greater or lesser degree. The efficiency also decreases as the actual feed tonnage falls below 80% of rated capacity because particles tend to bounce on the lightly loaded screen and make fewer contacts with the screen surface.

If W^F represents the actual feed tonnage, then the rating ratio is given by:

$$RR = \frac{W^F}{I_u \prod_i K_i \times \text{screen area}} \tag{4.12}$$

and the efficiency of transmission is given by

$$\begin{aligned} e &= 0.95 - 0.25(RR - 0.8) - 0.05(RR - 0.8)^2 && \text{for } RR \geq 0.8 \\ &= 0.95 - 1.67(0.8 - RR)^2 && \text{for } RR < 0.8 \end{aligned} \tag{4.13}$$

The actual tonnage passed to the undersize stream is

$$W^U = eP^F(h)W^F \tag{4.14}$$

Each size class smaller than the mesh size is subject to the same efficiency factor e so that the particle size distribution in the underflow stream is calculated in discrete form as

$$\begin{aligned} p_i^U &= \frac{ep_i^F W^F}{eP^F(h)W^F} = \frac{p_i^F}{P^F(h)} && \text{for } d_{pi} < h \\ &= 0 && \text{for } d_{pi} \geq h \end{aligned} \tag{4.15}$$

where p_i^U is the fraction of the underflow stream in the size class i and p_i^F that fraction in the feed stream.

The actual tonnage passed to the overflow stream is:

$$\begin{aligned} W^O &= (1 - P^F(h))W^F + (1 - e)P^F(h)W^F \\ &= W^F(1 - eP^F(h)) \end{aligned} \tag{4.16}$$

and the discrete size distribution in the overflow stream is given by:

$$p_i^O = \frac{(1 - e)p_i^F}{(1 - eP^F(h))} \qquad \text{for } d_p < h \tag{4.17}$$

$$= \frac{p_i^F}{1 - eP^F(h)} \qquad \text{for } d_p \geq h \tag{4.18}$$

The simple capacity model permits the calculation of the total screen area that is required for a specific duty. The aspect ratio of the screen (length/width ratio) must be determined by considering the depth of the bed of particles on the screen. An approximate guide is the restriction of the bed depth at the discharge end to no more than four times the mesh size. The thickness of the bed is determined by the total flowrate over the screen surface, the width of the screen, b, and the velocity of travel of the material along the screen.

$$t_b = \frac{W^O}{bu\rho_B} \tag{4.19}$$

where t_b is the bed thickness, W_d the mass flowrate across the discharge end, b the screen width, u the velocity of travel across the screen surface and ρ_B the bulk density. The velocity of travel depends primarily on the angle of inclination and the amplitude and mode of vibration.

4.2 The classification function

A more realistic description of the performance of a classification device is provided by the classification function. This function defines the probability that an individual particle will enter the oversize stream that leaves the classifier. This function is also known as the partition function, and, when shown graphically, as the partition curve. It is a complicated function of the particle properties and the classifying action of the particular device under consideration. The classification function $c(d_{pi})$ is defined as the mass fraction of material in size interval i in the feed which finally leaves in the oversize stream.

Once the partition function is known, the size distribution in both overflow and underflow can be calculated by a simple mass balance over the solids in size class i.

$$W^U p_i^U = (1 - c(d_{pi}))W^F p_i^F \tag{4.20}$$

$$W^O p_i^O = c(d_{pi})W^F p_i^F \tag{4.21}$$

where the superscripts O, U and F refer to oversize, undersize and feed respectively.

The total flowrates of solid in the oversize and the undersize are given by:

$$W^U = \sum_i W^U p_i^U = \sum_i (1 - c(d_{pi}))W^F p_i^F \tag{4.22}$$

$$W^O = \sum_i W^O p_i^O = \sum_i c(d_{pi})W^F p_i^F \tag{4.23}$$

$$p_i^U = [1 - c(d_{pi})]\frac{W^F}{W^U} p_i^F = \frac{[1 - c(d_{pi})]p_i^F}{\sum_i [1 - c(d_{pi})]p_i^F} \tag{4.24}$$

$$p_i^O = c(d_{pi})\frac{W^F}{W^O}p_i^F = \frac{c(d_{pi})p_i^F}{\sum_i c(d_{pi})p_i^F} \tag{4.25}$$

These formulas are often written in terms of the total yield of solids to the overflow:

$$Y_s = \frac{W^O}{W^F} \tag{4.26}$$

$$p_i^U = \frac{(1 - c(d_{pi}))p_i^F}{1 - Y_s} \tag{4.27}$$

$$p_i^O = \frac{c(d_{pi})p_i^F}{Y_s} \tag{4.28}$$

$$Y_s = \sum_i c(d_{pi})p_i^F \tag{4.29}$$

4.2.1 The Karra model

The approach described in the previous section is the traditional method used for sizing screens. Its chief limitation is that the screen must be sized according to the amount of feed that is presented to the screen. A more logical approach is based on the amount of material that must be actually transmitted by the screen to the underflow stream. This approach has been developed by V.K. Karra (1979) into an effective description of how a screen may be expected to perform during plant operation.

The approach is similar to the traditional approach but is based on the capacity of the screen to transmit undersize material proportional to the screen area. As in the traditional method this basic capacity is modified by a number of factors that allow for variations of the feed material and the screen from the standard test conditions.

Let A represent the basic capacity which is defined as the tonnage of undersize that a particular screen can transmit per unit of screen surface area. The basic capacity is increased or decreased depending on the nature of the feed and conditions on the screen. A number of capacity factors allow for the amount of oversize in the feed (factor B), the amount of half size in the feed (factor C), the location of the deck (factor D), factor for wet screening (factor E) and for material bulk density (factor F). These factors all have a value of unity at the nominal standard operating condition and move down or up as the screening duty becomes more or less arduous. The amount of near-size material also has a significant effect on the ability of the screen to transmit undersize material and an additional near-size capacity factor G_c is introduced.

The theoretical amount of undersize that can be transmitted by the screen is given by:

$$Th = A.B.C.D.E.F.G_c \times \text{screen area} \tag{4.30}$$

A screen will be well designed to handle its duty in the circuit if *Th* is approximately equal to the quantity of undersize in the feed. Each of the capacity factors is related to the quality of the feed and to the type of screen.

Karra bases the screen performance on the effective throughfall aperture of the screen defined by

$$h_T = (h + d_w) \cos \theta - d_w \tag{4.31}$$

where d_w is the diameter of the wire and θ is the angle of inclination of the deck. Equation 4.31 gives the effective aperture area projected onto the horizontal plane which is appropriate for particles that must pass through the screen under gravity.

The basic capacity A

The basic capacity is primarily determined by the mesh size of the screen with the effective throughfall aperture being used as the appropriate measure of screen mesh size.

$$A = 12.13h_T^{0.32} - 10.3 \qquad \text{for } h_T < 51 \text{ mm} \tag{4.32}$$

$$A = 0.34\, h_T + 14.41 \qquad \text{for } h_T \geq 51 \text{ mm} \tag{4.33}$$

with h_T in mm and A is in metric tons/hr m^2.

The basic capacity will also depend on the open area of the screen used. The basic capacity calculated from Equations 4.32 and 4.33 is applicable to standard industrial light-medium woven wire mesh. For other screen cloths and surfaces, A must be adjusted in proportion to the open area. The percent open area for light-medium wire mesh is related to the mesh size h by

$$OA = 21.5 \log_{10}h + 37 \tag{4.34}$$

with h in millimeters.

Thus capacity A must be adjusted to:

$$\frac{A \times \text{actual \% open area}}{OA} \tag{4.35}$$

The oversize factor B

$$\begin{aligned}
B &= 1.6 - 1.2\overline{P}^F(h_T) & \text{for } \overline{P}^F(h_T) \leq 0.87 \\
B &= 4.275 - 4.25\overline{P}^F(h_T) & \text{for } \overline{P}^F(h_T) > 0.87
\end{aligned} \tag{4.36}$$

The half-size factor C

$$\begin{aligned}
C &= 0.7 + 1.2P^F(0.5h_T) & \text{for } P^F(0.5h_T) \leq 0.3 \\
C &= 2.053P^F(0.5h_T)^{0.564} & \text{for } 0.3 < P^F(0.5h_T) \leq 0.55 \\
C &= 3.35P^F(0.5h_T)^{1.37} & \text{for } 0.55 < P^F(0.5h_T) \leq 0.8 \\
C &= 5.0P^F(0.5h_T) - 1.5 & \text{for } P^F(0.5h_T) > 0.8
\end{aligned} \tag{4.37}$$

The deck location factor D

$$D = 1.1 - 0.1S \qquad (4.38)$$

where $S = 1$ for top deck, $S = 2$ for 2nd deck and so on.

The wet screening factor E

Let $T = 1.26h_T$ (h_T in mm)

$E = 1.0$	for $T < 1$
$E = T$	for $1 \leq T < 2$
$E = 1.5 + 0.25T$	for $2 \leq T < 4$
$E = 2.5$	for $4 \leq T < 6$
$E = 3.25 - 0.125T$	for $6 \leq T < 10$
$E = 4.5 - 0.25T$	for $10 \leq T < 12$
$E = 2.1 - 0.05T$	for $12 \leq T < 16$
$E = 1.5 - 0.0125T$	for $16 \leq T < 24$
$E = 1.35 - 0.00625T$	for $24 \leq T < 32$
$E = 1.15$	for $T > 32$

whenever the material is sprayed with water.

The bulk density factor F

$$F = \frac{\rho_B}{1600} \qquad (4.39)$$

The near-size capacity factor G_c

The capacity of the screen is also affected by the presence of near-size material in the feed. The near-size material in the feed is in the size range from $0.75h_T$ to $1.25h_T$. Considerable quantities of near-size material will inhibit the passage of undersize material through the screen. The near-size capacity factor can be evaluated from:

$$G_c = 0.975(1 - \text{near-size fraction in feed})^{0.511}$$
$$G_c = 0.975(1 - P^F(1.25h_T) + P^F(0.75h_T))^{0.511} \qquad (4.40)$$

4.2.2 The screen classification function

In practice, not all of the undersize is transmitted because of various physical factors that impair the efficiency of the screen. This effect is described by the screen partition function. Several standard functional forms are available to describe this effect, and the Rosin-Rammler function is used here.

$$c(d_p) = 1 - \exp[-0.693(d_p/d_{50})^{5.9}] \qquad (4.41)$$

This gives the efficiency of transfer of particles of size d_p to oversize.

The parameter that will determine the screening efficiency is d_{50}. Values of d_{50} greater than the mesh size give high efficiencies and *vice-versa*.

The actual d_{50} achieved will depend primarily on the effective throughfall aperture of the wire mesh used on the screen, on a loading coefficient K defined by:

$$K = \frac{\text{tons of undersize in the feed/unit of screen area}}{ABCDEF}$$
$$= \frac{W^F P(h_T)/\text{screen area}}{ABCDEF} \tag{4.42}$$

and on the near-size factor G_c.

Experimental screening data are well represented by:

$$\frac{d_{50}}{h_T} = \frac{G_c}{K^{0.148}} \tag{4.43}$$

d_{50} can be substituted in Equations 4.41 and 4.20–25 to calculate the size distributions in the two product streams.

$$p_i^U = \frac{[1 - c(d_{pi})]p_i^F}{\sum_i [1 - c(d_{pi})]p_i^F} \tag{4.44}$$

$$p_i^O = \frac{c(d_{pi})p_i^F}{\sum_i c(d_{pi})p_i^F} \tag{4.45}$$

The model provides a simulation of the actual performance of the screen in the circuit. This performance can be compared with the design capacity of the screen and the screen performance evaluated. In particular, the actual operating efficiency can be calculated from

$$\text{simulated efficiency} = \frac{\text{tonnage transmitted to undersize}}{\text{tonnage of undersize in feed}} \tag{4.46}$$

$$= \frac{\sum_i [1 - c(d_{pi})]p_i^F}{P^F(h)} \tag{4.47}$$

The effective utilization of the screen area can be calculated from

AUF = area utilization factor

$$= \frac{\text{tonnage transmitted to undersize}}{\text{theoretical ability of the screen to pass undersize}} \tag{4.48}$$

$$= \frac{W^F P^F(h) \times \text{simulated efficiency}}{A.B.C.D.E.F.G_c \times \text{screen area}}$$

An AUF equal to unity indicates that the screen capacity is exactly balanced to the required duty. $AUF > 1$ indicates that the screen is overloaded while $AUF < 1$ indicates that the screen is underloaded.

A particular advantage of the Karra screen model is that no free parameters are required to be estimated from operating data.

4.3 A simple kinetic model for screening

Screening can be considered to be a kinetic process because the rate at which solid particulate material is transmitted through a screen is dependent on the nature of the screen, on the load that the screen is carrying and on the nature and size of particles. Obviously larger particles are transmitted more slowly than small particles and so on. The rate of transmission of particles through the screen will vary from point to point on an actual screen because the load that the screen carries varies in the direction of the material flow. In general the rate of transmission depends primarily on the particle size relative to the screen mesh size.

Observation of operating screens shows clearly that the screen action varies according to the load on the screen. Near the feed end, industrial screens are almost invariably heavily loaded so that the particulate material forms a multilayer of particles on the surface. As the particles move down the screen surface, the smaller material falls through and the load on the screen surface decreases steadily until only a monolayer of particles remain. These two conditions are referred to as the crowded and separated regimes respectively.

In the crowded condition, the undersize material must percolate through the upper layers of coarse material before it has a chance to contact the screen and thus fall through. Thus the rate of transmission of material is a function of the size distribution in the particle layer immediately above the screen surface. On the crowded feed end of the screen, this layer is replenished by downward percolation of undersize material through the upper layers of the bed. The net rate at which particles of a particular size are transmitted through the screen is a function of the percolation process as well as the screen transmission process. This model of the screening operation is illustrated in Figure 4.1.

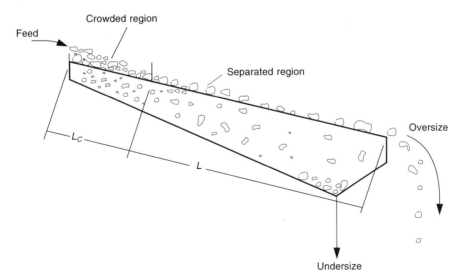

Figure 4.1 Varying rate of transmission of particles through a wire mesh screen

Let $w_i(l)$ = mass flowrate per unit width of screen of material in size interval i down the screen at distance l from the feed end (kg/ms). $w_i(l)$ is related to the discrete size distribution density function by $w_i(l) = W(l)p_i(l)$ where $W(l)$ is the total flow of material down the screen per unit width at position l and the dependence of these functions on position is shown explicitly by the argument l.

If the rate of transmission of particles in size interval i is represented by $r(d_{pi})$ kg/m^2 s, a differential mass balance gives

$$\frac{dw_i(l)}{dl} = -r(d_{pi}) \tag{4.49}$$

This equation can be integrated if a model for the rate of transmission can be developed.

The rate of transmission will be modeled differently under the crowded and separated conditions. Accurate models for the crowded condition are complex since they must take into account both the stratification and transmissions processes. A simple but useful model is used here but a more realistic model is developed in Section 4.3.1.

Unfortunately there is very little data available on the value of the transmission rate for crowded conditions, $r(d_p)$, and its variation with particle size and with conditions on the screen. It is also dependent on the size distribution in the feed.

If the rate of transmission under crowded conditions remains constant along the screen, Equation 4.49 can be integrated immediately to give

$$w_i(l) = w_i(0) - r(d_{pi})l \tag{4.50}$$

At the transition between the crowded regime and the separated regime $l = L_c$

$$w_i(L_c) = w_i(0) - r(d_{pi})L_c \tag{4.51}$$

with

$$r(d_{pi}) = k(d_{pi})M_o p_i^F \tag{4.52}$$

where M_o is the load on the screen in kg/m^2 at the feed end. There is some experimental evidence to support this assumption of a constant rate of transmission under crowded conditions.

$$w_i(L_c) = \frac{W^F}{b}p_i^F - k(d_{pi})M_o p_i^F L_c \tag{4.53}$$

Let $M(l)$ represent the load per unit area of screen surface at a distance l from the feed end in kg/m^2. The flowrate down the screen $W(l)$ is related to the load on the screen by

$$W(l) = M(l)u \tag{4.54}$$

where u is the velocity at which the material travels down the screen:

$$w_i(L_c) = M_o u p_i^F - k(d_{pi}) M_o p_i^F L_c \tag{4.55}$$

$$= M_o p_i^F (u - k(d_{pi}) L_c)$$

$$= \frac{W^F}{b} p_i^F \left(1 - \frac{k(d_{pi})}{u} L_c \right) \tag{4.56}$$

L_c can be defined as the position on the screen at which particles in the smallest size class just become depleted

$$L_c = \frac{u}{k(d_{pN})} \tag{4.57}$$

The simplest model for the variation of $k(d_{pi})$ with particle size is:

$$\frac{k(d_{pi})}{u} = \frac{k_0}{u} \left(1 - \frac{d_{pi}}{h} \right) m^{-1} \quad \text{for } d_{pi} < h \tag{4.58}$$

$$= 0 \quad \text{for } d_{pi} \geq h$$

where k_0/u is the operating parameter. k_0/u will depend on the mesh size of the screen, its fractional open area and on other operating parameters such as vibration amplitude and frequency.

Under separated conditions, the rate of transmission of particles if size d_p is assumed to be proportional to the amount of that material per unit area of screen surface

$$\frac{dw_i(l)}{dl} = -s(d_{pi}) M(l) p_i(l) \tag{4.59}$$

where

$p_i(l)$ = the discrete particle size density function of the material on the screen at distance l from the feed end.

$s(d_{pi})$ = the specific rate of transmission for particles of size d_{pi} (sec^{-1})

$w_i(l)$ is related to $M(l)$ by

$$w_i(l) = p_i(l) M(l) u \tag{4.60}$$

$$\frac{dw_i(l)}{dl} = -\frac{s(d_{pi})}{u} w_i(l) \tag{4.61}$$

This is easy to integrate and match the individual mass flowrates for each size interval at the boundary between the crowded and separated regions:

$$w_i(l) = w_i(L_c) \exp\left(-\frac{s(d_{pi})}{u}(l - L_c) \right) \tag{4.62}$$

$$= [w_i(0) - r(d_{pi}) L_c] \exp\left(-\frac{s(d_{pi})}{u}(l - L_c) \right) \tag{4.63}$$

The partition factor can be evaluated from Equation 4.63:

$$c(d_{pi}) = \frac{w_i(L)}{w_i(0)}$$

$$= \left[1 - \frac{r(d_{pi})L_c}{w_i(0)}\right] \exp\left(-\frac{s(d_{pi})}{u}(L - L_c)\right)$$

(4.64)

The specific rate constant $s(d_{pi})/u$ can be approximated by the probability of passage of a particle during a single contact with the screen. This is usually computed as the ratio of the area of the effective aperture to the total area of the screen. In order to pass cleanly through a square mesh opening a spherical particle must pass with its center within a square of side $h - d_p$ in the center of the mesh opening. Thus the probability of passage is given by

$$P_r = \frac{(h - d_p)^2}{(h + d_w)^2} = \frac{h^2}{(h + d_w)^2}\left(1 - \frac{d_p}{h}\right)^2$$

$$= f_0\left(1 - \frac{d_p}{h}\right)^2$$

(4.65)

where f_0 is the fraction open area of the screen. This is illustrated in Figure 4.2.

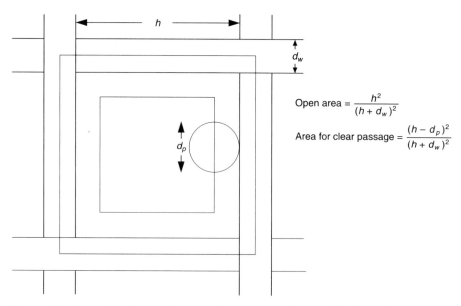

Figure 4.2 Geometrical restrictions on the passage of a particle through square-mesh screen

Ferrara *et al.* (1988) suggested an empirical functional form:

$$\frac{s(d_{pi})}{u} = nf_0^{\sigma/2}\left(1 - \frac{d_{pi}}{h}\right)^{\sigma} \tag{4.66}$$

where σ is a constant that should be close to 2. n is the number of presentations per unit length of bed. Equation 4.66 can be written

$$\frac{s(d_{pi})}{u} = s_{50}\, 2^{\sigma}\left(1 - \frac{d_{pi}}{h}\right)^{\sigma} \qquad \text{for } d_{pi} < h \tag{4.67}$$

$$= 0 \qquad \text{for } d_{pi} \geq h$$

where s_{50} is the value of $s(d_{pi})/u$ when the particle size is one half the mesh size. s_{50} should be proportional to $f_0^{\sigma/2}$ for similar screen types but will also vary with angle of inclination, vibration amplitude and frequency and the mode of vibration.

4.3.1 The Ferrara-Preti-Schena kinetic model

The modeling of the kinetics of the transmission process under crowded conditions given in the previous section does not represent the process particularly well. Ferrara *et al.* have analyzed an alternative model which is based on the assumption that the local rate of transmission under crowded conditions at size d_p is proportional to the mass fraction of material of size d_p on the screen. This reflects the obvious fact that particle sizes that are not abundant do not transmit large quantities of material.

Equation 4.49 can now be written

$$\frac{dw(d_p, l)}{dl} = -k(d_p)p(d_p, l) \tag{4.68}$$

where $k(d_p)$ is a kinetic transmission coefficient that is independent of the size distribution in the feed.

Using Equation 4.60

$$\frac{dw(d_p, l)}{dl} = -\frac{k(d_p)}{u}\frac{w(d_p, l)}{M(l)} \tag{4.69}$$

It is usual to write this in terms of the total mass of material flowing along the screen.

$$W(l) = uM(l) = \int_0^\infty w(d_p, l)\, dd_p \tag{4.70}$$

$$\frac{dw(d_p, l)}{dl} = -k(d_p)\frac{w(d_p, l)}{W(l)} \tag{4.71}$$

This equation is difficult to solve because $W(l)$ is related to the dependent variable $w(d_p, l)$ through the integral in Equation 4.70. It is best solved by noting the formal solution

$$E(d_p, l) = \frac{w(d_p, l)}{w(d_p, 0)} = \exp\left(-k(d_p)\int_0^l \frac{dl'}{W(l')}\right)$$

(4.72)

and Equation 4.71 becomes

$$\frac{W(l)}{E(d_p, l)}\frac{dE(d_p, l)}{dl} = -k(d_p)$$

(4.73)

Introduce a new variable

$$\chi(x) = \frac{k(x)}{k(d_p)}$$

(4.74)

where x can represent any particle size.
   Using Equation 4.72

$$\chi(x) = \frac{\ln E(x, l)}{\ln E(d_p, l)}$$

(4.75)

and noting that

$$W(l) = \int_0^\infty w(x, l)\, dx$$

$$= \int_0^\infty w(x, 0)E(x, l)dx$$

(4.76)

and

$$E(x, l) = E(d_p, l)^{\chi(x)}$$

(4.77)

Equation 4.73 becomes

$$\int_0^\infty w(x, 0)E(d_p, l)^{\chi(x)-1}\frac{dE(d_p, l)}{dl}\, dx = -k(d_p)$$

(4.78)

For every screen there will be a particle size D above which the rate of transmission is zero so that

$$\chi(x) = 0 \qquad \text{for } x \geq D$$

(4.79)

The integral in Equation 4.78 can be split into two parts which can be integrated with respect to l over the range 0 to L_c to give

$$\int_0^D \frac{w(x, 0)}{\chi(x)}[E(d_p, L_c)^{\chi(x)} - 1]dx + \int_D^\infty w(x, 0)\ln E(d_p, L_c)\, dx = -k(d_p)L_c$$

(4.80)

This equation can be written in terms of the size distribution in the feed

$$W^F \int_0^D \frac{p^F(x)}{\chi(x)} [E(d_p, L_c)^{\chi(x)} - 1] dx$$

$$+ W^F \overline{P}^F(D) \ln E(d_p, L_c) = - k(d_p) L_c b \tag{4.81}$$

where b represents the width of the screen.

In practice this equation is solved using a finite difference approximation

$$\frac{W^F}{L_c b} \sum_j \frac{p_j^F}{\chi_{ji}} [E(d_{pi}, L_c)^{\chi_{ji}} - 1] + W^F \overline{P}^F(D) \ln E(d_{pi}, L_c) = - k(d_{pi}) \tag{4.82}$$

where

$$\chi_{ji} = \frac{k(d_{pj})}{k(d_{pi})} \tag{4.83}$$

and the summation is taken only over those classes for which $d_{pj} < D$.

Equation 4.82 is solved for $E(d_{pi}, l)$ for each particle size class.

Ferrara *et al.* propose that $k(d_p)$ should be modeled by an equation of the form

$$k(d_p) = k_{50} 2^\sigma \left(1 - \frac{d_p}{D}\right)^\sigma \tag{4.84}$$

with σ approximately equal to 2. k_{50} is the rate of transmission of particles with size equal to $1/2$ the mesh size under crowded conditions in the absence of any other particle sizes. It must be determined by experiment. An approximate value can be obtained using the standard screen capacity factor for undersize transmission evaluated for the mesh size used and 0% oversize and 100% undersize. Using the Karra formulas this would be given by

$$k_{50} = A \times 1.6 \times 3.5 \times \frac{\rho_B}{1600} \qquad \text{tons/hr m}^2 \tag{4.85}$$

The partition coefficient at each size is given by Equation 4.64

$$c(d_{pi}) = E(d_{pi}, L_c) \exp\left(- \frac{s(d_{pi})}{u} (L - L_c)\right) \tag{4.86}$$

There is no known way of predicting a priori the value of the length L_c of the screen over which crowded screening conditions exist. It will be necessary to investigate by experiment how the kinetic constants are affected by operating conditions on the screen before this kinetic model finds any real application in practice.

4.4 Classification based on differential settling – the hydrocyclone

The technology of comminution is intimately connected with classification devices through the concept of closed-circuit milling. All comminution operations are not selective in that they can potentially reduce the size of all particles in the unit. It is undesirable to reduce the size of any particle beyond the desired product size for that unit since that consumes additional energy and further down-stream processing can be adversely effected. A classifier placed at an appropriate point in the circuit can selectively remove those particles that meet the product size criteria for the circuit and return coarse particles back to the comminution unit.

4.4.1 Interaction between fluids and particles

When a solid particle moves through a fluid it experiences a drag force that resists its motion. Drag force has its origin in two phenomena, namely the frictional drag on the surface and the increase in pressure that is generated in front of the particle as it moves through the fluid. The frictional drag is caused by the viscosity of the fluid as it flows over the surface of the particle. This component is called viscous drag and is illustrated in Figure 4.3.

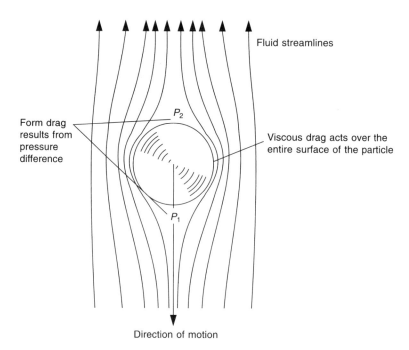

Figure 4.3 Streamlines that form around a particle that moves slowly through a fluid in the direction shown

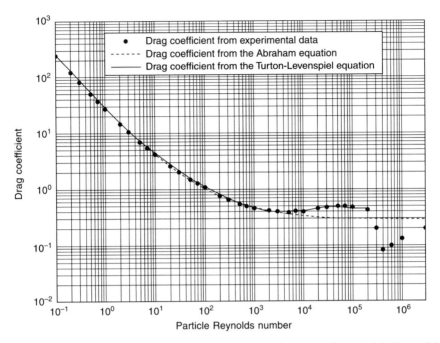

Figure 4.4 Drag coefficient for solid spheres plotted against the particle Reynolds number

A region of high pressure P_1 is formed immediately in front of the particle as it forces its way through the fluid. Likewise a region of relatively low pressure is formed immediately behind the particle in its wake. The pressure drop $P_1 - P_2$ gives rise to a force on the particle given by $(P_1 - P_2)A_c$ where A_c is the cross-sectional area of the particle measured perpendicular to the direction of motion. This is called the form drag. The total force on the particle is the sum of the viscous drag and the form drag.

Both the form drag and viscous drag vary with the relative velocity between particle and fluid and with the density of the fluid. Many experiments have revealed that particles of different size show very similar behavior patterns when moving relative to the surrounding fluid, be it air, water or any other viscous fluid. If the dimensionless groups

$$C_D = \frac{2 \times \text{Force on particle}}{\text{Cross-sectional area} \times \rho_f \times v^2} = \frac{2F}{A_c \rho_f v^2} \qquad (4.87)$$

and

$$Re_p = \frac{d_p v \rho_f}{\mu_f} \qquad (4.88)$$

are evaluated at any relative velocity v for any particle of size d_p, then all experimental data are described by a single relationship between C_D and Re_p.

Results from a large number of experimental studies are summarized in Figure 4.5. C_D is called the drag coefficient of the particle and Re_p the particle Reynolds number.

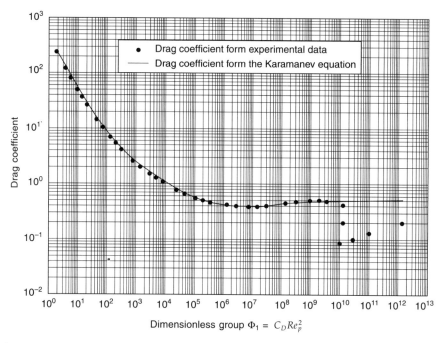

Figure 4.5 Drag coefficient for solid spheres plotted against the dimensionless group $\Phi_1 = C_D Re_p^2$. The line was plotted using equation (4.93). Use this graph to calculate the drag coefficient at terminal settling velocity when the particle size is known.

In the region $Re_p < 3 \times 10^3$ this data is described quite accurately by the Abraham (1970) equation:

$$C_D = 0.28 \left(1 + \frac{9.06}{Re_p^{1/2}} \right)^2 \tag{4.89}$$

In the region $Re_p < 2 \times 10^5$ the data is described well by the Turton-Levenspiel (1986) equation:

$$C_D = \frac{24}{Re_p}(1 + 0.173 Re_p^{0.657}) + \frac{0.413}{1 + 16300 Re_p^{-1.09}} \tag{4.90}$$

The Abrahams and Turton-Levenspiel equations are plotted in Figure 4.4.

Two alternative representations of the data are useful in practice and these are illustrated in Figures 4.5 and 4.6. Instead of using Re_p as the independent variable, the data is recomputed and plotted against the two dimensionless groups

$$\Phi_1 = C_D Re_p^2 \tag{4.91}$$

and

$$\Phi_2 = \frac{Re_p}{C_D} \tag{4.92}$$

as independent variables in Figure 4.5 and Figure 4.6 respectively. When the data are plotted in this way they can be seen to follow functional forms similar to the Turton-Levenspiel equation. This was pointed out by Karamanev who used the equation

$$C_D = \frac{432}{\Phi_1}(1 + 0.0470\Phi_1^{2/3}) + \frac{0.517}{1 + 154\Phi_1^{-1/3}} \tag{4.93}$$

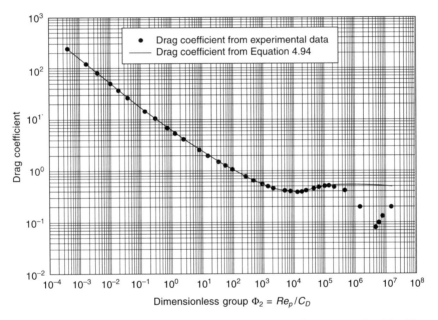

Figure 4.6 Drag coefficient plotted against the dimensionless group Re_p/C_D. The line is plotted using Equation 4.94. Use this graph to calculate drag coefficient at terminal settling velocity of a particle of unknown size when the terminal settling velocity is known

An equation of similar form describes the data in Figure 4.6:

$$C_D = \frac{4.90}{\Phi_2^{1/2}}(1 + 0.243\Phi_3^{1/3}) + \frac{0.416}{1 + 3.91 \times 10^4 \, \Phi_2^{-1}} \tag{4.94}$$

The drag coefficient can be used with Equation 4.87 to calculate the force on the particle as it moves through the fluid.

$$\text{Force} = \frac{C_D}{2}\rho_f v^2 A_c \tag{4.95}$$

4.5 Terminal settling velocity

If a particle falls under gravity through a viscous fluid it will accelerate for a short while but as the particle moves faster the drag force exerted by the fluid increases until the drag force is just equal to the net gravitational force less the buoyancy that arises from the immersion of the particle in the fluid.

The terminal settling velocity v_T can be evaluated by balancing the drag and buoyancy forces

$$v_p(\rho_s - \rho_f)g = \frac{C_D}{2}\rho_f v_T^2 A_c \tag{4.96}$$

where v_p is the volume of the particle.

4.5.1 Settling velocity of an isolated spherical particle

When the particle is spherical, the geometrical terms in Equation 4.96 can be written in terms of the particle diameter

$$\frac{\pi}{6}d_p^3(\rho_s - \rho_f)g = \frac{C_D}{2}\rho_f v_T^2 \frac{\pi}{4}d_p^2 \tag{4.97}$$

The drag coefficient at terminal settling velocity is given by

$$C_D^* = \frac{4}{3}\frac{(\rho_s - \rho_f)}{\rho_f v_T^2}gd_p \tag{4.98}$$

and the particle Reynolds number at terminal settling velocity is given by

$$Re_p^* = \frac{d_p v_T \rho_f}{\mu_f} \tag{4.99}$$

It is not possible to solve Equation 4.97 directly because C_D is a function of both v_T and the particle size d_p through the relationship shown in Figure 4.4 or either of the Abraham or Turton-Levenspiel equations. Two different solution procedures are commonly required in practice: the calculation of the terminal settling velocity for a particle of given size or the calculation of the size of the spherical particle that has a prespecified terminal settling velocity. These problems can be solved without recourse to trial and error methods by considering the two dimensionless groups Φ_1 and Φ_2 both evaluated at the terminal settling velocity.

$$\Phi_1^* = C_D^* Re_p^{*2} = \frac{4}{3} \frac{\rho_s - \rho_f}{\rho_f} \frac{g d_p}{v_T^2} \left(\frac{d_p v_T \rho_f}{\mu_f} \right)^2$$

$$= \left[\frac{4}{3} (\rho_s - \rho_f) \rho_f \frac{g}{\mu_f^2} \right] d_p^3 = d_p^{*3} \tag{4.100}$$

and

$$\Phi_2^* = \frac{Re_p^*}{C_D^*} = \left[\frac{3}{4} \frac{\rho_f^2}{(\rho_s - \rho_f)\mu_f g} \right] v_T^3 = V_T^{*3} \tag{4.101}$$

d_p^* and V_T^* are called the dimensionless diameter and the dimensionless terminal settling velocity of a sphere respectively. d_p^* can be evaluated from a knowledge of the properties of the fluid and the size of the particle. d_p^* is independent of the terminal settling velocity. V_T^* is independent of the particle size and can be evaluated if the terminal settling velocity is known. Φ_1^* is often called the Archimedes number which is usually represented by the symbol *Ar*.

From the definition of Φ_1^* and Φ_2^* given in Equations 4.100 and 4.101, the following relationships are easily derived.

$$Re_p^* = (\Phi_1^* \Phi_2^*)^{1/3} = d_p^* V_T^* \tag{4.102}$$

$$C_D^* = \frac{Re_p^*}{\Phi_2^*} = \frac{\Phi_1^*}{Re_p^{*2}} = \frac{d_p^*}{V_T^{*2}} \tag{4.103}$$

$$Re_p^* = C_D^* V_T^{*3} = \left(\frac{d_p^{*3}}{C_D^*} \right)^{1/2} \tag{4.104}$$

The drag coefficient at terminal settling velocity can be calculated easily from Figure 4.5 or Equation 4.93 if the dimensionless diameter of the particle is known. It is then a simple matter to evaluate the terminal settling velocity from Equation 4.97 or more conveniently from Equation 4.103.

The diameter of the particle that has known terminal settling velocity can be calculated using a similar procedure. In this case the dimensionless settling velocity V_T^* is calculated using Equation 4.101 and the drag coefficient at terminal settling velocity can then be calculated from Equation 4.94 or read from Figure 4.6. In either case the required variable can be calculated directly from Equation 4.97 once C_D^* is known or more conveniently from Equation 4.103.

A similar solution procedure can be obtained using the Abraham equation starting from Equation 4.102. The Abraham equation may be solved directly to give the dimensionless terminal settling velocity once d_p^* is known.

$$V_T^{*3} = \frac{Re_p^*}{C_D^*} = \frac{d_p^* V_T^*}{0.28\left(1 + \frac{9.06}{(d_p^* V_T^*)^{1/2}}\right)^2} \tag{4.105}$$

$$1 + \frac{9.06}{(d_p^* V_T^*)^{1/2}} = \frac{d_p^{*1/2}}{0.28^{1/2} V_T^*} \tag{4.106}$$

This equation can be re-arranged to show more clearly the relationship between V_T^* and d_p^*

$$0.28^{1/2} V_T^* d_p^{*1/2} + 9.06 \times 0.28^{1/2} V_T^{*1/2} - d_p^* = 0 \tag{4.107}$$

This equation can be solved for V_T^* in terms of d_p^* or for d_p^* in terms of V_T^*. Either way the solution procedure is straightforward because Equation 4.107 is quadratic in the variables $V_T^{*1/2}$ and $d_p^{*1/2}$.

The solutions are:

$$V_T^* = \frac{20.52}{d_p^*}[(1 + 0.0921 d_p^{*3/2})^{1/2} - 1]^2 \tag{4.108}$$

$$d_p^* = 0.070\left[\left(1 + \frac{68.49}{V_T^{*3/2}}\right)^{1/2} + 1\right]^2 V_T^{*2} \tag{4.109}$$

The drag coefficient at terminal settling velocity is related to d_p^* and V_T^* most directly by Equation 4.103.

These solutions are due to Concha and Almendra (1979).

This approach to the analysis of terminal settling velocity provides straightforward linear procedures for the calculation of terminal settling velocity or particle size. No inconvenient iterative calculations are required. Illustrative examples 4.1 and 4.2 demonstrate their application.

Illustrative example 4.1 Terminal settling velocity
Calculate the terminal settling velocity of a glass sphere of diameter 0.1 mm in a fluid having density 982 kg/m³ and viscosity 0.0013 kg/ms. Density of glass is 2820 kg/m³. Also calculate the drag coefficient at terminal settling velocity.

Solution
Since the particle size is known, calculate d_p^*.

$$\Phi_1^* = d_p^{*3} = \frac{4}{3} \frac{(\rho_s - \rho_f)\rho_f g \, d_p^3}{\mu_f^2}$$

$$= \frac{4}{3} \frac{(2820 - 982) \, 982 \times 9.81 \times (0.1 \times 10^{-3})^3}{(0.0013)^2}$$

$$= 13.97$$

Using the Karamanev modification to the Turton-Levenspiel equation,

$$C_D^* = \frac{432}{\Phi_1^*} (1 + 0.0470 \, \Phi_1^{*2/3}) + \frac{0.517}{1 + 154 \, \Phi_1^{-1/3}}$$

$$= 39.355 + 0.008 = 39.36$$

Check this on the graph of C_D vs Φ_1 (Figure 4.5)

$$Re_p^* = \left(\frac{\Phi_1^*}{C_D^*}\right)^{1/2} = \left(\frac{13.97}{39.36}\right)^{1/2} = 0.596$$

Check this on the C_D vs Re_p graph (Figure 4.4)

$$v_T = Re_p^* \frac{\mu_f}{d_p \rho_f} = \frac{0.596 \times 0.0013}{0.1 \times 10^{-35} \times 982} = 7.89 \times 10^{-3} \, \text{m/s}$$

Alternatively use the Concha-Almendra procedure

$$d_p^* = (13.97)^{1/3} = 2.408$$

$$V_T^* = \frac{20.52}{d_p^*} [(1 + 0.0921 d_p^{*3/2})^{1/2} - 1]^2$$

$$= 0.217$$

$$Re_p^* = d_p^* V_T^* = 2.408 \times 0.217 = 0.522$$

$$v_T = \frac{Re_p^* \mu_f}{d_p \rho_f} = \frac{0.522 \times 0.0013}{0.1 \times 10^{-3} \times 982} = 6.91 \times 10^{-3} \, \text{m/s}$$

The drag coefficient at terminal settling velocity is calculated from

$$C_d^* = \frac{d_p^*}{V_T^{*2}} = \frac{2.408}{0.217^2} = 51.1$$

Illustrative example 4.2 Size of settling particle
Find the size of a spherical particle of glass that has a terminal settling velocity of 7.4 mm/s in a fluid of density 982 kg/m^3 and viscosity 0.0013 kg/ms. Density of glass is 2820 kg/m^3.

Solution

Since the terminal settling velocity is known, start the calculation with V_T^*.

$$\Phi_2^* = V_T^{*3} = \frac{3}{4} \frac{\rho_f^2}{(\rho_s - \rho_f)\mu_f g} v_T^3$$

$$= \frac{3}{4} \frac{982^2 \times (7.4 \times 10^{-3})^3}{(2820 - 982) \times 0.0013 \times 9.81} = 0.0125$$

$$C_D^* = \frac{4.90}{\Phi_2^{*1/2}}(1 + 0.243\, \Phi_2^{*1/3}) + \frac{0.416}{1 + 3.91 \times 10^4\, \Phi_2^{*-1}}$$

$$= 46.29 + 1.33 \times 10^{-7} = 46.29$$

$$Re_p^* = C_D^* V_T^{*3} = 46.29 \times 0.0125 = 0.579$$

$$d_p = \frac{Re_p^* \times \mu_f}{v_T \times \rho_f} = \frac{0.579 \times 0.0013}{7.4 \times 10^{-3} \times 982} = 1.04 \times 10^{-4}\,\text{m}$$

Alternatively, the Concha-Almendra method can be used.

$$d_p^* = 0.070\left[\left(1 + \frac{68.49}{V_T^{*3/2}}\right)^{1/2} + 1\right]^2 \quad V_T^{*2} = 2.504$$

$$Re_p^* = d_p^* V_T^* = 2.504 \times (0.0125)^{1/3} = 0.581$$

$$d_p = \frac{Re_p^* \times \mu_f}{v_T \times \rho_f} = 1.04 \times 10^{-4}\,\text{m}$$

4.5.2 Stokes' law

When the relative velocity between the particle and the fluid is very low, the relationships that define the interaction between the fluid and the particle become very simple. Under this extreme condition

$$C_D = \frac{24}{Re_p} \tag{4.110}$$

The force experienced by the particle is given by Equation 4.87

$$\text{Force} = 12\frac{A_c}{d_p}\mu_f v \tag{4.111}$$

The terminal settling is obtained by substitution of this equation into Equation 4.96

$$\frac{\pi}{6} d_p^3 (\rho_s - \rho_f) g = \frac{24}{2 Re_p} \rho_f v_T^2 \frac{\pi}{4} d_p^2 \tag{4.112}$$

$$v_T = \frac{\rho_s - \rho_f}{18\mu_f} g d_p^2 \tag{4.113}$$

This is known as Stokes' law of settling. Stokes' law is particularly useful for the analysis of the motion of very small particles. It should not be used if the particle Reynolds number is greater than 1.0.

Illustrative example 4.3
Use Stokes' law to calculate the terminal settling velocity of a glass sphere of diameter 0.1 mm in a fluid having density 982 kg/m^3 and viscosity 0.0013 kg/ms. Density of glass is 2820 kg/m^3.

Solution
Stokes' law

$$v_T = \frac{\rho_s - \rho_f}{18\mu_f} g d_p^2$$

$$= \frac{2820 - 982}{18 \times 0.0013} 9.81 (0.1 \times 10^{-3})^2$$

$$= 7.71 \times 10^{-3} \text{ m/s}$$

Calculate the Reynolds number at terminal settling velocity to establish whether Stokes' law can be applied.

$$Re_p^* = \frac{d_p v_T \rho_f}{\mu_f}$$

$$= \frac{0.1 \times 10^{-3} \times 7.71 \times 10^{-3} \times 9.82}{0.0013} = 0.582$$

This satisfies the condition $Re_p < 1.0$ for the application of Stokes' law.

4.5.3 Isolated isometric particles of arbitrary shape

When the particles are not spherical but at least isometric (i.e. they have roughly equal dimensions in three orthogonal directions) useful correlations have been developed by Concha and Barrientos (1986). There are two factors that make the non-spherical particles differ from their spherical counterparts: the surface area per unit volume will be larger giving rise to larger surface drag and the irregular shape of the particle gives rise to oscillatory and vibratory motions as the particle settles through the fluid. The former effect is significant at all particle Reynolds numbers while the latter effect is significant only under turbulent flow conditions.

The particle size is now defined as the equivalent volume diameter.

$$d_e = \left(\frac{6v_p}{\pi}\right)^{1/3} \tag{4.114}$$

which is the diameter of the sphere having the same volume as the particle. Concha and Barrientos define a modified drag coefficient and a modified particle Reynolds number as

$$C_{DM} = \frac{C_D}{f_A(\psi)f_C(\lambda)} \tag{4.115}$$

and

$$Re_M = \frac{Re_p}{[f_B(\psi)f_D(\lambda)]^2} \tag{4.116}$$

In these equations ψ represents the particle sphericity and λ the density ratio

$$\psi = \frac{\text{Surface area of sphere with equal volume}}{\text{Surface area of particle}} \tag{4.117}$$

$$= \frac{\pi d_e^2}{a_p}$$

$$\lambda = \frac{\rho_s}{\rho_f} \tag{4.118}$$

The functions f_A, f_B, f_c and f_D account for the effect of sphericity and density ratio on the drag coefficient and the particle Reynolds number.

The Abrahams equation is modified in a simple way to allow for non-spherical particles.

$$C_{DM} = 0.284\left(1 + \frac{9.06}{Re_M^{1/2}}\right)^2 \tag{4.119}$$

The empirical functions proposed by Concha and Barrientos are given by

$$f_A(\psi) = \frac{5.42 - 4.75\psi}{0.67} \tag{4.120}$$

$$f_B(\psi) = \left(0.843 f_A(\psi)\log\frac{\psi}{0.065}\right)^{-1/2} \tag{4.121}$$

$$f_C(\lambda) = \lambda^{-0.0145} \tag{4.122}$$

$$f_D(\lambda) = \lambda^{0.00725} \tag{4.123}$$

Note that

$$f_A(1) = f_B(1) = f_C(1) = f_D(1) = 1.0 \tag{4.124}$$

These functions allow the definition of modified dimensionless particle size and dimensionless settling velocity as follows:

$$d_{eM}^* = d_e^* \, (\beta(\psi))^{2/3} \, (\eta(\lambda))^{2/3} \tag{4.125}$$

and

$$V_m^* = \frac{V_T^*}{\alpha(\psi)\gamma(\lambda)\beta(\psi)^{2/3}\,\eta(\lambda)^{2/3}} \tag{4.126}$$

where d_e^* and V_T^* are evaluated from Equations 4.100 and 4.101 using d_e rather than d_p in Equation 4.100. The new functions α, β, γ and η are related to f_A, f_B, f_c and f_D as follows:

$$\alpha(\psi) = f_B^2(\psi) \tag{4.127}$$

$$\beta(\psi) = (f_A^{1/2}(\psi) f_B^2(\psi))^{-1} \tag{4.128}$$

$$\gamma(\lambda) = f_D^2(\lambda) \tag{4.129}$$

$$\eta(\lambda) = (f_c(\lambda)^{1/2} f_D(\lambda)^2)^{-1} \tag{4.130}$$

With these definitions of $\alpha(\psi)$, $\beta(\psi)$, $\gamma(\lambda)$ and $\eta(\lambda)$, it is easy to show that the modified variables satisfy the same consistency relationships 4.102 and 4.103 at terminal settling velocity.

$$Re_M^* = (\Phi_{1M}^* \Phi_{2M}^*)^{1/3} = d_{eM}^* V_M^* \tag{4.131}$$

and

$$C_{DM} = \frac{Re_M^*}{\Phi_{2M}^*} = \frac{\Phi_{1M}^*}{Re_M^2} = \frac{d_{eM}^*}{V_M^{*2}} \tag{4.132}$$

With these definitions the modified dimensionless terminal settling velocity is related to the modified dimensionless particle diameter by

$$V_M^* = \frac{20.52}{d_{eM}^*}[(1 + 0.0921d_{em}^{*3/2})^{1/2} - 1]^2 \tag{4.133}$$

and

$$d_{eM}^* = 0.070\left[\left(1 + \frac{68.49}{V_M^{*3/2}}\right)^{1/2} + 1\right]^2 V_M^{*2} \tag{4.134}$$

which are identical in form to Equations 4.108 and 4.109. In practice it may be more convenient to work with V_T^* and d_e^* and then the relationship between them varies when the particle shape factor varies. This is shown in Figure 4.7 for a number of values of ψ.

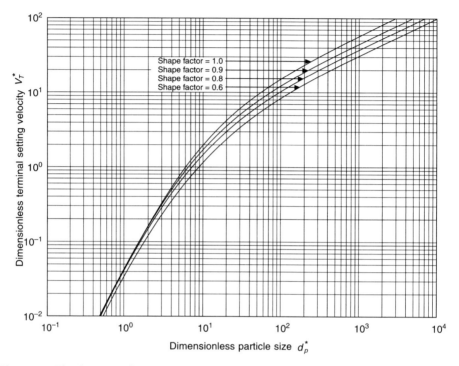

Figure 4.7 The dimensionless terminal settling velocity as a function of the dimensionless effective particle diameter shown for different values of the particle shape factor

---

Illustrative example 4.4
Calculate the terminal settling velocity of a glass cube having edge dimension 0.1 mm in a fluid of density 982 kg/m³ and viscosity 0.0013 kg/ms. The density of the glass is 2820 kg³.
   Calculate the equivalent volume diameter and the sphericity factor.

Solution

$$d_e = \left(\frac{6v_p}{\pi}\right)^{1/3} = \left(\frac{6 \times 10^{-12}}{\pi}\right)^{1/3} = 1.241 \times 10^{-4} \text{ m}$$

$$\psi = \frac{\pi d_e^2}{a_p} = \frac{\pi(1.241 \times 10^{-4})^2}{6 \times 10^{-8}} = 0.806$$

$$\lambda = \frac{\rho_s}{\rho_f} = \frac{2820}{982} = 2.872$$

$$f_A(\psi) = \frac{5.42 - 4.75\psi}{0.67} = 2.375$$

$$f_B(\psi) = \left(0.843 f_A(\psi) \log \frac{\psi}{0.065}\right)^{-1/2}$$

$$= 0.676$$

$$f_c(\lambda) = 0.985$$

$$\alpha(\psi) = f_B^2(\psi) = 0.457$$

$$\beta(\psi) = (f_A^{1/2}(\psi) f_B^2(\psi))^{-1} = 1.421$$

$$\gamma(\psi) = f_D^2 = 1.015$$

$$\eta(\psi) = (f_c(\lambda)^{1/2} f_D(\lambda)^2)^{-1} = 0.992$$

$$d_e^* = \left[\frac{4}{3}(\rho_s - \rho_f)\, \rho_f \frac{g}{\mu_f^2}\right]^{1/3} \quad d_e = 2.989$$

$$d_{eM}^* = d_e^* (\beta(\psi))^{2/3} (\eta(\lambda))^{2/3} = 3.757$$

$$V_m^* = \frac{20.52}{d_{eM}^*} [(1 + 0.0921 d_{eM}^{*3/2})^{1/2} - 1]^2$$

$$= 0.467$$

$$V_T^* = V_M^* \, \alpha(\psi)\, \gamma(\lambda)\, \beta(\psi)^{2/3}\, \eta(\lambda)^{2/3}$$

$$= 0.273$$

This combination of d_e^* and V_T^* can be checked on Figure 4.7.

$$v_T = V_T^* \left[\frac{3}{4} \frac{\rho_f^2}{(\rho_s - \rho_f)\mu_{fg}}\right]^{-1/3}$$

$$= 8.70 \times 10^{-3} \text{ m/s}$$

4.5.4 *General principles of the operation of the hydrocyclone*

The principle of operation of the hydrocyclone is based on the concept of the terminal settling velocity of a solid particle in a centrifugal field. The conditions in an operating hydrocyclone can be described by reference to Figures 4.8 and 4.9. The feed enters tangentially into the cylindrical section of the hydrocyclone and follows a circulating path with a net inward flow of fluid from the outside to the vortex finder on the axis. The high circulating velocities generate large centrifugal fields inside the hydrocyclone. The centrifugal field is usually high enough to create an air core on the axis that usually extends from the spigot opening at the bottom of the conical section through the vortex finder to the overflow at the top. In order for this to occur the centrifugal force field must be many times larger than the gravitational field.

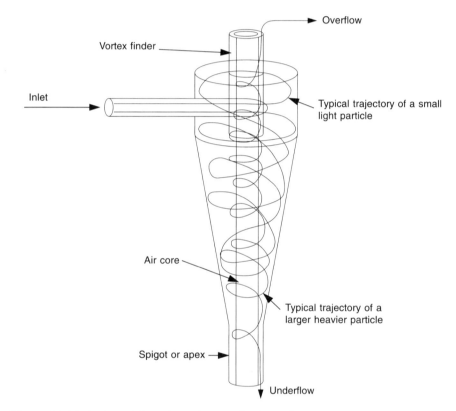

Figure 4.8 Typical particle trajectories in a hydrocyclone

Particles that experience this centrifugal field will tend to move outwards relative to the carrier fluid because of their relatively greater density. The larger, heavier particles will migrate rapidly to the outside walls of the cylindrical section and will then be forced to move downward on the inside of the conical wall. Small, light particles, on the other hand will be dragged inwards by the fluid as it moves toward the vortex finder. The drag force experienced by any particle will be a complex function of the hydrodynamic conditions inside the hydrocyclone and the shape and size of the particle.

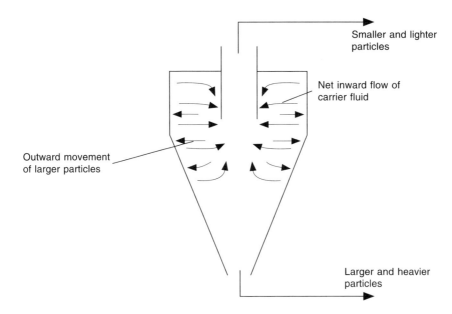

Smaller and lighter
particles

Net inward flow of
carrier fluid

Outward movement
of larger particles

Larger and heavier
particles

Figure 4.9 Schematic representation of the net flow of water and counter flow of
larger particles in the hydrocyclone

4.5.5 The equilibrium orbit hypothesis

The classification action of the hydrocyclone is determined by the net effect
of the two competing forces that act on every particle – the outward centrifugal
force and the inward drag force. A rough guide to the effect of various operating
variables on the performance of the device can be established using the so-
called equilibrium orbit hypothesis. Any particle that experiences an
equilibrium between the drag and centrifugal forces inside the hydrocyclone
will have an equal chance to exit through either the underflow or the overflow
because they will tend to circulate on a circular orbit inside the hydrocyclone.
They will be moved toward one or other outlet only by random impacts with
other particles and the random eddy motion in the highly turbulent flow
field inside the hydrocyclone body. An orbit on which a particle experiences
a balance between the centrifugal and drag forces is called an equilibrium
orbit.

The conditions that define an equilibrium orbit can be defined as follows:
if v_r is the radial velocity of the fluid at a point in the hydrocyclone and u_r the
radial velocity of the particle, the drag force is given by Equation 4.87.

$$\text{Drag force} = 0.5C_D(v_r - u_r)^2 \, \rho_f A_c \qquad (4.135)$$

where ρ_f is the fluid density, A_c the cross-sectional area of the particle and C_D
is the drag coefficient. It is not difficult to show that, within the time taken for
a particle to make a single orbit in the cyclone, the drag force is balanced by
the centrifugal force due to the circulating motion and the particles move at

their terminal settling velocities relative to the inward moving fluid. The centrifugal force is given by

$$\text{Centrifugal force} = \frac{v_\theta^2}{r} v_p (\rho_s - \rho_f) \tag{4.136}$$

where v_θ is the tangential component of the particle velocity vector, and r the radius of the tangential motion. v_p the volume of the particle and ρ_s the density of the solid. Balancing these forces

$$0.5C_D (v_r - u_r)^2 \rho_f A_c = \frac{v_\theta^2}{r} v_p (\rho_s - \rho_f) \tag{4.137}$$

The particles that have a 50% chance of passing to overflow will establish an equilibrium orbit somewhere within the cyclone. The position of this equilibrium orbit is not precisely defined although some authors claim that it is at the point where the locus of zero vertical velocity meets the spigot opening. The uncertainty regarding the actual position of the equilibrium orbit of the 50% particle is not important since we are interested only in establishing a functional form for the relationship among the particle properties that give it a 50% chance of leaving in either the overflow or underflow. All particles that have combinations of density and size that produce a 50% split in the cyclone are assumed to have equilibrium orbits at the same location in the cyclone and this assumption allows a useful correlation to be developed for the cut point as shown in the following analysis.

On an equilibrium orbit $u_r = 0$ and Equation 4.137 can be written for a particle having a 50% chance of passing to overflow.

$$\frac{v_p}{A_c} = \frac{0.5C_D\rho_f}{(\rho_s - \rho_f)} \cdot \frac{rv_r^2}{v_\theta^2} \tag{4.138}$$

Equation 4.138 defines the so-called cut point for the hydrocyclone. This is the size of the particle that has a 50% chance of leaving in either the underflow or overflow. The cut point is normally represented by the symbol d_{50} and for spherical particles

$$\frac{v_p}{A_c} = \frac{2}{3}d_p \tag{4.139}$$

so that

$$d_{50_c} = 0.75 \frac{\rho_f C_D}{(\rho_s - \rho_f)} \cdot \frac{rv_r^2}{v_\theta^2} \tag{4.140}$$

For a particular cyclone the particles that have a 50% chance of passing to overflow will satisfy Equation 4.140 at a particular value of rv_r^2/v_θ^2. These particles define the midpoint of the partition curve as shown in Figure 4.10, and

Equation 4.140 provides a correlation for the cut point as a function of particle properties.

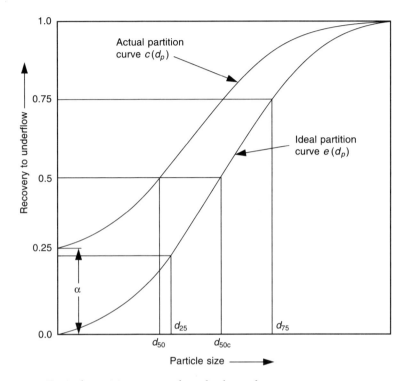

Figure 4.10 Typical partition curves for a hydrocyclone

The drag force experienced by a particle due to the relative motion between it and the fluid is the key to using the equilibrium orbit theory as a basis for modeling the behavior of the hydrocyclone. It is known that these relative velocities are small and the particle Reynolds number is usually so low that it is commonly assumed that the drag force can be calculated from the formula for the slow relative motion between a sphere and a Newtonian fluid. This is the well-known Stokes formula. This approach neglects two important phenomena: the highly turbulent nature of the fluid inside the hydrocyclone and the relatively high concentration of particles. Thus the use of Stokes' flow theories which assumes that only a single isolated particle is present is not very realistic.

Four limiting cases can be considered: isolated particles in a laminar flow field (Stokes regime), isolated particles in a turbulent flow field (Newton regime), interacting particles in a laminar flow field (Blake-Kozeny regime) and interacting particles in a turbulent flow field (Burke-Plummer regime).

The drag coefficient in Equation 4.140 is a function of the particle size and shape and the environment that the particle finds on its equilibrium orbit. Each of the four environments defined above give different expressions for C_D.

Stokes regime

In this regime the particles do not interact with each other and are surrounded by fluid in laminar motion. The drag coefficient is given by Equation 4.110.

$$C_D = \frac{24}{Re_p}$$

$$= \frac{24\mu_f}{d_p(v_r - u_r)\rho_f} \tag{4.141}$$

$$= \frac{24\mu_f}{d_p v_r \rho_f} \quad \text{on an equilibrium orbit}$$

Substituting in Equation 4.140 and noting that $d_p = d_{50c}$ on an equilibrium orbit:

$$d_{50c} = \frac{18.0\mu_f}{(\rho_s - \rho_f)d_{50}} \cdot \frac{rv_r}{v_\theta^2}$$

$$d_{50c}^2 = \frac{K_1}{(\rho_s - \rho_f)} \cdot \frac{rv_r}{v_\theta^2} \tag{4.142}$$

Newtonian regime

In this regime isolated particles are surrounded by fluid in turbulent motion and from the Abraham equation:

$$C_D = 0.28 \tag{4.143}$$

Substituting in Equation 4.141:

$$d_{50c} = 0.22 \frac{\rho_f}{(\rho_s - \rho_f)} \cdot \frac{rv_r^2}{v_\theta^2}$$

$$d_{50c} = \frac{K_2}{(\rho_s - \rho_f)} \cdot \frac{rv_r^2}{v_\theta^2} \tag{4.144}$$

The two limiting cases for the equilibrium orbit hypothesis (Equations 4.142 and 4.144) indicate that the variation of d_{50c} with the density difference and with the volumetric feed rate should take the form:

$$d_{50c} = \frac{KD_c^k}{(\rho_s - \rho_f)^m Q_c^n} \tag{4.145}$$

where m and k are constants having a values between 0.5 and 1, and n is a constant between 0 and 0.5. The lower limit of 0.5 for m and k represents laminar flow conditions and turbulence inside the hydrocyclone will give rise to higher values of m and k and lower values of n.

Experimental observations have shown that the ratio v_r/v_θ is independent of total flow rate at every point inside the hydrocyclone and the assumption of turbulent flow (Equation 4.144) indicates that d_{50c} should be at most a weak function of the total flow rate while the assumption of laminar flow conditions (Equation 4.142) indicates that d_{50c} should be inversely proportional to the square root of the volumetric flow rate. Under turbulent conditions, d_{50c} may be expected to increase linearly with the cyclone diameter.

No one set of assumptions is likely to describe the operating behavior of the hydrocyclone under all conditions. Consequently only comparatively crude empirical correlations are currently available for the prediction of the variation of d_{50c} with hydrocyclone geometry and operating conditions.

4.5.6 *Empirical performance models for hydrocyclones*

Unfortunately no classifier operates perfectly and classifiers will not divide a population of particles into two groups separated at a definite and particular size. All classifiers are characterized by a distribution function which gives the efficiency of separation at any size and the distribution function is more or less sharp depending on the efficiency of separation. A typical classification curve is shown in Figure 4.10. This is called the partition curve (sometimes called the Tromp curve) and it shows the fraction of particles at a particular size that will be partitioned to the coarser fraction. The S shaped curve is typical of all practical classifiers and a variety of quantitative expressions have been used to describe the shape of the curve.

A characteristic of virtually all practical classifiers is the phenomenon of short circuiting. All classifiers exploit some physical process to separate particles on the basis of size. This will be the differential settling velocity in a viscous fluid in spiral, rake and hydrocyclone classifiers or the physical sieving action in a screening operation. Some particles pass through the equipment without being subjected to the physical separation action. In practice it is only the short circuiting of the fine particles to the coarser product stream that is significant. This shows up as a non-zero intersection on the partition axis at zero size. In the hydrocyclone this is due to the water carrying fine particles into the boundary layer on the outer wall of the conical section and discharging them with the underflow. In other classifiers such as spiral, rake and the various screening operations, fine particles are physically carried with the large particles into the coarse product stream.

This short circuit effect can easily be accounted for by reference to Figure 4.11.

If the ideal classification action of the unit is described by classification function $e(d_p)$ and a fraction α of the feed short circuits directly to the

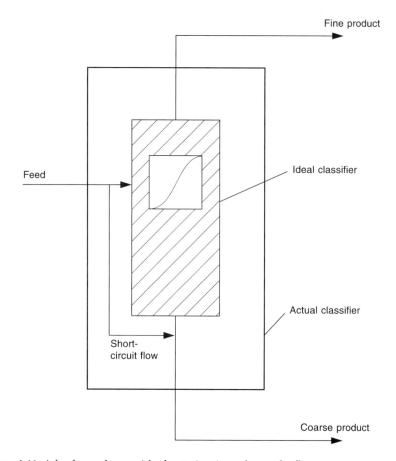

Figure 4.11 A hydrocyclone with short circuit to the underflow

coarser product then a simple mass balance gives the actual classification curve as

$$c(d_p) = \alpha + (1 - \alpha)e(d_p) \tag{4.146}$$

$e(d_p)$ is called the corrected classification function and $c(d_p)$ the actual classification function.

The important parameter that characterizes the operation of the classifier is the size at which $e(d_p)$ is 0.5. This size is usually called the corrected d_{50c} (corrected since the effect of the short circuit flow has been eliminated as shown in Figure 4.10). d_{50c} and d_{50} are defined by the equations

$$e(d_{50c}) = 0.5 \tag{4.147}$$

and

$$c(d_{50}) = 0.5 \tag{4.148}$$

d_{50} is clearly an indicator of the size at which the classifier cuts the particle population.

Although phenomenological models have been developed for the main classifier types, no completely satisfactory procedure exists for the calculation of the classification function in any of the main types of classifier. In the hydrocyclone the function is determined primarily by the turbulent dispersion within the centrifugal velocity field in the conical section and this has been a fertile field of research for many years. On the various types of screening devices, the classification function is determined by the kinetics of the transmission process of particles through the apertures of the screen. These processes have been comprehensively studied but it is not yet possible to make accurate calculations of the classification function in terms of the dimensions of and operating load on the classifier. However, a number of useful empirical functional forms are available to describe the ideal classification function. The most commonly used are:

1. Rosin-Rammler

$$e(d_p) = 1 - \exp(-0.693x^\lambda) \qquad (4.149)$$

2. Exponential sum

$$e(d_p) = \frac{\exp(\lambda x) - 1}{\exp(\lambda x) + \exp(\lambda) - 2} \qquad (4.150)$$

3. Logistic

$$e(d_p) = 1/(1 + x^{-\lambda}) \qquad (4.151)$$

In these equations $x = d_p/d_{50c}$ and λ is a parameter that quantifies the sharpness of the classification. It is relatively easy to appreciate the sharpness of classification in terms of the sharpness index defined by

$$SI = d_{25}/d_{75} \qquad (4.152)$$

with $e(d_{25}) = 0.25$ and $e(d_{75}) = 0.75$. SI has a value between 0 and 1 with low values indicating poor and inefficient separation while a value of 1.0 indicates perfect classification at the cut size d_{50c}.

The parameter λ is related to the sharpness index for each case given above as follows:

1. Rosin-Rammler

$$(SI)^{-\lambda} = 4.819 \qquad (4.153)$$

$$SI = \exp(-1.572/\lambda) \qquad (4.154)$$

$$\lambda = \frac{-1.572}{\ln(SI)} \qquad (4.155)$$

2. Exponential sum

$$SI = \frac{\ln[\exp \lambda + 2)/3]}{\ln[3 \exp \lambda - 2]} \qquad (4.156)$$

λ is usually quite large and Equation 4.156 can be solved approximately for λ in terms of SI as follows:

$$\lambda = 1.099 \frac{(1 + SI)}{(1 - SI)} \tag{4.157}$$

3. Logistic

$$(SI)^{-\lambda} = 9 \tag{4.158}$$

$$SI = \exp(-2.1972/\lambda) \tag{4.159}$$

$$\lambda = \frac{-2.1972}{\ln(SI)} \tag{4.160}$$

These functions are useful for the quantitative description of the behavior of the various classification units but it is necessary to evaluate the three parameters α, d_{50} and λ in terms of the physical dimensions and the actual operating conditions relevant to the unit in question.

Illustrative example 4.5 Effect of particle density
A hydrocyclone was found to have a corrected classification curve given by

$$e(d_p) = 1/(1 + x^{-\lambda})$$

where $x = d_p/d_{50c}$.
   When processing a quartz slurry, $d_{50c} = 16.5$ μm and $SI = 0.72$.
   Calculate the recovery of 10 μm particles to overflow when this hydrocyclone treats a magnetite slurry under comparable conditions. Density of quartz is 2670 kg/m^3 and that of magnetite is 5010 kg/m^3. Recovery of water to underflow is 18.5% in both cases.

Solution

$$\text{For magnetite} \quad d_{50c} = 16.5 \times \left(\frac{2670 - 1000}{5010 - 1000} \right)^m$$

Evaluate this for the two limiting cases: Stokes regime and fully turbulent regime.

$$\text{Stokes regime: } m = 0.5 \quad d_{50c} = 10.65 \text{ } \mu\text{m}$$
$$\text{Turbulent regime } m = 1.0 \quad d_{50c} = 6.87 \text{ } \mu\text{m}$$

Classification function $C(d_{pi}) = \alpha + (1 - \alpha) \dfrac{1}{1 + x^{-\lambda}}$

$$= 0.185 + (1 - 0.185) \frac{1}{1 + \left(\dfrac{10}{10.65} \right)^{-6.686}}$$

$$= 0.185 + (1 - 0.185) \times 0.396$$

$$= 0.5079$$

For the logistic function $\lambda = -\dfrac{2.1972}{\ln(SI)} = -\dfrac{2.1972}{\ln(0.72)}$

$$= 6.689$$

Recovery to overflow $= 1 - 0.5079 = 49.2\%$

4.5.7 The Plitt model for the hydrocyclone

The classification action of the hydrocyclone depends on a balance between the hydrodynamic drag forces that tend to convect the particle toward the axis and the centrifugal force that tends to move the particle outward towards the wall of the cone. The d_{50} size will be determined for the particle that finds these two forces in balance and this gives rise to the equilibrium orbit hypothesis that is discussed in Section 4.5.5. This hypothesis allows some general conclusion to be drawn for predicting d_{50} as a function of hydrocyclone geometry and operating conditions.

The performance of the hydrocyclone is strongly influenced by the short circuit to underflow and this is determined by the volumetric flow split between over- and underflow. The volumetric flow split, S, is a function primarily of the ratio of spigot to vortex finder diameters but is also a strong function of the total feed rate. As the flow rate through the hydrocyclone increases, the diameter of the air core increases choking off the underflow. This effect has not been comprehensively studied and we use a correlation developed by Plitt (1976) based on experimental data.

$$S = \frac{a(D_u/D_o)^b(D_u^2 + D_o^2)^c h^d \exp(0.54\varphi)}{D_c^f H^g} \tag{4.161}$$

where

S = volumetric flow rate in underflow/volumetric flow rate in overflow
D_u = spigot diameter
D_o = vortex finder diameter
D_c = cyclone diameter
φ = volume fraction solids in the feed
h = vortex finder to spigot distance
H = slurry feed head

The following values are recommended for hydrocyclones operating with free discharge: $a = 3.79, b = 3.31, c = 0.36, d = 0.54, f = 1.11, g = 0.24$. The inverse dependence on the feed head H should be noted. As the head increases the flow through the hydrocyclone increases, the centrifugal field increases and the air core expands choking off the discharge from the spigot.

The following recoveries are defined:

R_v = volumetric flowrate in the underflow/volumetric flowrate in the feed
R_f = recovery of fluid phase to the underflow
R_s = recovery of solid phase to the underflow.

The volumetric recovery to underflow is related to S by:

$$R_v = \frac{S}{S+1} \tag{4.162}$$

The bypass fraction α is assumed to be equal to the fraction of water that reports to the underflow R_f, and this is related to the recovery of solids R_s by

$$R_v = R_s\varphi + R_f(1-\varphi) \tag{4.163}$$

$$\alpha = R_f = \frac{R_v - R_s\varphi}{1-\varphi} \tag{4.164}$$

R_s is determined from Equation 4.29 by the actual classification function which in turn is itself a function of R_f

$$R_s = \sum_i c(d_{pi})p_i^F \tag{4.165}$$

where p_i^F is the particle size distribution in the feed.

Using Equation 4.146

$$R_s = \sum_i [R_f + (1-R_f)\, e(d_{pi})]\, p_i^F$$

$$= R_f + (1-R_f)\sum_i e(d_{pi})\, p_i^F \tag{4.166}$$

Substitution into Equation 4.164 and simplifying gives

$$R_f = \frac{R_v - \varphi \sum_i e(d_{pi})p_i^F}{1 - \varphi \sum_i e(d_{pi})p_i^F} \tag{4.167}$$

Thus the hydrocyclone performance is completely determined if the corrected classification function $e(d_p)$ is known together with the volumetric flow split S.

The Rosin-Rammler model 4.149 is used to describe the partition function, which requires two parameters, d_{50} and λ. Plitt has correlated these parameters in terms of the cyclone geometry and the operating variables as follows:

$$d_{50c} = \frac{aD_c^b D_i^c D_o^d \exp[6.3\varphi]}{D_u^f h^g Q^i (\rho_s - \rho_f)^{0.5}} \tag{4.168}$$

D_i is the inlet diameter and Q the volumetric flowrate to the cyclone.

Recommended values for the constants are $a = 2.69 \times 10^3$ and $b = 0.46$, $c = 0.6$, $d = 1.21$, $f = 0.71$, $g = 0.38$, $i = 0.45$. With this value of a, d_{50c} from Equation 4.168 will be in microns. Comparison of Equation 4.168 with Equation 4.145 indicates that Plitt's model for d_{50c} is consistent with the main conclusions drawn from the equilibrium orbit hypothesis in that d_{50c} varies roughly in proportion to the cyclone diameter (cyclone size to the power $b + c + d - f - g = 1.18$) and inversely with feed rate to a power less than 1. The 0.5 power dependence on $\rho_s - \rho_f$ indicates that the interaction between particles and fluid is governed by Stokes' law but slightly higher values can be used.

The parameter λ in Equation 4.149 is correlated by:

$$\lambda = a \left(\frac{D_c^2 h}{Q} \right)^b \exp(-1.58 R_v) \tag{4.169}$$

Recommended values for the constants are $a = 2.96$ and $b = 0.15$. This equation reveals that the efficiency of separation depends on the volumetric recovery R_v but is a comparatively weak function of the cyclone size particularly when the feed rate Q is properly matched to the cyclone size.

It is useful to convert Equation 4.169 to a consistent basis for all classification functions using Equation 4.155 as follows:

$$\ln(SI) = -1.24 \left(\frac{Q}{D_c^2 h} \right)^{0.15} \exp(1.58 R_v) \tag{4.170}$$

The value of the selectivity index calculated from Equation 4.170 can be used in Equations 4.157 and 4.160 to evaluate λ for exponential sum and logistic partition functions.

The values of the parameters a in Equations 4.161, 4.168 and 4.169 are often estimated from experimental data obtained from an existing cyclone installation in order to make the model correspond to the actual operating performance and each of these parameters can be multiplied by a separate calibration factor.

In spite of the empirical nature of the Plitt hydrocyclone model it has proved to be robust for practical work. The chief source of uncertainty is in the prediction of the flow split S.

The particle size distribution in the overflow stream is given by:

$$p_i^O = \frac{(1 - c(d_{pi}))p_i^F}{\sum_i (1 - c(d_{pi}))p_i^F} \tag{4.171}$$

$$p_i^U = \frac{c(d_{pi})p_i^F}{\sum_i c(d_{pi})p_i^F} \tag{4.172}$$

It often happens that the different minerals present in the solid have different densities. Then the classification function will be a function of the particle size and the particle composition and the size distribution density in the overflow will be given by:

$$p_i^O = \sum_j p_{ij}^O (d_p, g) \tag{4.173}$$

$$= \frac{\sum_j (1 - c(d_{pi}, g_i))p_{ij}^F (d_p, g)}{\sum_i \sum_j (1 - c(d_{pi}, g_i))p_{ij}^F (d_p, g)} \tag{4.174}$$

The recovery of solids is given by:

$$R_s = R_f + (1 - R_f)\,\bar{\rho}_s^F \sum_i \sum_j \frac{e(d_{pi},\, g_j)}{\rho_s(g_j)}\, p_{ij}^F(d_{pi},\, g_j) \tag{4.175}$$

where $\bar{\rho}_s^F$ is the average density of the solids in the feed and $\rho_s(g)$ is the density of a particle of composition g.

Equation 4.167 becomes:

$$\alpha = R_f = \frac{R_v - \varphi\bar{\rho}_s^F \sum_i \sum_j \dfrac{e(d_{pi},\, g_j)}{\rho_s(g_j)}\, p_{ij}^F(d_{pi},\, g_j)}{1 - \varphi\bar{\rho}_s^F \sum_i \sum_j \dfrac{e(d_{pi},\, g_j)}{\rho_s(g_j)}\, p_{ij}^F(d_{pi},\, g_j)} \tag{4.176}$$

4.6 Capacity limitations of the hydrocyclone

The capacity of the hydrocyclone is essentially limited by the ability of the spigot opening to discharge solids. If the solids load is too high the solids content of the underflow increases and the viscosity of the discharged pulp becomes high. The tangential velocity of the slurry at the spigot decreases and the usual umbrella-shaped discharge becomes a rope-like stream of dense pulp. Under these conditions the cyclone is said to be roping and the classifier performance is significantly degraded with the discharge of significant quantities of oversize material through the overflow. Hydrocyclones should not normally be operated in the roping condition and they are often installed in clusters to spread the feed over several units to ensure that no one cyclone is overloaded. When a cyclone is operating properly the discharge should flare out in an umbrella shape.

Preliminary selection of hydrocyclones is usually done using manufacturers' charts. These show regions of operation of each cyclone in the manufacturer's range as a function of the inlet pressure and the volumetric flow rate that must be processed. Typically a range of cut sizes is associated with each hydrocyclone. It is generally not possible to use these selection charts effectively until an estimate of the circulating load is known. It is however a simple matter to make a tentative selection of an approximate cyclone size and then determine the number of cyclones in a cluster to handle the required load by simulating the operation of the entire circuit including any comminution and other processing units.

4.7 Symbols used in this chapter

AUF Area utilization factor.
b Width of screen.
$c(d_p)$ Classification function.

C_D Particle drag coefficient.
d_{pi} Representative size of particles in size class i.
d_w Wire diameter.
e Screen efficiency.
G_c Near-size factor.
h Mesh size of screen.
h_T Throughfall aperture of screen.
L_c Length of crowded section of screen.
M Load on screen per unit area.
OA   Open area of light-medium wire mesh.
p_i Mass fraction of particle population in size class i.
P Cumulative distribution of particle size.
SI   Sharpness index.
W Mass flowrate.
$W(l)$ Mass flow down screen per unit width of screen.
θ Angle of inclination of screen.
ρ_B Bulk density.

References

Abraham, F.F., (1970) Functional dependence of drag coefficient of a sphere on Reynolds number. *Phys. Fluids,* Vol. 13, pp. 2194–2195.

Concha, F. and Almendra, E.R. (1979) Settling velocities of particulate systems. *International Journal of Mineral Processing,* Vol. 5, pp. 349–367.

Concha, F. and Barrientos (1986) Settling velocities of particulate systems: 4. Settling of nonspherical isometric particles. *International Journal of Mineral Processing,* Vol. 18, pp. 297–308.

Ferrara, G. Preti, U. and Schena, G.D. (1988) Modelling of screening operations. *International Journal of Mineral Processing,* Vol. 22, pp. 193–222.

Karamanev, D.G., (1996) Equations for calculation of terminal velocity and drag coefficient of solid spheres and gas bubbles. *Chem. Eng. Communications,* Vol. 147, pp. 75–84.

Karra, V.K. (1979) Development of a model for predicting the screening performance of vibrating screens. *CIM Bulletin,* pp. 168–171.

Plitt, V.R. (1976) A mathematical model of the hydrocyclone classifier. *CIM Bulletin,* Dec. 1976, pp. 114–123.

Turton, R. and Levenspiel, O. (1986) A short note on drag correlation for spheres. *Powder Technology,* Vol. 47, pp. 83–86.

5
Comminution operations

5.1 Fracture of brittle materials

5.1.1 Fracture toughness and critical stress

Particulate materials are fractured primarily by the imposition of compressive stress applied fairly rapidly by impact. Subsidiary fracture is caused by high shearing stress particularly at the surface of the particle. The former leads to cleavage and shattering while the latter leads to attrition and wear. All materials resist fracture to a greater or lesser extent and energy must be expended to break a solid.

It is important to understand the difference between hardness and toughness. Hardness of a material represents its ability to withstand indentation and therefore deformation. Mineralogical materials are generally quite hard. Toughness on the other hand is the ability to withstand stress without fracture or failure. It is the latter characteristic that is primarily of interest in comminution. There is usually a definite correlation between hardness and toughness. Most mineralogical materials are brittle. They resist the tendency to deform plastically and they show a linear relationship between stress and strain. However, they are generally not always tough and many can be fractured relatively easily.

The reason for the comparatively low toughness is the presence of micro and macro flaws in the materials. Under stress, a flaw leads to an increase in stress intensity at the flaw tip or edge. Thus the material can be stressed locally beyond its strength at the flaw edge. The energy required for bond breakage and crack propagation into the material is readily available as stored elastic energy when the particle as a whole is stressed and this energy can be released rapidly at the flaw edge when the crack grows.

A theory of brittle fracture was developed by Griffith during the 1920s and this led ultimately to the modern theory of fracture mechanics. In essence Griffith was able to explain why the actual strength of brittle materials like glass was considerably less than the strength expected from theoretical calculations assuming that the interatomic bonds between pairs of atoms need to be ruptured. Griffith postulated that the low observed strengths of brittle elastic materials were due to the presence of small cracks or flaws in the material. When a material containing flaws is stressed, the flaws act as stress raisers so that the stress along the edge of the flaws is considerably larger than the average stress across the cross-section of the material.

The stress distribution in the vicinity of a crack that has been opened to a shallow elliptical cross-section can be calculated. These calculations indicate that the stress at the edge of the crack is given by:

$$\sigma_e = 2\sigma\left(\frac{L}{r}\right)^{1/2} \tag{5.1}$$

where σ is the average stress, L is half the crack width and r the radius of curvature of the crack opening at the crack edge.

The theoretical stress required to fracture the interactomic bonds can be shown to be

$$\sigma_{th} = \left(\frac{\gamma Y}{\alpha}\right)^{1/2} \tag{5.2}$$

where γ is the surface tension of the solid, Y is Young's modulus and α is the mean interatomic distance across the fracture plane at zero stress.

Equation 5.2 can be used to calculate the theoretical strength of some typical solids. In many crystalline solids α is about 3 Å (3×10^{-10} m), Y about 10^{10} N/m^2 and γ about 1 N/m. This makes σ_{th} approximately 6000 Mpa, which is considerably larger than the strength of most crystalline solids.

The crack will grow if $\sigma_e \geq \sigma_{th}$ so that the critical failure stress σ_f can be found from

$$2\sigma_f\left(\frac{L}{r}\right)^{1/2} = \left(\frac{\gamma Y}{\alpha}\right)^{1/2} \tag{5.3}$$

$$\sigma_f = 1/2\left(\frac{\gamma Y}{L}\frac{r}{\alpha}\right)^{1/2} \tag{5.4}$$

or

$$\sigma_f = \text{constant}\left(\frac{\gamma Y}{L}\right)^{1/2} \tag{5.5}$$

Equation 5.5 suggests that the group $\sigma_f L^{1/2}$ will have a critical value for each material. It is customary to define the stress intensity factor, K, as the critical value of K with $\sigma = \sigma_f$ as an indicator of rapid crack growth. The critical value of K is called the fracture toughness of the material K_c.

$$K_c = \sigma_f(\pi L)^{1/2} \tag{5.6}$$

and this parameter measures the resistance of a material to crack extension. A crack will grow if $K > K_c$. Unfortunately this criterion cannot be used directly to describe fracture of irregular and inhomogeneous particles because it is impossible to evaluate the stress distribution inside the particle when it is stressed. Experimental measurements show that the conditions for unstable crack growth normally exist at the crack tip under conditions where the particle is subject to impact loading such as would be experienced in an industrial comminution machine (Anderson, 1995). This means that the crack grows very rapidly. The energy that is required to drive this rapid crack growth comes from the elastic energy that is stored in the particle due to deformation during the rapid increase in stress prior to fracture. The energy that is stored during impact prior to fracture represents the minimum energy that is required to break the particle under the particular external loading and stress pattern that the particle experiences. This energy is called the particle fracture energy and its measurement and use is described later.

Unfortunately only few measurements of fracture toughness have been made on rocks and similar mineralogical materials. Values of the fracture toughness ranging from 0.8 Mpa m$^{1/2}$ for limestone to 2.8 Mpa m$^{1/2}$ for diorite have been measured. The value for quartzite has been measured at about 2.4 Mpa m$^{1/2}$.

The Griffith theory indicates that the larger the flaw size the lower the stress required for fracture. It is tensile stress that is required to cause fracture and it is difficult to impose tensile stress directly on a particle because it cannot be readily gripped and pulled. Tensile stress is induced indirectly by non uniform compressive stress.

Consider a spherical particle compressed by a force P as shown in Figure 5.1. The normal stresses in the z and x directions are shown. The principal stress in the x–y plane is thus tensile throughout a considerable portion of the particle. Thus local compressive stresses on the microscale can produce local tensile stresses and cause fracture.

5.2 Patterns of fracture when a single particle breaks

Shatter

This mechanism of fracture is induced by rapid application of compressive stress. A broad spectrum of product sizes is produced and this process is unselective. Multiple fracture processes occur so that progeny particles are immediately subject to further breakage by successive impacts. In reality the shattering process consists of a series of steps in which the parent particle is fractured and this is followed immediately by the sequential fracturing of successive generations of daughter fragments until all of the energy available for fracture is dissipated. These successive fractures take place in very rapid succession and on the macro time scale they appear to be one event.

This fracture mechanism is illustrated in Figure 5.2, which shows the fracture pattern that can be expected, the range of daughter fragments that are formed and the shape of the size distribution density function for the daughter fragments of a single fracture event. The population of progeny particles is made up of a number of sub populations – those that originate from primary

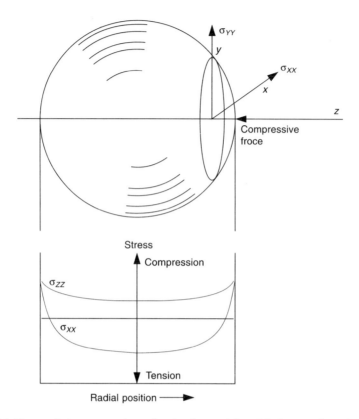

Figure 5.1 Two-point compression of a single particle with the resultant internal stress pattern

fracture processes and those that result from successive rebreakage. The size distribution of the progeny population will reflect this mixture of sub-populations and forms the basis for models of the breakage function. Shattering is the most common mode of fracture that occurs in industrial autogenous, rod and ball mills.

Cleavage

When the original solid has some preferred surfaces along which fracture is likely to occur, cleavage results. If multiple fracture of the daughter fragments does not occur, this mechanism of fracture tends to produce several relatively large fragments together that reflect the grain size of the original material with much finer particles that originate at the points of application of the stress. The size distribution of the product particles is relatively narrow but will often be bimodal as illustrated in Figure 5.3 or even multi-modal.

Attrition and chipping

Attrition occurs when the particle is large and the stresses are not large

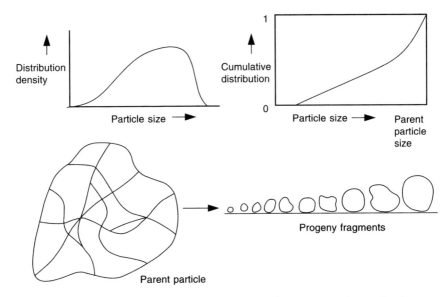

Figure 5.2 The shattering process produces a broad spectrum of particle sizes

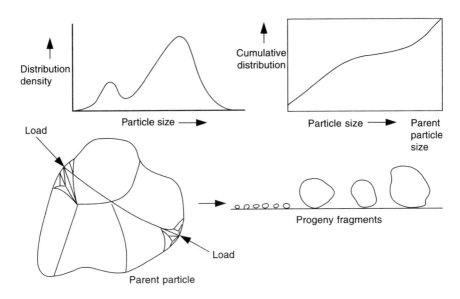

Figure 5.3 Fracture by cleavage without successive fracture of progeny

enough to cause fracture. This often occurs in autogenous mills where large particles are present to act as media. Notice that the parent particle hardly changes size but the attrition process generates a significant number of particles that are much smaller than the parent size. This is called a birth process because of the appearance of small particles. However, there is no corresponding death process since the parent particle is not destroyed and at most moves across the lower boundary of its size class to the size class below. This behavior is reflected in the population balance models for autogenous mills. This comminution process is illustrated in Figure 5.4. The particle size distribution density for the progeny particles shows up as a distinct peak at the small sizes. This peak is well separated from the peak generated by the residual parent particles which distribute themselves in a narrow size range just below the parent size. The two peaks are separated by a range of sizes that contains no particles. This gives rise to a flat plateau in the cumulative distribution function as shown in Figure 5.4. This is a characteristic of attrition breakage.

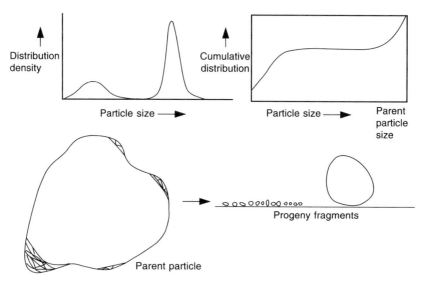

Figure 5.4 Attrition and chipping produce products with a narrow size spectrum at the parent size and another hump at small sizes

By contrast shatter and cleavage give rise to both birth and death processes. Sometimes it is convenient to regard attrition as a wear process although there is no clear dividing line between wear and attrition.

5.3 Breakage probability and particle fracture energy

Fracture of particulate material is a random process. Many factors contribute to the fracture of any single particle and the result is never exactly predictable. It is never certain whether a particular particle that is involved in an impact will actually break but it is known that the probability of fracture is related to the amount of energy that is absorbed by the particle during the impact event. When a particle is subject to an impact it is able to absorb only a certain amount of energy before it fractures. This is known as the particle fracture energy and it is a property of the particle. The particle fracture energy of a particle is a material property that is a function of its size and shape and also of the orientation that the particle presents to the impact stress. Obviously different materials also show significant differences in particle fracture energy.

Particles of the same material, similar in size and shape, show a wide variation in individual fracture energies. This reflects the inherent randomness of the fracture process and the wide distribution of microscopic flaws that exist in particles found in any naturally occurring particle population. An individual particle does not have a unique particle fracture energy. The value will depend on the orientation of the particle relative to the impacting forces and relative to the stress field external to the particle such as when it is stressed as part of a bed of particles. Measurements of the fracture energy are usually made using an experimental device such as the fast load cell. Two typical sets of data are shown in Figure 5.5, which shows the measured distribution of particle fracture energies for two populations of quartz particles. These are typical of the measurements made on many different materials. The data shows clearly that smaller particles are stronger than larger particles of the same material – a fact that has been known for many years.

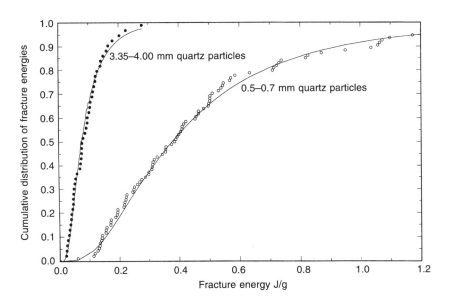

Figure 5.5 Measured distribution of particle fracture energies for two sizes of quartz particles. Data from Bourgeois (1993)

Most materials that have been tested follow a log-normal distribution of particle fracture energies. Results for four different materials are shown in

Figure 5.6 and the variation in the strength of the different materials is immediately obvious. The data in this figure are plotted in the log-normal coordinate system which produces straight lines when the data are log-normally distributed. The data presented in Figure 5.5 also show the strong linear characteristic when plotted in the log-normal coordinate system.

The fracture energy is not only a function of the material and of the particle size but it also reflects the presence of flaws in the material. Flaws in the material prior to impact are referred to as material damage and they are generated during the many natural and artificial processes that influence the particles before they are impacted and broken in a comminution machine. The role that flaws play in particle fracture is described by the Griffith theory of brittle fracture. When a particle is subject to rapid impact stress it initially deforms elastically absorbing impact energy from the impacting devices as stored elastic strain. The internal stresses in the particle increase until the stress intensity at the most sensitive flaw exceeds the Griffith criterion for unstable crack growth. A period of exceedingly rapid crack growth follows with the crack tip accelerating and often bifurcating as well. The energy that drives the growing crack comes from the stored elastic strain energy in the particle, which is available to the growing crack at the requisite high rate. The crack continues to grow until it intersects the edge of the particle at which time the particle is fractured. Any remaining elastic stored energy is dissipated as kinetic energy of the progeny, which may suffer further fracture as a result. This model of particle fracture provides a clear explanation of the particle fracture energy and the role that it plays in the fracture process. It is the energy that is stored as elastic strain up to the instant of first fracture and this represents the minimum energy that must be supplied to the particle before fracture will occur.

In general no one particle will absorb all of the energy that is available in any one impact event and the residual energy that remains after a particle breaks will be available to break progeny particles in the manner described above. This model of particulate fracture suggests that it is the ratio of impact energy to particle fracture energy that will determine the size distribution of the progeny that are produced as a result of a single impact event.

The distribution of fracture energies has another interesting interpretation. The cumulative distribution function also gives the probability that a specific particle will fracture if it receives a given amount of energy during an impact. This follows from the fact that a particle will not break if it is impacted with less than its fracture energy. Consider a representative sample of say 100 particles of quartz in the size range 0.5–0.7 mm. If each of these particles is subjected to an impact of say, 0.5 J/g the data shown in Figure 5.5 indicate that about 35 of the particles will not break because they have particle fracture energies greater than 0.5 J/g. Thus the probability of breakage for these particles is 65% at an impact energy of 0.5 J/g. The same argument can be made for any other impact energy. Thus the curve in Figure 5.5 has two equally valid interpretations. On the one hand it represents the cumulative distribution of particle fracture energies and on the other it represents the breakage probability of an individual particle as a function of impact energy.

The straight lines in Figure 5.6 show that the data are well represented by the log-normal distribution and the breakage probability of a particle that suffers an impact of E J/kg is given by

$$P(E; d_p) = G\left(\frac{\ln(E/E_{50})}{\sigma_E}\right) \tag{5.7}$$

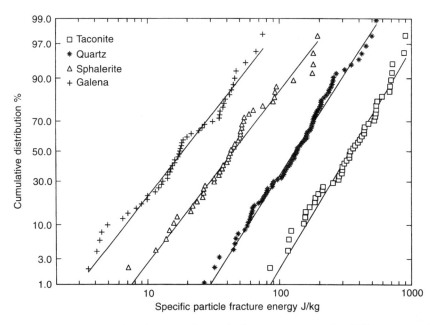

Figure 5.6 Cumulative distribution of particle fracture energies for different materials measured in single-particle fracture tests. The data are plotted in the log-normal coordinate system (Tavares and King, 1998)

with

$$G(x) = \frac{1}{\sqrt{2\pi}} \int_{-\infty}^{x} e^{-t^2/2} dt \tag{5.8}$$

E_{50} is the median fracture energy, which varies with particle size according to

$$E_{50} = E_{\infty}\left(1 + \frac{d_0}{d_p - d_{p\,min}}\right)^{\varphi} \tag{5.9}$$

Values of the parameters d_0 and φ are material specific and have been measured for several common minerals and these are shown in Table 5.1. $d_{p\,min}$ is the lower limit of brittle fracture. Particles smaller than $d_{p\,min}$ are deformed during impact and do not fracture. This limit is about 0.3 μm for quartz. E_∞ is the median particle fracture energy for large particles, usually about 1 cm or more.

Table 5.1 Parameters that relate the mean mass-specific particle fracture energy to particle size for some common minerals. From Tavares and King (1998)

Material	E_∞ J/kg	d_{p0} mm	φ	Size range, mm
Apatite	1.05	19.3	1.62	0.25–8.00
Galena	3.19	7.31	1.03	0.70–7.60
Gilsonite	5.50	7.03	1.60	1.18–10.0
Quartz	43.4	3.48	1.61	0.25–4.75
Sphalerite	7.00	8.24	1.16	0.35–10.0
Magnetite	9.56	3.93	1.96	0.25–7.20
Copper ore	96.1	1.17	1.26	0.25–15.8
Iron ore	47.3	1.08	2.30	0.25–15.0
Limestone	114.2	0.490	2.05	0.35–5.60
Marble	45.9	0.882	2.66	0.50–15.0
Taconite	235.9	0.803	1.42	0.35–6.00
Taconite	163.3	0.856	1.76	0.35–10.0

Equation 5.9 reflects the fact that the measured median particle fracture energy is independent of particle size for particles larger than a few millimeters, but increases quite sharply as the particle size decreases below approximately 1 mm.

5.4 Progeny size distribution when a single particle breaks – the breakage function

It is clear that the breakage of a single particle will produce an entire spectrum of progeny sizes. When dealing with comminution processes, the distribution of breakage product sizes must be considered on a number of levels.

(a) Breakage of a specific single particle only once generating a few daughter fragments. (This is referred to as single-particle primary fracture or cleavage).

(b) Breakage of a single particle followed by sequential breakage of some of the progeny during a single breakage event. This is referred to as single-particle, single-event fracture. All the impact energy is dissipated in a single impact event. Some of the energy is consumed by the fracture of the parent particle and by some members of successive generations of the progeny, some is absorbed by progeny particles without fracture and the remainder is dissipated by steel-on-steel impact. A single-impact event has total duration of a few milliseconds and the successive progeny breakage events follow each other at intervals of a few tens of microseconds.

(c) The breakage of many single particles all of the same size in single-particle, single-impact events that produces statistically many daughter fragments. In other words repeated events of type (b).

(d) The breakage of many particles and the immediately succeeding progeny during a single impact on a bed of particles. Particles in the bed suffer single-particle, single-event impact each one using only a fraction of the total impact energy that is available for the event. This is referred to as a particle-bed single-impact event.

(e) Concurrent breakage of many particles in a comminution machine by particle-bed single impact events. The progeny from these events immediately become available for fracture in later impact events.

Each of these conceptual situations produces a different size distribution and it is important to distinguish clearly between them. (a) is not of much interest except as a basic event to produce the statistically meaningful (c). The product size distribution produced by a test of type (c) is known as the single-particle breakage function and it can be measured in a sequence of single-particle fracture tests all starting with particles of the same size.

The single-particle breakage function is defined by $B(x; y)$ = fraction of daughter particles smaller than size x that result from fracture of a single particle of size y as described in paragraph (c) above.

Single-particle breakage tests are easy to do in the laboratory and a number of ingenious apparatus designs have been developed. The simplest is the drop weight test apparatus which is illustrated in Figure 5.7. A weight, usually spherical, is dropped from a known height on to a single particle that is supported on a hard surface. The kinetic energy of the sphere just prior to impact is equal to mgh where m is the mass of the drop weight. This kinetic energy is transferred to the particle, which may or may not fracture. The test is repeated many times on single particles all having the same size. The fracture products are collected for the determination of the product size distribution.

In spite of the use of a single particle for each test, each fracture event consists of a sequence of individual breakage events. Usually the energy in the drop weight is sufficient to cause a primary fracture to occur and the residual energy in the drop weight is then absorbed by the daughter fragments which in turn are fractured forming their own progeny. The process of breaking successive generations of progeny continues until all of the energy in the drop weight is dissipated or until all of the progeny have been expelled from the impact zone and the drop weight collides with the hard support surface and dissipates all of its remaining energy as metal deformation or as elastic rebound.

The successive breakage and rebreakage of progeny particles in the single-particle, single-impact test means that the breakage function determined from such a test is itself a composite of many individual primary breakage functions

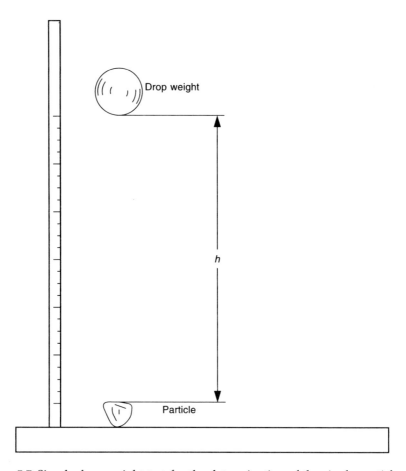

Figure 5.7 Simple drop weight test for the determination of the single-particle breakage function

each one of which results from the breakage of a single progeny fragment. These breakage events follow each other in rapid succession (separated by a few microseconds) and the intermediate products cannot normally be observed. Special experimental techniques have been developed to arrest the impacting drop weight immediately after the first fracture, thus effectively preventing the subsequent breakage of progeny particles. The resulting breakage function is called the primary breakage function and it has been determined for a few situations (Bourgeois, 1993). The primary breakage function is not relevant to conditions in any real comminution equipment and is not considered any further here.

   The number of sequential breakage events that occur during a single drop weight single-particle experiment is determined primarily by the energy of the impacting drop weight. The higher the energy the more sequential breakage and *vice versa*. Consequently the size distribution of the final progeny will be determined to a large extent by the impact energy and to a lesser extent by the nature of the material. A model for the single-particle breakage function that is based on these ideas is given below.

The breakage function plays a pivotal role in the quantitative description of comminution processes. It is not the final product size distribution from any comminution operation because the progeny particles are themselves broken and rebroken in many successive breakage events. In any comminution machine, many particles are involved in each impact event so that the single-particle breakage function that is measured in a single-particle test cannot be applied directly to the breakage processes that occur inside the machine. Single-impact test can be conducted on particle beds and the resultant breakage function can be measured. It is important to distinguish among the primary breakage event, the single-particle impact event and the single-impact particle bed event. Each kind of event is characterized by its own distinctive breakage function. It is only the single-impact particle-bed event that is directly relevant to conditions inside a grinding mill.

A single impact on a bed of particles is regarded as the elementary breakage event that occurs in a grinding mill. Such an event itself consists of many subevents such as the capture of energy by individual particles and the fracture of parent particles and the subsequent immediate rebreakage of some of their progeny. It is the combination of the results of all of these subevents that constitute the elementary breakage event. It is only the result of all the subevents together that can be observed because the subevents occur too quickly and cannot be distinguished. The total event typically lasts no more than a few milliseconds.

5.4.1 *Empirical models for the breakage function that are based on mixtures of progeny populations*

Mixtures of populations with logarithmic distributions

There have been many attempts to determine the form of the function $B(x; y)$ for each kind of breakage event on theoretical grounds but no reliable theoretical model has been found that will reproduce the size distribution that is produced in single-impact fracture tests. However some empirical models have been found to provide good descriptions of the experimental data.

The most popular of these models is based on the idea that the progeny population is made up of a mixture of two separate populations as illustrated in Figures 5.3 and 5.4. Each has a cumulative distribution that can be modeled by

$$B(x; y) \text{ is proportional to } \left(\frac{x}{y}\right)^n \qquad (5.10)$$

Different values of n describe the larger products produced by tensile stress and the smaller products produced by the intense compressive stress at the points of application. The two distributions are added using an appropriate fractional weighting factor.

$$B(x;y) = K\left(\frac{x}{y}\right)^{n_1} + (1 - K)\left(\frac{x}{y}\right)^{n_2} \tag{5.11}$$

The first term in Equation 5.11 describes the size distribution of the fine fraction in the population of progeny particles and K is the fraction of daughter products that contribute to the finer fraction.

The geometry of this function provides an insight into the role that each parameter plays and also provides a convenient method for the estimation of the parameters K, n_1 and n_2 when experimental data is available for the breakage function. The data is plotted on log-log coordinates as shown in Figure 5.9. If $n_2 > n_1$ then

$$\left(\frac{x}{y}\right)^{n_1} > \left(\frac{x}{y}\right)^{n_2} \quad \text{as} \quad \frac{x}{y} \to 0 \tag{5.12}$$

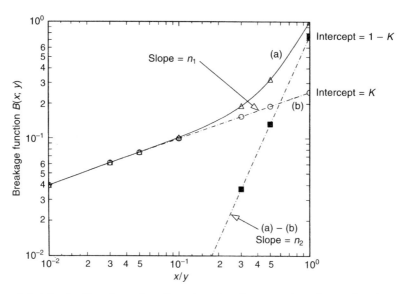

Figure 5.8 A typical breakage function showing the procedure for evaluation of the parameters k, n_1 and n_2

Then at low values of $\dfrac{x}{y}$

$$B(x;y) = K\left(\frac{x}{y}\right)^{n_1} \tag{5.13}$$

which plots as a straight line with slope n_1 on the log-log coordinates as shown in Figure 5.8. The straight line will intersect the ordinate $x/y = 1$ at the point $B(x; y) = K$ as shown. Thus the parameters n_1 and K can be established easily by drawing the tangent from the lower left of the curve and reading the intercept at $x/y = 1$. Once this line has been established, the difference

$$B(x; y) - K\left(\frac{x}{y}\right)^{n_1} = (1 - K)\left(\frac{x}{y}\right)^{n_2} \tag{5.14}$$

can be plotted as shown in the figure. This will produce another straight line of slope n_2 and intercept $(1 - K)$ at $x/y = 1$. The value of n_2 can be determined from the slope of this line.

It sometimes happens that the difference between the original curve (a) and the first straight line (b) is itself a curve rather than a straight line. Such a curve can be resolved into two terms using the same procedure and the entire function $B(x; y)$ will then consist of three power function terms.

The breakage function described by Equation 5.11 is independent of the size of the parent particle and depends only on the ratio x/y. The same function can be used for all parent sizes and the distribution is said to be normalized with respect to the parent size.

If the experimental data do not normalize, the fraction K must be determined

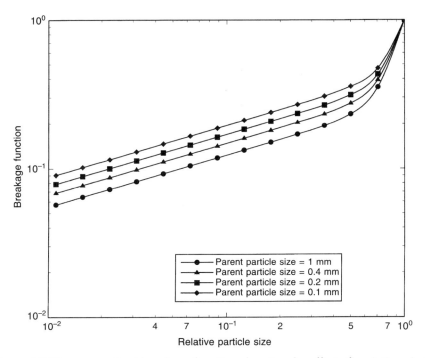

Figure 5.9 Non-normalized breakage function showing the effect of variations in the parent size

as a function of the parent size. This is illustrated in Figure 5.9. Usually this effect is modeled using a simple inverse power function

$$K(y) = K_0\left(\frac{y_0}{y}\right)^\delta \tag{5.15}$$

where y_0 is a reference parent size usually taken to be 5 mm.

As the parent particles become too large to be properly nipped between the media particles, the breakage function becomes bimodal reflecting the greater tendency of the particles to chip rather than to shatter. Under these circumstances the breakage function becomes bimodal as shown in Figure 5.10.

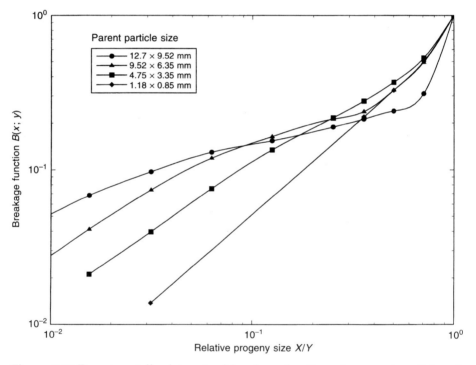

Figure 5.10 Experimentally determined breakage functions for parent particles of different size. 26.6 mm diameter balls were used. The distribution is distinctly bimodal at larger particle sizes. Data from Austin *et al.* (1987)

This behavior can be modeled approximately in the following way.

$$B(x;y) = K\left(\frac{x}{y_0}\right)^{n_3}\left(\frac{x}{y}\right)^{n_1} + (1-K)\left(\frac{x}{y}\right)^{n_2} \quad \text{for } x < y_0 \tag{5.16}$$

and

$$B(x; y) = K \left(\frac{x}{y} \right)^{n_1} + (1 - K) \left(\frac{x}{y} \right)^{n_2} \quad \text{for } x \geq y_0 \qquad (5.17)$$

y_0 is the largest chip size which must obviously be smaller than the parent size y.

The four parameters K, n_1, n_2, and n_3 are functions of the impact energy but the exact dependence has never been systematically investigated. They can also vary with ball size in a ball mill and a weighted average breakage function can be used to increase the accuracy of calculation.

Models for the breakage function that are based on truncated standard distributions

Any of the truncated distributions that are described in Section 2.2.2 are useful models for the breakage function. Typical representations of the truncated Rosin-Rammler, truncated logistic and truncated log-normal distributions are shown in Figure 5.11.

Measurements indicate that the truncated distributions are not always flexible enough to describe the breakage behavior under all conditions, for

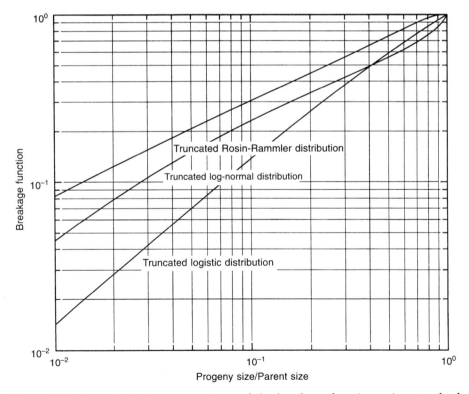

Figure 5.11 Some typical representations of the breakage function using standard truncated distribution functions

example during slow compression of constrained beds and the size distributions measured in single particle breakage tests have been correlated assuming that the progeny are made up from several populations each of which has a truncated log-normal size distribution.

$$B(x; y, E) = \sum_{n=1}^{N} \mu_n(E)F_n(x, y) \tag{5.18}$$

where the functions $F_n(x, y)$ are truncated log-normal distributions given by

$$F_n(x, y) = G\left[\dfrac{\ln\left(\dfrac{x(x_0^{(n)} - x_{50}^{(n)})}{x_{50}^{(n)}(x_0^{(n)} - x)}\right)}{\sigma^{(n)}}\right] \tag{5.19}$$

and G is the normal distribution function defined by

$$G(x) = \frac{1}{\sqrt{2\pi}} \int_{-\infty}^{x} \exp(-t^2/2)dt \tag{5.20}$$

and $x_0^{(n)}$ is the upper truncation size for the nth population. The coefficients $\mu_n(E)$ are the weighting fractions for the separate distributions and these depend quite strongly on the impact energy. The parameters $x_0^{(n)}, x_{50}^{(n)}$ and $\sigma^{(n)}$ of Equation 5.19 are largely insensitive to the impact energy but depend on the parent size y. In Equation 5.19, $F_n(x, y)$ is defined to be 1 whenever $x \geq x_0^{(n)}$. See Dan and Schubert (1990).

5.4.2 Models for the breakage function that are based on the impact energy

Single-impact tests have shown clearly that the breakage function is determined primarily by the amount of energy that is available to break the material during the impact. The larger the amount of energy in the impact, the finer the progeny that result from that impact. This is due largely to the successive breakage and rebreakage of the progeny until the total impact energy is dissipated. Data from a series of single-particle, single-impact tests at different impact energies are shown in Figure 5.12. The data are plotted using a relative size scale $n = $ (parent size/progeny size). The effect of the increasing impact energy is apparent in the data. It is postulated that the same relationship between breakage function and energy will be obtained when the impact involves a bed of particles. In that case however the progeny that results from any one particle in the bed will be governed by the fraction of the total impact energy that is absorbed by that particle. The breakage function that results from the impact as a whole is the sum total of all the individual breakage functions from each particle that absorbs enough energy to break.

Thus the impact event is defined by both the total amount of energy that is available in the impact and the way in which this energy is partitioned among all the particles in the impact zone.

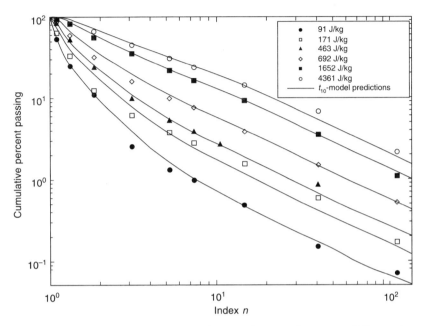

Figure 5.12 Breakage functions measured in single-particle tests at different impact energies. 3.35–4.00 mm quartz. Bourgeois (1993)

Many sets of data of the type shown in Figure 5.12 reveal that the measured size distributions can be characterized effectively in terms of the parameter t_{10} which is defined as the fraction of the progeny that are smaller than 1/10th of the parent size. The breakage function at any other size can be represented in terms of t_n, the fraction smaller than $1/n$ of the parent size. Each curve shown in Figure 5.12 has a unique value of t_{10} and the values of t_n can be uniquely related to t_{10}. Replotting the data in this way produces curves of the type shown in Figure 5.13. It is remarkable that data plotted in this way for a number of different materials that are subject to impact have curves that are similar. Thus this model can be considered to be a single-parameter representation of the breakage function which makes it attractive for practical modeling work. When the measured breakage function is not available for a specific material, a general plot such as that shown in Figure 5.13 can be used to regenerate the breakage function provided that the value of t_{10} is known. The data shown in the figure were obtained from single-particle breakage tests on three different mechanical devices and can be considered representative.

t_{10} is determined primarily by the energy that is absorbed during the impact and is modeled by

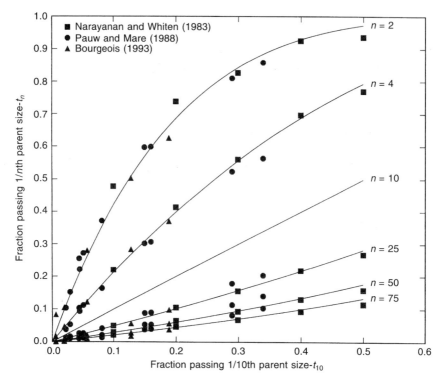

Figure 5.13 Reconstruction plot for the breakage function that is defined in terms of the parameter t_{10}. The solid lines were calculated using Equation 5.25 with $\alpha = 0.75$

Table 5.2 Parameters that relate t_{10} to the specific impact energy for some common minerals

Material	t_{10max}	β
Magnetite	47.1	0.0098
Apatite	45.4	0.0115
Quartz	38.8	0.0176
Hematite	45.6	0.0164
Galena	44.5	0.0176
Limestone	54.5	0.0176
Copper ore 1	44.8	0.0263
Copper ore 2	58.9	0.0204
Basalt	52.0	0.0252
Titanium ore	51.0	0.0269
Cement clinker 1	69.2	0.0276
Cement clinker 2	60.5	0.0437
Marble	76.3	0.0792
Iron ore	65.4	0.0932

$$t_{10} = t_{10max}(1 - e^{-\beta E/E_{50}})$$ (5.21)

The parameters t_{10max} and β are material specific and they can be determined easily in the laboratory using standard drop weight apparatus. The test can be performed on single particles or on beds of particles. Values for some common minerals are given in Table 5.2.

The entire breakage function can be regenerated from the single value of t_{10} using a reconstruction plot such as that shown in Figure 5.13. A number of models have been investigated to provide a suitable structure for the behavior shown in Figure 5.13. Models based on the standard truncated distribution functions are particularly useful because they are characterized in terms of only one or two parameters and these are described in the next section.

The t_{10} model based on truncated distribution functions
The truncated Rosin-Rammler distribution is given by Equation 2.18.

$$P(D) = 1 - \exp\left[-\left(\frac{\eta}{\eta_{63.2}}\right)^{\alpha}\right] \quad \text{for } D \leq d'_p$$ (5.22)

with

$$\eta = \frac{\xi}{1 - \xi} \quad \text{and} \quad \xi = \frac{D}{d'_p}$$

From the definition of t_{10}

$$t_{10} = P\left(\frac{d'_p}{10}\right) = 1 - \exp\left[-\left(\frac{1}{(10-1)\eta_{63.2}}\right)^{\alpha}\right]$$ (5.23)

and from the definition of t_n

$$t_n = P\left(\frac{d'_p}{n}\right) = 1 - \exp\left[-\left(\frac{1}{(n-1)\eta_{63.2}}\right)^{\alpha}\right]$$ (5.24)

When $\eta_{63.2}$ is eliminated from these two equations

$$t_n = 1 - (1 - t_{10})^{\left(\frac{10-1}{n-1}\right)^{\alpha}}$$ (5.25)

which shows the unique relationship between t_n and t_{10}. The parameter α is material specific but representative values for different materials have not yet been investigated.

Similar manipulations give the following relationships for the truncated logistic and truncated log-normal distributions respectively.

$$t_n = \frac{1}{1 + \left(\frac{1 - t_{10}}{t_{10}}\right)\left(\frac{n-1}{9}\right)^{\lambda}}$$ (5.26)

$$t_n = G\left[\frac{1}{\sigma}\ln\left(\frac{9}{n-1}\right) + G^{-1}(t_{10})\right] \tag{5.27}$$

Illustrative example 5.1

Plot a graph of the single-particle breakage function when 4 mm particles of apatite are subject to an impact of 874 J/kg.

Use Equation 5.9 to get E_{50} using parameter values from Table 5.1.

$$E_{50} = 1.50\left[1 + \frac{19.3}{4}\right]^{1.62} = 26.1 \text{ J/kg}$$

Use Equation 5.21 to calculate t_{10} using parameter values from Table 5.2.

$$t_{10} = 0.454\left[1 - \exp\left(-0.0115\left(\frac{874}{26.2}\right)\right)\right] = 0.145$$

Use Equation 5.25 to reconstruct the breakage function from this value of t_{10}.

$$t_n = 1 - (1 - t_{10})^{\left(\frac{9}{n-1}\right)^{0.75}}$$

The calculated values are given in the table

D/d'_p	1.0	0.9	0.8	0.7	0.5	0.2	0.1	0.05	0.01
n	1	1.11	1.25	1.429	2	5	10	20	100
t_n	1.0	0.985	0.900	0.785	0.557	0.250	0.145	0.086	0.026

These values are plotted in Figure 5.14 together with similarly calculated breakage functions at other impact energies. Similar data is shown for a

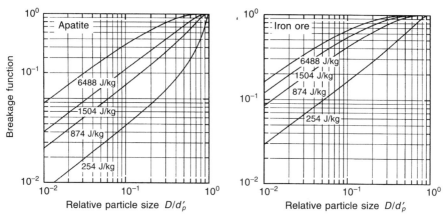

Figure 5.14 Calculated single-particle breakage functions for materials at various impact energies

typical iron ore which is more brittle than apatite and therefore produces more fines at the same impact energy.

The breakage function is an important descriptor of the comminution process and is critically important for modeling purposes. It is material specific and varies according to the precise nature of the comminution process. It is a strong function of the impact energy that is delivered during any impact.

The breakage function is difficult to measure experimentally in an operating mill because it essentially requires an observation of the fracture products that result from a single-impact event that occurs in the environment of the comminution machine. This cannot be done directly since the environment is determined by all particles that are breaking. Also daughter fragments are continually broken and it is impossible to collect the particles of single breakage events. The most acceptable procedure is to assume a functional form for the breakage function and use this within a model for the comminution operation and estimate the parameters by regression to experimentally determined size distributions in the product from the mill. Such a procedure requires test work to be performed on an operating industrial mill or at least on a laboratory mill that operates under conditions that are equivalent to the industrial operations. This is a time-consuming and costly task so prediction of the breakage function from a single-impact laboratory test is an attractive alternative.

5.4.3 A generalized model

Models for the breakage function that are based on mixtures of progeny populations all have a common feature: the set of weighting coefficients that are used to combine the distribution functions for each population. A logical scheme is to weight the distributions according to the relative rates of formation of the various populations during the fracture event (Menacho, 1986). This is usually difficult to do because these rates are not easy to model *a priori*. An alternative method is developed here based on the probabilistic nature of the fracture process.

The fracture probability provides the appropriate averaging of the single-particle breakage functions to yield the breakage function in the mill as a whole.

$$B(x; y) = \int_0^\infty \int_0^1 B(x; y, eE) P(eE, y) p(e) p(E) \, de \, dE \qquad (5.28)$$

where $P(E, y)$ is the probability of fracture of a particle of size y when it receives an amount of energy E from an impact in the mill and $B(x; y, E)$ is the breakage function that results from the single impact event having impact energy E. $p(E)$ is the distribution density for impact energies in the mill and e is the fraction of the energy in the impact that is captured by an individual particle. $p(e)$ describes how e is distributed over the particle population. In practice the available impact energy of an event is not shared equally among all particles that are involved with the event. Consequently the available impact energy is shared among the particles. The fraction of the available energy that a particular particle captures during the event is primarily a function of the position that the particle occupies relative to the impact axis among those particles that are in the impact zone. Particles that are on or close to the impact axis are likely to capture more of the available energy than particles that are off the axis. Larger particles tend to capture more of the energy than smaller particles. There will be considerable rearrangement of particles in the bed during the course of a single impact. At present little is known about these processes. $p(E)$ is the distribution density for impact energies in the mill as a whole and it has been extensively investigated in recent years using the discrete element method (Rajamani *et al.*, 1995). Equation 5.28 provides a link between single-particle fracture tests and the behavior of the material in an operating mill.

5.5 Energy requirements for comminution

Obviously it takes energy to fracture a particle. The energy absorbed by single-particle fracture is a random variable. Each particle will require a different amount of energy depending on the distribution of flaw sizes and on the fracture process (rate of stress application, orientation of particle in the stress field etc.).

Single-particle impact fracture data show that greater amounts of energy absorbed by the particle leads to finer average size of the product population. This leads to the concept of energy absorption by a particle population and the continuing absorption of energy as the size distribution shifts to finer and finer sizes.

A rough empirical expression for energy absorption during comminution can be obtained by following some representative size of the particle population as it steadily decreases with the amount of comminution energy that is put into the particulate material.

Let d_r be a representative size. It could be the mean size on either a number-weighted or mass-weighted basis, but is often the 80% passing size (mass basis). There is nothing magical about the choice of representative size but it is important to maintain consistency once a choice has been made.

The relationship between comminution energy absorbed per unit mass and the representative size is defined by a differential equation:

$$\frac{dE}{dd_r} = f(d_r) \qquad (5.29)$$

$f(d_r)$ is a decreasing function of d_r reflecting the fact that more energy per unit mass is required as the particles get smaller.

A variety of formulations for the functional form of $f(d_r)$ have been proposed and the three most common approaches are ascribed to Kick (1883), Rittinger (1857) and Bond (1952). Each of these suggest a functional form of the type

$$f(d_r) = -Kd_r^{-n} \tag{5.30}$$

with n having the value

$n = 1$ (Kick)
$n = 1.5$ (Bond)
$n = 2$ (Rittinger)

A value of n greater than 1 reflects an increase in energy per unit mass fractured as the particle size decreases.

The energy equation can be integrated from the initial condition that $E = 0$ when $d_r = d_{rl}$ for the parent population

$$\frac{dE}{dd_r} = -Kd_r^{-n}$$

$$E = -\frac{K}{1-n}d_r^{(1-n)} + C \qquad n \neq 1$$

$$0 = -\frac{K}{1-n}d_{rl}^{(1-n)} + C \tag{5.31}$$

$$E = \frac{K}{n-1}\left(\frac{1}{d_r^{(n-1)}} - \frac{1}{d_{rl}^{(n-1)}} \right)$$

when $n = 1$ Kick

$$E = K \ln \frac{d_{rl}}{d_r} \tag{5.32}$$

when $n = 1.5$ Bond

$$E = 2K\left(\frac{1}{d_r^{1/2}} - \frac{1}{d_{rl}^{1/2}} \right) \tag{5.33}$$

when $n = 2$ Rittinger

$$E = K\left(\frac{1}{d_r} - \frac{1}{d_{rl}} \right) \tag{5.34}$$

These equations can be given approximate physical interpretations:

In the case of Kick's law energy consumed is proportional to the size reduction ratio, i.e. it takes the same amount of energy to reduce a population of particles from a representative size of 100 cm to a representative size of 1 cm as it takes to reduce from 1 mm to 10 microns.

In the case of Rittinger's law, the reciprocal of representative size can be regarded as being proportional to the average surface area per unit volume in the particle population. Thus energy consumed is proportional to the increase in surface area per unit volume.

This simple analysis forms the basis of the use of the Bond work index to assess energy requirements for rod and ball milling. This method is described in Section 5.15.1.

5.6 Crushing machines

Crushers usually precede SAG, FAG, rod and ball mills in mineral processing plants. A popular and effective model that applies to a variety of different crusher types is discussed in this section. This model has been widely used to simulate the operation of crushing plant circuits.

5.6.1 Jaw and gyratory crushers

Jaw and gyratory crushers are used mostly for primary crushing. They are characterized by wide gape and narrow discharge and are designed to handle large quantities of material. The capacity of these crushers is determined primarily by their sizes.

The gape determines the maximum size of material that can be accepted. Primary crushers are designed so that the maximum size that can be presented to the crusher is approximately 80% of the gape. Jaw crushers are operated to produce a size reduction ratio between 4 : 1 and 9 : 1. Gyratory crushers can produce size reduction ratios over a somewhat larger range of 3 : 1 to 10 : 1.

The primary operating variable available on a crusher is the set and on jaw and gyratory the open-side set (OSS) is specified. This reflects the fact that considerable portions of the processed material fall through the crusher at OSS and this determines the characteristic size of the product. The set of a crusher can be varied in the field and some crushers are equipped with automatically controlled actuators for the automatic control of the set. However the set of gyratory and jaw crushers is not customarily changed during operation except to compensate for wear in the machine.

The open- and closed-side sets and the gape are identified in Figure 5.15.

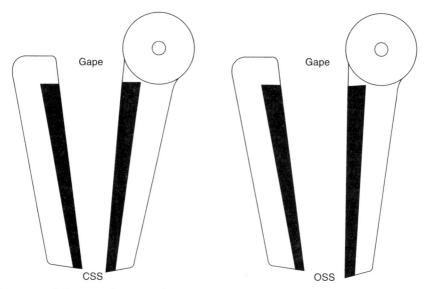

Figure 5.15 Schematic diagram of a crusher showing the open- and closed-side settings

The throw of the crusher is the distance that moving jaw moves in going from OSS to CSS.

Throw = OSS – CSS.

The capacity is a function of size and OSS. Manufacturers publish tables of capacity for their crushers of various size as a function of the open-side set.

Product size distributions in gyratory and jaw crusher products

Primary crushers usually receive feed that contains only small amounts of material that is smaller than the open-side set of the crusher. Thus almost all of the feed material is crushed in the machine.

The size distribution of the material in the product is independent of the size distribution in the feed and is a function primarily of the setting of the crusher and to a lesser extent of the nature of the material that is crushed. Product size distributions are often of Rosin-Rammler type with the parameters of the distribution varying with the product particle size.

The size distribution of the product is characterized by the size relative to the open-side setting

$$r_i = \frac{D_i}{OSS} \tag{5.35}$$

and to the 'product type' P_T which is specified in terms of the fraction of the product that is smaller than the open-side setting. This parameter is related to the nature of the material, as shown in Table 5.3.

Table 5.3 Characteristics that determine the product type

Crusher work index, kW hr/tonne	Material characteristic	Product type, P_T
5–10	Soft	90
	Soft spongy	85
10–13	Average	90
	Average spongy	85
>13	Hard brittle	90
	Hard tough	82
	Hard slabby	75

Two derived parameters are derived from P_T as follows:

$$K_U = \left[\ln\left(\frac{1}{1 - P_T} \right) \right]^{-0.67} \tag{5.36}$$

and

$$K_L = 0.5\left[\ln\left(\frac{1}{1-P_b}\right)\right]^{-1.18} \tag{5.37}$$

with

$$P_b = 1 - \exp\left[-\left(\frac{0.5}{K_U}\right)^{1.5}\right] \tag{5.38}$$

The size distribution is given by

$$P(D) = 1 - \exp\left[-\left(\frac{r}{K_U}\right)^{1.5}\right] \quad \text{for } r > 0.5$$

$$= 1 - \exp\left[-\left(\frac{r}{K_L}\right)^{0.85}\right] \quad \text{for } r \le 0.5 \tag{5.39}$$

These equations can lead to inconsistent results since they take no account of the distribution of particle sizes in the feed to the crusher and should be applied only when the open-side setting of the crusher is smaller than the d_{50} size in the feed.

The model that is described above is based on data from a single crusher manufacturer and is probably specific to the machine design. An alternate model that has been reported to describe the distribution of particle sizes in the product from gyratory crushers is the simple normalized logarithmic distribution

$$P(D_i) = \left(\frac{r}{1.25}\right)^{0.843} \quad \text{for } r \le 1.25 \tag{5.40}$$

5.6.2 *Cone crushers*

Cone crushers are commonly used for secondary, tertiary and quaternary crushing duties. Two variations of cone crusher are used – standard and short head. The chief difference between cone and gyratory or jaw crushers is the nearly parallel arrangement of the mantle and the cone at the discharge end of the cone crusher. This is illustrated in Figure 5.16. Reduction ratios in the following ranges are common for cone crushers: 6 : 1 – 8 : 1 for secondaries and 4 : 1–6 : 1 for tertiary and quaternary crushing.

The size distribution of the products tends to be determined primarily by the CSS since no particle can fall through during a single open-side period and all particles will experience at least one closed-side nip during passage through the crusher. The CSS is adjusted by screwing the bowl of the crusher up or down. It is not unusual for cone crushers to be fitted with hydraulic controls so that the setting on the crusher can be changed during operation with relative ease and as part of an automatic control loop if necessary.

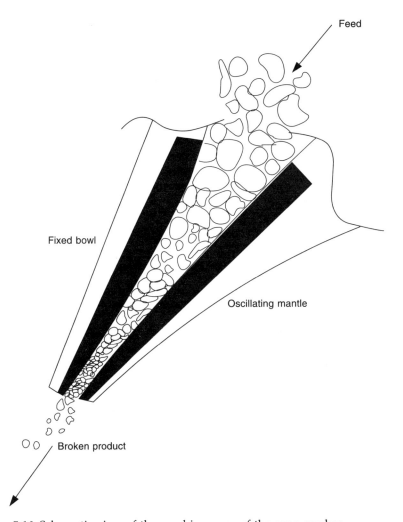

Figure 5.16 Schematic view of the crushing zone of the cone crusher

5.6.3 Crushing mechanisms and product size distributions

The crushing action of a crushing machine is described most usefully through the classification-breakage cycle model. The operation of a crusher is periodic with each period consisting of a nipping action and an opening action. During the opening part of the cycle material moves downward into the crusher and some material falls through and out. A certain amount of fresh feed is also taken in. This is illustrated in Figure 5.17.

This behavior can be described quantitatively and the model will be developed in terms of a discrete size distribution. This makes the model immediately useful for simulation calculations.

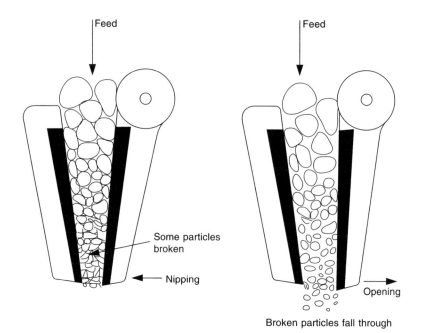

Figure 5.17 The opening and nipping cycles in the crusher on which the model is based

The following symbols are defined for use in the model:

p_i^F = fraction of the feed in size class i,
p_i = fraction of the product in size class i,
M = mass of material held in the crusher,
b_{ij} = fraction of particles breaking in size class j that end up in size class i,
m_i = fraction of material in the crusher in size class i,
c_i = $c(d_i)$
    = fraction of material in size class i that is retained for breakage during the next nip of the crusher,
W = mass of total feed that is accepted during a single opening
    = mass of product discharged.

As always, size class 1 contains the largest particles.

The fortunes of material in the largest size class can be followed through a complete opening and nipping cycle in the crusher, starting with an amount Mm_1 in the crusher.

During an opening phase of the cycle:

Material discharged from the crusher = $(1 - c_1)Mm_1$

Material positioned for breakage in the breakage zone during next nip = $c_1 Mm_1$

Material accepted from feed = Wp_1^F

After the next nip the crusher must again have an amount Mm_1 in the crushing zone since the overall operation is at steady state:

$$Mm_1 = Wp_1^F + c_1 Mm_1 b_{11}$$

$$\frac{Mm_1}{W} = \frac{p_1^F}{1 - c_1 b_{11}} \tag{5.41}$$

Product discharged during the opening phase = $Wp_1 = (1 - c_1)Mm_1$ so that

$$p_1 = (1 - c_1)\frac{Mm_1}{W} \tag{5.42}$$

which gives the fraction of size class 1 in the crusher product.

Now consider the next size down:

During an opening phase of the cycle:

Material discharged = $(1 - c_2)Mm_2$

Material positioned for breakage during next nip = $c_2 Mm_2$

Material accepted from feed = Wp_2^F

After next nip:

$$Mm_2 = Wp_2^F + c_2 Mm_2 b_{22} + c_1 Mm_1 b_{21}$$

$$\frac{Mm_2}{W} = \frac{1}{1 - c_2 b_{22}}\left(p_2^F + c_1 \frac{Mm_1}{W} b_{21}\right) \tag{5.43}$$

Product discharged during the opening phase which gives the fraction of this size class in the crusher product = $Wp_2 = (1 - c_2)Mm_2$

$$p_2 = (1 - c_2)\frac{Mm_2}{W} \tag{5.44}$$

The next size down can be handled in exactly the same way to give

$$Mm_3 = Wp_3^F + c_3 Mm_3 b_{33} + c_2 Mm_2 b_{32} + c_1 Mm_1 b_{31}$$

$$\frac{Mm_3}{W} = \frac{1}{1 - c_3 b_{33}}\left(p_3^F + c_2 \frac{Mm_2}{W} b_{32} + c_1 \frac{Mm_1}{W} b_{31}\right) \tag{5.45}$$

This procedure can be continued from size to size in an obvious way.

Thus for any size class i

$$\frac{Mm_i}{W} = \frac{1}{1 - c_i b_{ii}}\left(p_i^F + \sum_{j=1}^{i-1} c_j \frac{Mm_j}{W} b_{ij}\right) \tag{5.46}$$

The series of Equations 5.46 can be easily solved sequentially for the group Mm_i/W starting from size class number 1. The size distribution in the product can then be calculated from

$$p_i = \frac{(1 - c_i)Mm_i}{W} = \frac{1 - c_i}{1 - c_i b_{ii}}\left(p_i^F + \sum_{j=1}^{i-1} c_j \frac{Mm_j}{W} b_{ij}\right) \tag{5.47}$$

and the distribution of sizes in the product is completely determined from the size distribution in the feed and a knowledge of the classification and breakage functions c_i and b_{ij}. This model is completely general and for a specific crusher reduces to the specification of the appropriate classification and breakage functions.

The classification function is usually of the form shown schematically in Figure 5.18. d_1 and d_2 are parameters that are characteristic of the crusher. They are determined primarily by the setting of the crusher. Data from operating crusher machines indicate that both d_1 and d_2 are proportional to the closed side setting. d_1 is the smallest size particle that can be retained in the crushing zone during the opening phase of the cycle. d_2 is the largest particle that can fall through the crushing zone during the opening phase of the cycle.

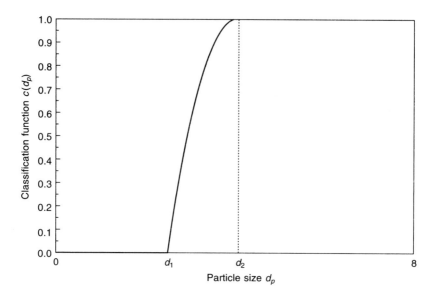

Figure 5.18 A typical internal classification function for a crusher

A useful form of the classification function is

$$c_i = 1 - \left(\frac{d_{pi} - d_2}{d_1 - d_2} \right)^n \qquad \text{for } d_1 < d_{pi} < d_2$$
$$= 0 \qquad \text{for } d_{pi} \leq d_1$$
$$= 1 \qquad \text{for } d_{pi} \leq d_2$$

(5.48)

For both standard and short-head Symons cone crushers, the parameters are related to CSS as follows:

$$d_1 = \alpha_1 \text{ CSS}$$

(5.49)

$$d_2 = \alpha_2 \, \text{CSS} + d^* \tag{5.50}$$

α_1 varies from about 0.5 to 0.95 and α_2 varies from about 1.7 to 3.5. n is usually approximately 2 but can be as low as 1 and as high as 3. Higher values of n usually require higher values of α_2. d^* is usually set to 0.

Breakage functions of the type

$$B(x; y) = K\left(\frac{x}{y}\right)^{n_1} + (1 - K)\left(\frac{x}{y}\right)^{n_2} \tag{5.51}$$

are normally used to describe crusher behavior.

The values of b_{ij} are obtained from the cumulative breakage function by

$$b_{ij} = B(D_{i-1}; d_{pj}) - B(D_i; d_{pj}) \tag{5.52}$$

and

$$b_{jj} = 1 - B(D_j; d_{pj}) \tag{5.53}$$

represents the fraction of material that remains in size interval j after breakage. These relationships are illustrated in Figure 5.19. n_1 is approximately 0.5 for both standard and short-head crushers and n_2 is approximately 2.5 for short-head and 4.5 for standard crushers.

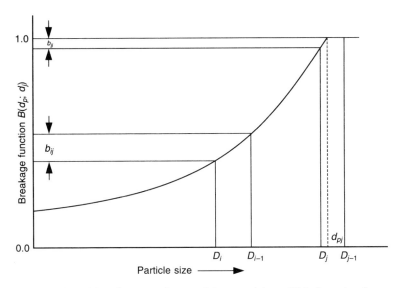

Figure 5.19 The breakage function for crushing machines. This function has a value 1 at the representative size of the parent class. Compare this with the breakage function used for grinding machines

---

Illustrative example 5.2
The breakage function in a crusher machine was determined to be given by

$$B(x:y) = 0.3\left(\frac{x}{y}\right)^{0.45} + 0.7\left(\frac{x}{y}\right)^{3.2}$$

If size class 2 has mesh sizes boundaries 13.4 cm and 9.5 cm and size class 5 has boundaries 7.50 cm and 5.60 cm, calculate b_{52}.

$$b_{ij} = B(D_{i-1}; dp_j) - B(D_i; dp_j)$$

$$d_{p2} = \sqrt{13.4 \times 9.4} = 11.28 \text{ cm}$$

$$b_{52} = 0.3\left(\frac{7.5}{11.28}\right)^{0.45} + 0.7\left(\frac{7.5}{11.28}\right)^{3.2} - 0.3\left(\frac{5.6}{11.28}\right)^{0.45} - 0.7\left(\frac{5.6}{11.28}\right)^{3.2}$$

$$= 0.4393 - 0.2934$$

$$= 0.1459$$

The parameters in the classification and breakage functions are obviously specific to the type and size of crusher. Unfortunately not many studies have been done to establish their values under a range of actual operating conditions using predictive equations. In practice it is usual to estimate them from measured particle size distributions in the products from operating crushers. Once established for a particular material in a particular crusher, they remain independent of the closed side setting. This allows the crusher performance to be simulated as the CSS is varied.

Approximate capacities of some common crushers are given in Tables 5.4–5.7. These are average values from a number of manufacturers' tables and should be used only to obtain approximate values of sizes and settings for simulation calculations.

5.7 Grinding

5.7.1 Grinding action

Industrial grinding machines used in the mineral processing industries are mostly of the tumbling mill type. These mills exist in a variety of types – rod, ball, pebble autogenous and semi-autogenous. The grinding action is induced by relative motion between the particles of media – the rods, balls or pebbles. This motion can be characterized as collision with breakage induced primarily by impact or as rolling with breakage induced primarily by crushing and attrition. In autogenous grinding machines fracture of the media particles also occurs by both impact (self breakage) and attrition.

The relative motion of the media is determined by the tumbling action which in turn is quite strongly influenced by the liners and lifters that are always fixed inside the shell of the mill. Liners and lifters have two main purposes:

Table 5.4 Approximate capacities of jaw crushers in tonnes/hr. The size designation used here is the traditional one in which the feed opening is specified as gape × length in inches

Size Feed opening	Motor kW	Open-side setting mm								
		25	32	38	51	63	76	102	127	152
10×20	15	12.7	15.4	18.2	23.0	31.0				
10×24	11	14.5	17.3	20.0	31.0	30.0				
15×24	22		20.9	24.5	30.0	38.1	45.4			
14×24	19			23.6	70.0	37.2	45.4			
24×36	56					86.3	103	136		
30×42	75					113	136	182	227	272

		Open-side setting mm								
		63	76	102	127	152	178	203	229	254
32×42	75				227	263	300	327	363	
36×48	93		189	245	300	354	409			
42×48	110				345	381	426	463	490	527
48×60	1580					436	481	517	554	600
56×72	186						454	500	567	617
66×84	225						700	772	863	950

Table 5.5 Approximate capacities of gyratory crushers in tonnes/hr. Size is specified as gape × lower mantle diameter in inches

Size	Motor kW	Open-side setting mm									
		51	63	76	89	102	114	127	140	152	178
30×60	150	313	381	450	508	567	630	695	760		
30×55	300		381	463	518	590	663	735	817		
36×60	186		458	540	604	680	755	830	900	970	
42×65	400						800	908	1017	1317	1500

		Open-side setting mm									
		127	140	152	178	190	203	216	229	241	254
42×70	300	708	790	863	944	1017	1090				
48×74	500	1544	1680	1816	1952	2088	2452				
48×80	500			1376	1462	1562	1662	1770	1870		
54×74	500		1634	1771	1907	2043	2180	2315			
54×80	500			1307	1394	1490	1580	1680	1770		
60×89	600			2270	2424	2580	2815	2960	3270		

		Open-side setting mm									
		190	203	216	229	241	254	267	279	292	305
60×102	800		2542	2760	2974		3396		3827		4254
60×109	800			3904	4195	4485	4776	5067	5357	5675	5993

Table 5.6 Approximate capacities of standard Symons cone crushers in tonnes/hr

Size (Max. power kW)	Type of cavity	Feed opening on the closed side* with minimum CSS mm	Closed-side setting mm										
			6	9	13	16	19	22	25	31	38	51	64
2 ft (22)	Fine	57	16	18	23	27	32	36	41	45	54		
	Coarse	83		18	23	27	32	41	45	54	68		
	Extra coarse	100			23	27	36	45	50	63	72		
3 ft (56)	Fine	83		45	59	72	81	91	118	136	163		
	Coarse	119			59	72	91	99	118	136	163		
	Extra coarse	163					99	109					
4 ft (93)	Fine	127		63	91	109	127	140	154	168			
	Medium	156			99	118	136	145	163	181	199		
	Coarse	178					140	154	181	199	245	308	
	Extra coarse	231							190	208	254	317	
4½ ft (112)	Fine	109			109	127	145	154	163	181			
	Medium	188				131	158	172	199	227	264		
	Coarse	216					172	195	217	249	295	349	
	Extra coarse	238							236	272	303	358	
5½ ft (150)	Fine	188				181	204	229	258	295	326		
	Medium	213						258	290	335	381	417	
	Coarse	241							290	354	417	453	635
	Extra coarse	331									431	476	680
7 ft (224)	Fine	253					381	408	499	617	726		
(260)	Medium	303							607	726	807	998	
	Coarse	334								789	843	1088	1270
(EHD)	Extra coarse	425									880	1179	1380
10 ft (450)	Fine	317						934	1179	1469	1632		
	Medium	394							1570	1633	1814	2267	
	Coarse	470									1905	2449	2857
	Extra coarse	622									1995	2630	3084

EHD – extra heavy duty

Table 5.7 Approximate capacities of short head Symons cone crushers in tonnes/hr

Size (Max. power kW)	Type of cavity	Recommended minimum CSS mm	Feed opening with minimum CSS mm		Closed-side setting mm							
			Closed side	Open side	3	5	6	10	13	16	19	25
2 ft (22)	Fine	3	19	35	9	6	18	27	36			
	Coarse	5	38	51		16	22	29	41			
3 ft (56)	Fine	3	13	41	27	41	54	68	91			
	Medium	3	33	60	27	41	54	68	91	99		
	Extra coarse	5	51	76		50	59	72	95	113	127	
4 ft (93)	Fine	5	29	57			77	86	122	131		
	Medium	8	44	73				91	131	145		
	Coarse	13	56	89					140	163	181	
	Extra coarse	16	89	117					145	168	190	217
4¹/₂ ft (112)	Fine	5	29	64		59	81	104	136	163		
	Medium	6	54	89			81	104	136	163		
	Coarse	8	70	105				109	158	181	199	227
	Extra coarse	13	98	133					172	190	254	238
5¹/₂ ft (150)	Fine	5	35	70		91	136	163	208	254		
	Medium	6	54	89			136	163	208	281	281	
	Coarse	10	98	133				190	254	281	308	598
	Extra coarse	13	117	133					254		308	653
7 ft (224)	Fine	5	51	105		190	273	326	363	408		
(260)	Medium	10	95	133				354	408	453	506	
	Coarse	13	127	178					453	481	544	598
EHD)	Extra coarse	16	152	203						506	589	653
10 ft (450)	Fine		76	127			635	735	816	916	1106	
	Medium		102	152				798	916	1020	1224	
	Coarse		178	229						1125	1324	1360
	Extra coarse		203	254								1478

EHD = extra heavy duty.

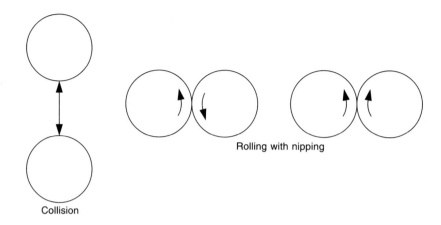

Rolling with nipping

Collision

Figure 5.20 Different types of grinding action by the grinding media

1. Liners protect the outer shell of the mill from wear – liners are renewable.
2. Lifters prevent slipping between the medium and slurry charge in the mill and the mill shell. Slippage will consume energy wastefully but more importantly it will reduce the ability of the mill shell to transmit energy to the tumbling charge. This energy is required to cause grinding of the material in the mill. The shape and dimensions of the lifters control the tumbling action of the media.

The tumbling action is difficult to describe accurately but certain regions in the mill can be characterized in terms of the basic pattern of motion of material in the mill.

The motion of an individual ball in the charge is complicated in practice and it is not possible to calculate the path taken by a particular particle during rotation of the charge. However the general pattern of the motion of the media can be simulated by discrete element methods, which provide valuable information about the dynamic conditions inside the mill.

Some of the terms that are often used to describe the motion of the media in a tumbling mill are shown in Figure 5.21.

5.7.2 Critical speed of rotation

The force balance on a particle against the wall is given by

Centrifugal force outward

$$F_c = m_p \omega^2 \frac{D_m}{2} \qquad (5.54)$$

ω is the angular velocity, m_p is the mass of any particle (media or charge) in the mill and D_m is the diameter of the mill inside the liners.

Gravitational force

$$F_g = m_p g \tag{5.55}$$

The particle will remain against the wall if these two forces are in balance, i.e.

$$F_c = F_g \cos \theta \tag{5.56}$$

where θ is shown in Figure 5.21.

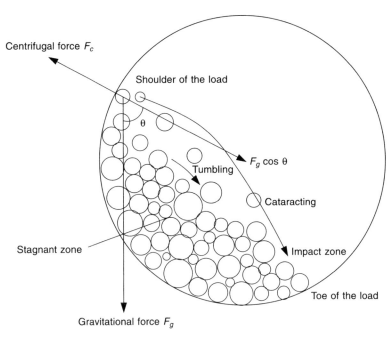

Figure 5.21 Media motion in the tumbling mill

Thus a particle will separate from the wall at the point where

$$\cos \theta = \frac{F_c}{F_g} \tag{5.57}$$

The critical speed of the mill, ω_c, is defined as the speed at which a single ball will just remain against the wall for a full cycle. At the top of the cycle $\theta = 0$ and

$$F_c = F_g \tag{5.58}$$

$$m_p \frac{\omega_c^2 D_m}{2} = m_p g \tag{5.59}$$

$$\omega_c = \left(\frac{2g}{D_m} \right)^{1/2} \tag{5.60}$$

The critical speed is usually expressed in terms of the number of revolutions per second

$$N_c = \frac{\omega_c}{2\pi} = \frac{1}{2\pi}\left(\frac{2g}{D_m}\right)^{1/2} = \frac{(2 \times 9.81)^{1/2}}{2\pi D_m^{1/2}}$$

$$= \frac{0.705}{D_m^{1/2}} \text{ revs/sec} \tag{5.61}$$

$$= \frac{42.3}{D_m^{1/2}} \text{ revs/min}$$

The liner profile and the stickiness of the pulp in the mill can have a significant effect on the actual critical velocity. Mills usually operate in the range 65–82% of critical but values as high as 90% are sometimes used.

A parameter that defines the performance of a mill is the energy consumption. The power supplied to the mill is used primarily to lift the load (medium and charge). Additional power is required to keep the mill rotating.

5.7.3 *Power drawn by ball, semi-autogenous and autogenous mills*

A simplified picture of the mill load is shown in Figure 5.22 and this can be used to establish the essential features of a model for mill power.

The torque required to turn the mill is given by:

$$\text{Torque} = T = M_c g d_c + T_f \tag{5.62}$$

where M_c is the total mass of the charge in the mill and T_f is the torque required to overcome friction.

$$\text{Power} = 2\pi NT \tag{5.63}$$

For mills of different diameter but running at the same fraction of critical speed and having the same fractional filling

$$\text{Net power} = 2\pi N M_c d_c$$

$$= \alpha N_c M_c d_c$$

$$= \alpha \frac{1}{D_m^{0.5}} LD_m^2 D_m \tag{5.64}$$

$$= \alpha LD_m^{2.5}$$

The exponent 2.5 on D_m has been variously reported to have values as low as 2.3 and as high as 3.0.

The effect of varying mill speed on the power drawn by the mill is shown graphically in Figure 5.26. The speed of rotation of the mill influences the power draft through two effects: the value of N and the shift in the center of gravity with speed. The center of gravity first starts to shift to the left as the

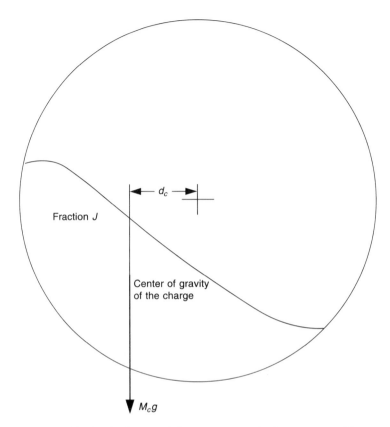

Fraction *J*

Center of gravity
of the charge

$M_c g$

Figure 5.22 Simplified calculation of the torque required to turn a mill

speed of rotation increases but as critical speed is reached the center of gravity
moves towards the center of the mill as more and more of the material is held
against the shell throughout the cycle. Since critical speed is larger at smaller
radii the centrifuging layer gets thicker and thicker as the speed increases
until the entire charge is centrifuging and the net power draw is zero.

The effect of mill charge is primarily through the shifting of the center of
gravity and the mass of the charge. As the charge increases the center of
gravity moves inward. The power draft is more or less symmetrical about the
50% value.

A simple equation for calculating net power draft is:

$$P = 2.00 \; \varphi_c D_m^{2.5} L K_l \quad \text{kW} \tag{5.65}$$

K_l is the loading factor which can be obtained from Figure 5.23 for the popular
mill types. φ_c is the mill speed measured as a fraction of the critical speed.

More reliable models for the prediction of the power drawn by ball, semi-
autogenous and fully autogenous mills have been developed by Morrell (1996)
and by Austin (1990). The relationship between gross power and net power
is given by

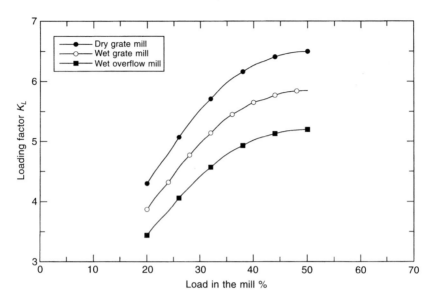

Figure 5.23 Effect of mill filling on power draft for ball mills. The data is taken from *Rexnord Process Machinery Reference Manual*, Rexnord Process Machinery Division, Milwaukee, 1976

Gross power = No-load power + Net power drawn by the charge   (5.66)

The net power is calculated from

$$\text{Net power} = KD_m^{2.5}L_e\rho_c\alpha\delta \quad \text{Watts} \tag{5.67}$$

In Equation 5.67, D_m is the diameter inside the mill liners and L_e is the effective length of the mill including the conical ends. ρ_c is the specific gravity of the charge and α and δ are factors that account for the fractional filling and the speed of rotation respectively. K is a calibration constant that varies with the type of discharge. For overflow mills $K = 7.98$ and for grate mills $K = 9.10$. This difference is ascribed to the presence of a pool of slurry that is present on the bottom of overflow-discharge mills but which is not present to the same extent in grate-discharge mills. This pool is situated more or less symmetrically with respect to the axis of the mill and therefore does not draw significant power. Austin recommends $K = 10.6$ for overflow semi-autogenous mills, however a value of $K = 9.32$ gives a better fit between Austin's formula and Morrell's data and this value is recommended.

The no-load power accounts for all frictional and mechanical losses in the drive system of the mill and can be calculated from Morrell (1996):

$$\text{No-load power} = 1.68D_m^{2.05}[\varphi_c(0.667L_d + L)]^{0.82} \quad \text{kW} \tag{5.68}$$

L_d is the mean length of the conical ends and is calculated as half the difference between the center-line length of the mill and the length of the cylindrical section.

The geometry of a mill with conical ends is shown in Figure 5.24. The total volume inside the mill is given by

$$V_m = \frac{\pi}{4} D_m^2 L \left(1 + \frac{2(L_c - L)}{L} \frac{1 - (D_t/D_m)^3}{1 - D_t/D_m} \right) \tag{5.69}$$

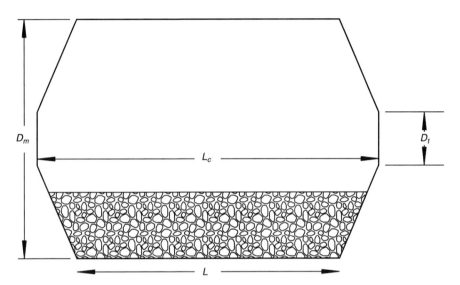

Figure 5.24 Geometry of a mill with cylindrical ends. All dimensions are inside liners. L_c = centerline length, L = belly length, D_m = mill diameter, D_t = trunnion diameter

The density of the charge must account for all of the material in the mill including the media which may be steel balls in a ball mill, or large lumps of ore in an autogenous mill or a mixture in a semi-autogenous mill, as well as the slurry that makes up the operating charge. Let J_t be the fraction of the mill volume that is occupied by the total charge, J_b the fraction of the mill volume that is occupied by steel balls and E the voidage of the balls and media. U is the fraction of the voidage that is filled by slurry. φ_v is the volume fraction of solids in the slurry. Let V_B be the volume of steel balls in the mill, V_{Med} be the volume of autogenous media and V_S the volume of slurry.

$$V_B = J_b(1 - E)V_m$$

$$V_S = J_t U E V_m \tag{5.70}$$

$$V_{Med} = (J_t - J_b)(1 - E)V_m$$

The charge density is calculated from

$$\rho_c = \frac{V_B \rho_b + V_{Med} \rho_m + V_S(1 - \varphi_v)1000 + V_S \varphi_v \rho_s}{J_t} \tag{5.71}$$

where ρ_b is the density of the balls and ρ_m the density of the media. The effective length of the mill is dependent on the load and is calculated from

$$L_e = L\left(1 + 2.28J_t(1 - J_t)\frac{L_d}{L}\right) \tag{5.72}$$

according to Morrell and from

$$L_e = L(1 + f_3)$$

$$f_3 = \frac{0.092}{J_t(1 - 1.03J_t)}\left(\frac{L_d}{L(1 - D_t/D)}\right)\left[\left(\frac{0.625}{0.5 - J_t}\right)^{0.1} - \left(\frac{0.625}{0.5 - J_t}\right)^{-4}\right] \tag{5.73}$$

according to Austin.

The functions α and δ account for the effects of mill filling and rotation speed respectively. However each of these factors is a function of both mill filling and rotation speed. Morrell recommends the following formulas:

$$\alpha = \frac{J_t(\omega - J_t)}{\omega^2} \tag{5.74}$$

$$\omega = 2(2.986\varphi_c - 2.213\varphi_c^2 - 0.4927)$$

and

$$\delta = \varphi_c(1 - [1 - \varphi_{max}]\exp(-19.42(\varphi_{max} - \varphi_c)))$$
$$\varphi_{max} = 0.954 - 0.135J_t \tag{5.75}$$

Austin recommends the following formulas for the factors α and δ:

$$\alpha = J_t(1 - 1.03J_t)$$
$$\delta = \varphi_c\left(1 - \frac{0.1}{2^{9-10\varphi_c}}\right) \tag{5.76}$$

The formulas proposed by Austin and Morrell give substantially the same estimates of the net power that is drawn by the mill. Some typical cases are shown in Figure 5.25.

5.7.4 Power drawn by rod mills

The power drawn by a rod mill is given by:

$$P = 1.752D^{0.33}(6.3 - 5.4V_p)\varphi_c \quad \text{kW hr/tonne of rods charged} \tag{5.77}$$

where V_p = fraction of the mill volume that is loaded with rods. Rod mills operate typically at φ_c between 0.64 and 0.75 with larger diameter mills running at the lower end and smaller mills at the upper end. Typical rod loads are 35% to 45% of mill volume and rod bulk densities range from about 5400 kg/m^3 to 6400 kg/m^3 (Rowland and Kjos, 1980).

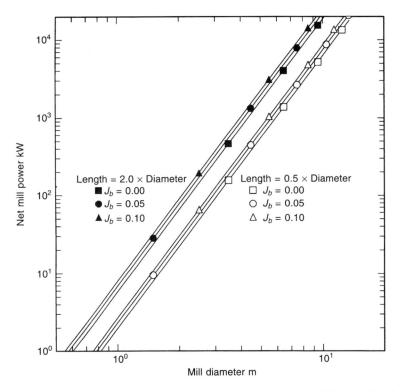

Figure 5.25 Variation of net power drawn by semi-autogenous mills calculated using the formulas of Austin and Morrell. Mill conditions used: $J_t = 0.35$, $E = 0.4$, $U = 1.0$, $\varphi_c = 0.72$, $D_o = 0.1D_m$, $L_d = 0.25D_m$, $\rho_B = 7800$, $\rho_S = 2750$, $\varphi_v = 0.459$

5.7.5 The impact energy spectrum in a mill

The energy that is required to break the material in the mill comes from the rotational energy that is supplied by the drive motor. This energy is converted to kinetic and potential energy of the grinding media. The media particles are lifted in the ascending portion of the mill and they fall and tumble over the charge causing impacts that crush the individual particles of the charge. The overall delivery of energy to sustain the breakage process is considered to be made up of a very large number of individual impact or crushing events. Each impact event is considered to deliver a finite amount of energy to the charge which in turn is distributed unequally to each particle that is in the neighborhood of the impacting media particles and which can therefore receive a fraction of the energy that is dissipated in the impact event. Not all impacts are alike. Some will be tremendously energetic such as the impact caused by a steel ball falling in free flight over several meters. Others will result from comparatively gentle interaction between media pieces as they move relative to each other with only little relative motion. It is possible to calculate the distribution of impact energies using discrete element methods to simulate

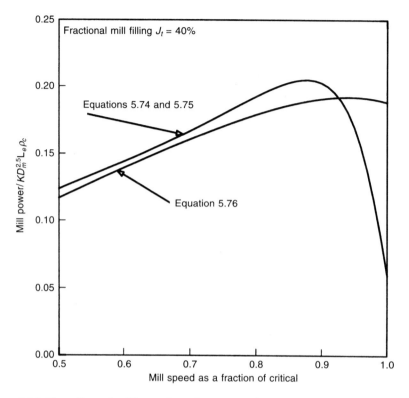

Figure 5.26 The effect of mill speed on the power drawn by a rotating mill

the motion of the media particles including all the many collisions in an operating mill. The distribution of impact energies is called the impact energy spectrum of the mill and this distribution function ultimately determines the kinetics of the comminution process in the mill.

5.7.6 *Kinetics of breakage*

The process engineering of milling circuits is intimately linked with the kinetic mechanisms that govern the rate at which material is broken in a comminution machine. The discussion of breakage phenomena in previous sections has been based on the concept of individual impact events in which a collection of parent particles are involved in a single rapid application of compressive stress in a mechanical event that has been loosely described as an impact. During each impact event each parent particle captures some of the available impact energy and as a result it may break. If it does break its immediate progeny will also absorb some of the residual impact energy and these progeny particles may or may not break. Any progeny particle that breaks will produce progeny that in turn may break. This repeated fracture of successive generations of progeny continues until all the energy of the original impact has been dissipated. Each parent has a unique set of progeny particles that result from the impact and each progeny comes from one and only one original parent. The distribution of sizes of the set of progeny particles is defined by the breakage function of the parent. These discrete impact events happen with high frequency in a grinding mill and it is convenient to consider the breakage as a continuous process. This provides an effective modeling method that is based on the population balance approach.

In any region of a comminution machine, the rate at which material of a particular size is being broken is a strong function of the amount of that size of material present. This is the basis of the model. The rate of breakage varies with size. It is convenient to use discrete size classes that provide an adequate approximation for practical computation.

The fundamental postulate that supports the kinetic model is that the rate of breakage of material out of size class i is proportional to the amount of material of size i in the mill. Thus

$$\text{Rate of breakage} = k_i M m_i$$

k_i is called the specific rate of breakage. It is also called the selection function. Note that k_i is a function of the representative size in the class. It may also be a function of the composition of the parent particle.

Material breaking out of class i distributes itself over all other classes according to the breakage function which is specific to the comminution operation and the material being broken.

It is common practice to normalize the breakage function to the lower mesh size of the interval so that breakage means breakage out of a size class. If the breakage function is described by the standard form:

$$B(x; y) = K \left(\frac{x}{y} \right)^{n_1} + (1 - K) \left(\frac{x}{y} \right)^{n_2} \tag{5.78}$$

then

$$b_{ij} = B(D_{i-1}; D_j) - B(D_i; D_j) \tag{5.79}$$

$$b_{jj} = 0$$

The relationship between the cumulative breakage function and b_{ij} is shown in Figure 5.27. The breakage function has the value 1.0 at the lower boundary of the parent size interval. This reflects the convention that breakage is assumed to occur only when material leaves the parent size interval. This convention has been adopted because it is only possible to detect breakage in a laboratory experiment if the material leaves the sieve interval. The convention is not an essential feature of the population balance method but it is convenient because most of the breakage data that appears in the literature has it as the base. At all events it is not difficult to convert the breakage function data to the alternative convention if required. The differences between Equation 5.79 and the equivalent expressions for crushing machines (5.52 and 5.53) should be noted.

If normalization is satisfactory and if a strictly geometric progression is

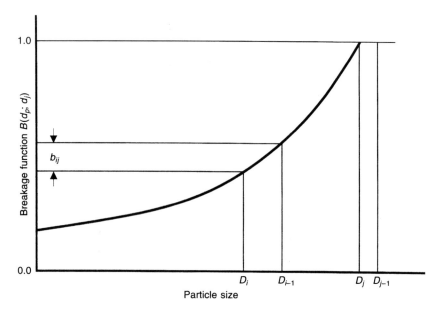

Figure 5.27 The breakage function used for the description of milling operations in rotating mills. Compare with the equivalent diagram for crushing machines (Figure 5.19)

used for the mess size then b_{ij} is a function only of the difference $i - j$. For example if a root 2 series is used for the size classes.

$$\frac{D_i}{D_j} = \frac{D_i}{(\sqrt{2})^{i-j} D_i} = (\sqrt{2})^{j-i} \tag{5.80}$$

$$b_{ij} = K\left(\frac{D_{i-1}}{D_j}\right)^{n_1} + (1 - K)\left(\frac{D_{i-1}}{D_j}\right)^{n_2} - K\left(\frac{D_i}{D_j}\right)^{n_1} + (1 - K)\left(\frac{D_i}{D_j}\right)^{n_2} \tag{5.81}$$

$$b_{ij} = K[(\sqrt{2})^{n_1 (j-i+1)} - (\sqrt{2})^{n_1 (j-i)}] + (1 - K)[(\sqrt{2})^{n_2 (j-i+1)} - \sqrt{2}^{n_2 (j-i)}]$$

$$= K(\sqrt{2})^{n_1 (j-i)}(\sqrt{2} - 1) + (1 - K)(\sqrt{2})^{n_2 (j-i)}(\sqrt{2} - 1) \tag{5.82}$$

$$= 0.414 K(\sqrt{2})^{n_1 (j-i)} + 0.414(1 - K)(\sqrt{2})^{n_2 (j-i)}$$

5.8 The continuous mill

Industrial grinding mills always process material continuously so that models must simulate continuous operation. Suitable models are developed in this section.

5.8.1 The population balance model for a perfectly mixed mill

The equations that describe the size reduction process in a perfectly mixed ball mill can be derived directly from the master population balance equation that was developed in Chapter 2. However, the equation is simple enough to derive directly from a simple mass balance for material in any specific size class.

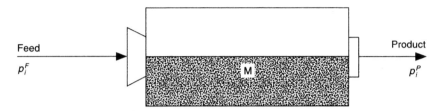

Figure 5.28 The perfectly mixed continuous mill

This simple equation is derived below using the following definitions:

p_i^F = fraction feed in size class i
p_i^P = fraction of product in size class i
m_i = fraction of mill contents in size class i
M = mass of material in the mill
W = mass flowrate through the mill.

A mass balance on size class i is developed by noting that the contents of the mill receives material in size class i from the feed and from the breakage of material in the mill that is of size greater than size i. Material of size i is destroyed in the mill by fracture.

$$Wp_i^P = Wp_i^F + M \sum_{j=1}^{i-1} b_{ij}k_j m_j - Mk_i m_i$$

$$p_i^P = p_i^F + \tau \sum_{j=1}^{i-1} b_{ij}k_j m_j - k_i \tau m_i \qquad (5.83)$$

where $\tau = M/W$ is the average residence time of the material in the mill.
If the contents of the mill are assumed to be perfectly mixed $m_i = p_i^P$

$$p_i^P = p_i^F + \sum_{j=1}^{i-1} b_{ij}k_j \tau p_j^P - k_i \tau p_i^P$$

$$p_i^P = \frac{p_i^F + \sum_{j=1}^{i-1} b_{ij}k_j \tau p_j^P}{1 + k_i \tau} \quad \text{for all } i \qquad (5.84)$$

These can be solved by a straightforward recursion relationship starting with size class 1.

$$p_1^P = \frac{p_1^F}{1 + k_1 \tau}$$

$$p_2^P = \frac{p_2^F + b_{21} \tau k_1 p_1^P}{1 + k_2 \tau}$$ (5.85)

$$p_3^P = \frac{p_3^F + b_{31} \tau k_1 p_1^P + b_{32} \tau k_2 p_2^P}{1 + k_3 \tau}$$

etc.

5.8.2 The mill with post classification

In practice the material in the mill does not have unrestricted ability to leave in the outlet stream. Larger particles are prevented from leaving by the discharge grate if one is present and even in overflow discharge mills, the larger particles do not readily move upward through the medium bed to the discharge. On the other hand very small particles move readily with the water and are discharged easily. Thus the discharge end of the mill behaves as a classifier which permits the selective discharge of smaller particles and recycles the larger particles back into the body of the mill. This is illustrated schematically in Figure 5.29. The operation of such a mill can be modeled as a perfectly mixed mill with post classification as shown in Figure 5.30.

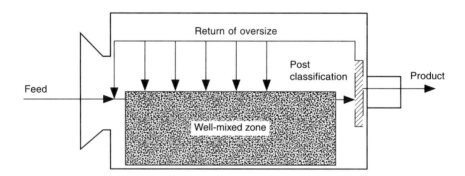

Figure 5.29 The classification mechanism at the mill discharge end recycles larger particles into the body of the mill

Applying Equation 5.84 to the perfectly mixed milling section:

$$m_i = \frac{f_i' + \sum_{j=1}^{i-1} b_{ij} k_j \tau' m_j}{1 + \tau' k_i}$$ (5.86)

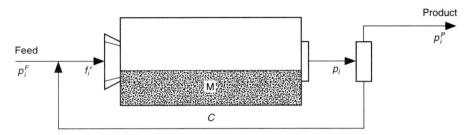

Figure 5.30 The perfectly mixed mill with post classification. The classification mechanism illustrated in Figure 5.29 is equivalent to a perfectly mixed mill with external classifier

In this equation, τ' is the effective residence time in the mixed section of the mill:

$$\tau' = \frac{M}{W(1 + C)} = \frac{\tau}{(1 + C)} \tag{5.87}$$

where C is the ratio of recirculation rate to feed rate.

A mass balance on size class i at the point where the post classified material re-enters the mill gives

$$(1 + C)f_i' = c_i m_i(1 + C) + p_i^F \tag{5.88}$$

where c_i is the classification constant for the particles of size i at the mill discharge.

Substituting the expression for f_i' into Equation 5.84 gives, after some simplification,

$$m_i(1 + C) = \frac{p_i^F + \sum\limits_{j=1}^{i-1} b_{ij} k_j \tau' m_j (1 + C)}{1 + \tau' k_i - c_i} \tag{5.89}$$

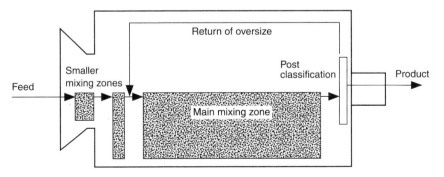

Figure 5.31 Schematic representation of the mill with three perfectly mixed segments of unequal size. Discharge from the last segment is restricted by a post classifier

If a new variable $m_i^* = m_i(1 + C)$ is defined, this equation takes a form identical to Equation 5.84

$$m_i^* = \frac{p_i^F + \sum\limits_{j=1}^{i-1} b_{ij} k_j \tau' m_j^*}{1 + \tau' k_i - c_i} \tag{5.90}$$

This equation can be solved using the same recursive procedure as that used for Equation 5.84 starting from the largest size.

The size distribution in the product can be obtained from the calculated values of m_i^* using the properties of the classifier.

$$p_i^P = (1 - c_i)(1 + C)m_i = (1 - c_i)m_i^* \tag{5.91}$$

This solution is complicated a little because the modified residence time is not usually known. It can be calculated only after the size distribution in the mixed section has been evaluated. This requires an iterative solution for convenient implementation starting with an assumed value of τ' to calculate m_i^* from Equation 5.90. C is calculated from

$$C = \sum_i c_i (1 + C)m_i = \sum_i c_i m_i^* \tag{5.92}$$

and the assumed value for τ' can be checked using Equation 5.87 and modified until convergence is obtained.

The actual residence time can be obtained from the load of solid in the mill and the mass flowrate of the solid through the mill or from a dynamic tracer experiment. It is usually easier to trace the liquid phase than the solid but the residence time of the liquid will be considerably shorter because of the holdback of the solid by the classification mechanism of the discharge.

The precise form of the classification function can be determined by measuring the size distribution of the material in the mill contents and in the product stream

$$1 - c_i = \frac{p_i^P}{(1 + C)m_i} \tag{5.93}$$

The presence of post classification in a mill can be detected by noting the difference in particle-size distribution between the mill contents and the discharged product. An example is shown in Figure 5.32.

5.8.3 Mineral liberation in ball mills

When mineral liberation is important in the mill, the population must be distributed over the two variables size and composition and the population balance equation becomes:

$$p_{ij}^P = \frac{p_{ij}^F + \sum\limits_{k=k_{ij}^L}^{k=k_{ij}^U} \sum\limits_{l=1}^{j-1} b_{ijkl} k_{kl} \tau p_{kl}^P}{1 + k_{ij}\tau} \tag{5.94}$$

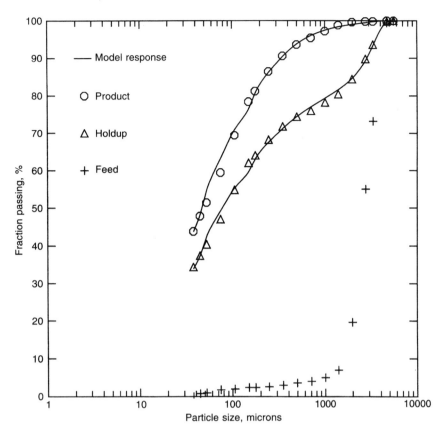

Figure 5.32 Experimentally determined particle-size distributions in the contents of the mill and the discharged product (Schneider, 1995)

where k_{ij}^L and k_{ij}^U are the lower and upper boundaries of the feeder region for progeny particles in grade class i and size class j as described in Chapter 3. Equation 5.94 can be solved recursively in the same way as Equation 5.84 provided that the full breakage function b_{ijkl} is known. This requires the full internal and boundary structure of the Andrews-Mika diagram to be modeled. The model that is described in Chapter 3 has been found to be effective for describing liberation in both laboratory and industrial mills.

5.9 Mixing characteristics of operating mills

In practice operating mills do not conform particularly well to the perfectly mixed pattern because there is considerable resistance to the transport of material, both solids and water, longitudinally along the mill. The actual behavior of the mill can be modeled quite well by dividing it into several perfectly mixed segments in series, with discharge from the last segment being restricted by a post classifier. The size distribution in the material that leaves each segment can be calculated by repeated application of Equation 5.84 successively to each mixed segment in turn. The product from the first segment becomes the feed to the next and so on down the mill. The number of mixed sections and their relative sizes can be determined from the residence-time distribution function in the mill. Residence distribution functions have been measured in a number of operating ball, pebble and autogenous mills and it is not unusual to find that three unequal perfectly mixed segments are adequate to describe the measured residence-time distribution functions. Usually the last segment is significantly larger than the other two. This is consistent with the behavior of a post classifier that holds up the larger particles at the discharge end of the mill, which are then thrown quite far back into the body of the mill.

It has been suggested in the literature that a further refinement to the mixed-region model can be achieved by the use of a classification action between each pair of segments, but since it is impossible to make independently verifiable measurements of such inter-stage classifications, this refinement cannot be used effectively.

The structure of a mill with three perfectly-mixed segments with post classification on the last is illustrated in Figure 5.31.

The residence-time distribution function for a mill can be measured experimentally by means of a dynamic tracer test. It is considerably easier to trace the liquid phase than the solid phase and this is usually done in practice but the two phases have significantly different residence-time distributions. The water is not restricted during its passage through the mill as much as the solid phase is. Consequently the hold-up of solid in the mill is greater than that of the liquid and the solid has a higher mean residence time. Although the two phases have significantly different mean residence times, the behavior of each phase is consistent with the three-stage model that is described above. This is demonstrated in the two measured residence-time distributions that are shown in Figures 5.33 and 5.34. The data shown in Figures 5.33 and 5.34 were collected from the same mill. The solid phase was traced by adding a sample of silica rapidly to the feed stream, which was normally limestone. The mill product was sampled frequently and the concentration of SiO_2 in each sample by X-ray fluorescence analysis. The liquid was traced by rapidly adding a small sample of water-soluble colored tracer to the mill feed and analyzing the liquid phase in the product samples by UV absorbance.

5.10 Models for rod mills

The physical arrangement of rods in the rod mill inhibits the effective internal mixing that is characteristic of ball mills. The axial mixing model for the mixing pattern is more appropriate than that based on the perfectly-mixed region. When axial mixing is not too severe in the mill, an assumption of plug flow is appropriate. In that case the population balance model for the batch

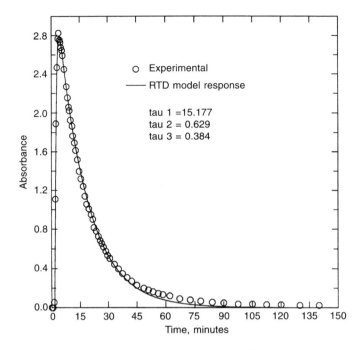

Figure 5.33 Residence-time distribution function for the water in a continuous ball mill. The line shows the small-small-large segment model response to this experimental data. Note the smaller residence times than for the solid phase in the same mill shown in Figure 5.34 (Schneider, 1995)

mill can be used to simulate the behavior of the rod mill with the time replaced by the average residence time in the mill which is equal to the holdup divided by the throughput.

5.11 The population balance model for autogenous mills

Four distinct mechanisms of size reduction have been identified in fully autogenous mills: attrition, chipping, impact fracture and self-breakage. Attrition is the steady wearing away of comparatively smooth surfaces of lumps due to friction between the surfaces in relative motion. Chipping occurs when asperities are chipped off the surface of a particle by contacts that are not sufficiently vigorous to shatter the particle. Attrition and chipping are essentially surface phenomena and are commonly lumped together and identified as wear processes. Impact fracture occurs when smaller particles are nipped between two large particles during an impact induced by collision or rolling motion. Self-breakage occurs when a single particle shatters on impact after falling freely in the mill. Rates of breakage and the progeny

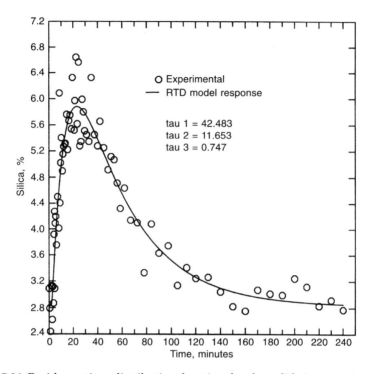

Figure 5.34 Residence-time distribution function for the solids in a continuous ball mill. The line is the small-small-large segment model response to the experimental data. Note the larger residence time than for the liquid phase in the same mill shown in Figure 5.33 (Schneider, 1995)

spectrum formed during these processes differ considerably from each other and each should be modeled separately.

A fifth breakage mechanism occurs in a semi-autogenous mill when particles are impacted by a steel ball. Breakage and selection functions that describe this mechanism can be modeled in a manner similar to those used for the ball mill.

In practice three distinct fracture subprocesses are modeled: wear, impact fracture, and self-breakage. Each of these produces essentially different progeny size distributions and the appropriate breakage function must be used for each. The breakage function for attrition and chipping $A(x; y)$ can be measured in the laboratory in mills that contain only large lumps that are abraded under conditions similar to those found in operating mills. No general models for the attrition breakage function have been developed. The breakage function $B(x; y)$ for the impact fracture process can be modeled using the same model structures that were used for ball milling. The breakage function for self-breakage $C(x; y)$ can be modeled using the t_{10} method that describes single-particle impact fracture using the impact energy equal to the kinetic energy of the particle immediately before impact. Particles will have a wide distribution of free-fall impact energies in a real mill and the breakage function for self-breakage is obtained by integration over the impact energy spectrum

$$C(x; y) = \int_0^\infty C(x; y, h)P(y, h)p(h)dh \qquad (5.95)$$

where $C(x; y, h)$ is the single particle breakage function for self-breakage of a particle of size y in free fall from height h, $P(y, h)$ is the probability that a particle of size y will shatter when falling a vertical height h and $p(h)$ is the distribution density for effective drop heights in the mill.

The attrition and wear processes that occur in autogenous, semi-autogenous, and pebble mills produce a significant fraction of the final production of fine particles in the mill. These processes are not satisfactorily described by the kinetic breakage models that are useful for ball and rod mills. It is necessary to invoke the full generalized population balance model that was developed in Section 2.14 in order to describe the autogenous milling operations adequately.

Various finite difference representations of Equation 2.116 are available but those based on the predefined size classes as described in Chapter 2 are generally used. It is usual and useful to interpret the rate of breakage process $\Re$ in such a way that

$$\int_{D_i}^{D_{i-1}} \Re'(p(x), x, F[p(x)]) \, dx = \bar{R}(p_i, x_i, F[p(x)]) \qquad (5.96)$$

is the rate of breakage out of the interval i. This convention, which was used to develop the simple finite difference models for ball and rod mills, is useful particularly because it is comparatively easy to measure the rate of breakage out of a screen interval in the laboratory and to correlate the measured rates with the representative size for a size class.

Equation 2.116 is integrated over a typical size class i to give

$$-\tau\Delta_i\kappa p(x) + 3\tau \int_{D_i}^{D_{i-1}} \kappa(x)\frac{p(x)}{x}dx + \tau\bar{R}(p_i, x_i, F_1[p(x)])$$

$$-\tau \sum_{j=1}^{i-1} \bar{R}(p_j, x_j, F_1[p(x)]\Delta_i B(x, D_j)$$

$$-3\tau \int_{D_i}^{D_{i-1}} \int_{R''(x)} \kappa(x')\frac{p(x')}{x'} a(x; x')dx'dx \qquad (5.97)$$

$$= P_{i\,in} - P_{i\,out}$$

where

$$\Delta_i\kappa p(x) = \kappa(D_{i-1})p(D_{i-1}) - \kappa(D_i)p(D_i) \qquad (5.98)$$

and

$$\Delta_i B(x; D_j) = B(D_{i-1}; D_j) - B(D_i, D_j)$$

$$= \Delta B_{ij} \qquad (5.99)$$

Finite difference approximations are required for the remaining terms in Equation 5.97 in terms of the representative size for each size class. The following approximations have been used successfully:

$$p(D_i) \simeq \frac{p_i}{D_i - D_{i+1}} \qquad (5.100)$$

so that

$$\Delta_i \kappa p(x) \simeq \frac{\kappa_{i-1} p_{i-1}}{D_{i-1} - D_i} - \frac{\kappa_i p_i}{D_i - D_{i+1}} \qquad (5.101)$$

$$\int_{D_i}^{D_{i-1}} \kappa(x) \frac{p(x)}{x} dx \simeq \frac{\kappa_i p_i}{d_{pi}} \qquad (5.102)$$

$$\int_{D_j}^{D_{j-1}} \kappa \frac{p(x')}{x'} \int_{D_i}^{D_{i-1}} a(x; x') dx\, dx' = \frac{\kappa_j p_j}{d_{pj}} \Delta A_{ij} \qquad (5.103)$$

which implies that the products of wear and attrition must leave the size class of the parent particle, i.e., must be smaller than D_j.

$$\Delta A_{ij} = A(D_{i-1}; D_j) - A(D_i; D_j) \qquad (5.104)$$

If the total region is perfectly mixed then $p_i = p_{i\,out}$ and Equation 5.97 becomes

$$p_i - \frac{\tau \kappa_{i-1} p_{i-1}}{D_{i-1} - D_i} + \frac{\tau \kappa_i p_i}{D_i - D_{i+1}} + \frac{3 \tau \kappa_i p_i}{d_{pi}} + k_i p_i \tau$$

$$- \tau \sum_{j=1}^{i=1} \left(k_j \Delta B_{ij} + \frac{3 \kappa_j \Delta A_{ij}}{d_{pj}} \right) p_j = p_{i\,in} \qquad (5.105)$$

This equation can be simplified somewhat by defining the ratio

$$\gamma_i = \frac{3 \kappa_i}{d_{pi}} \qquad (5.106)$$

and

$$\sigma_i = \frac{d_{pi}}{3(D_i - D_{i+1})} \qquad (5.107)$$

$$p_i + \sigma_i \gamma_i \tau p_i + \gamma_i \tau p_i = p_{i\,in} + \sigma_{i-1} \gamma_{i-1} \tau p_{i-1} + \tau \sum_{j=1}^{i-1} (k_j \Delta B_{ij} + \gamma_j \Delta A_{ij}) p_j \qquad (5.108)$$

$$p_i = \frac{p_{i\,in} + \sigma_{i-1} \gamma_{i-1} \tau p_{i-1} + \tau \sum_{j=1}^{i-1} (k_j \Delta B_{ij} + \gamma_j \Delta A_{ij}) p_j}{1 + (\sigma_i \gamma_i + \gamma_i + k_i) \tau} \qquad (5.109)$$

This equation can be simplified further by making the following definitions:

$$\zeta_i = \sigma_i \gamma_i \tau \qquad (5.110)$$

$$\xi_i = (k_i + \gamma_i)\tau \qquad (5.111)$$

Both ζ_N and ξ_N are zero because there is no breakage out of the smallest size class.

$$(k_j \Delta B_{ij} + \gamma_j \Delta A_{ij})\tau = \Delta_{ij} \qquad (5.112)$$

Using the definitions of ΔA_{ij} and ΔB_{ij}

$$\sum_{i=j+1}^{N} \Delta_{ij} = (k_j + \gamma_j)\tau = \xi_j \qquad (5.113)$$

Equation 5.109 can be written

$$p_i = \frac{P_{i\,in} + \zeta_{i-1} p_{i-1} + \sum_{j=1}^{i-1} \Delta_{ij} p_j}{\zeta_i + \xi_i + 1} \qquad (5.114)$$

This is the fundamental discrete population balance equation for autogenous mills and it should be compared with the equivalent Equation 5.84 for the ball mill. The appearance of the term p_{i-1} on the right hand side of Equation 5.114 should be particularly noted. This term arises from the wear process that reduces the size of large lumps in the mill charge which is an essential feature of autogenous milling. Equation 5.114 can be solved recursively in a straightforward manner starting at p_0. However it is necessary to choose a value of $\zeta_0 p_0$ to ensure that

$$\sum_{i=1}^{N} p_i = 1 \qquad (5.115)$$

is satisfied.

Using Equation 5.114

$$\sum_{i=1}^{N} p_i (1 + \xi_i + \zeta_i) = \sum_{i=1}^{N} p_{i\,in} + \sum_{i=1}^{N} \zeta_{i-1} p_{i-1} + \sum_{i=1}^{N} \sum_{j=1}^{i-1} \Delta_{ij} p_j$$

$$= 1 + \zeta_0 p_0 + \sum_{i=2}^{N} \zeta_{i-1} p_{i-1} + \sum_{j=1}^{N-1} p_j \sum_{i=j+1}^{N} \Delta_{ij} \qquad (5.116)$$

$$= 1 + \zeta_0 p_0 + \sum_{j=1}^{N-1} \zeta_j p_j + \sum_{j=1}^{N-1} p_j \xi_j$$

$$= 1 + \zeta_0 p_0 + \sum_{j=1}^{N-1} (\zeta_j + \xi_j) p_j$$

which leads to

$$\sum_{i=1}^{N} p_i + (\zeta_N + \xi_N)p_N = 1 + \zeta_0 p_0 \tag{5.117}$$

Thus the condition

$$\zeta_0 p_0 = 0 \tag{5.118}$$

guarantees the consistency of the solution.

5.11.1 The effect of the discharge grate

In practice it is always necessary to maintain some classification action at the mill discharge to ensure that large pebbles do not escape from the mill. Autogenous and semi-autogenous mills are always equipped with a steel or rubber grate to hold back the grinding media. The classification action of the grate can be described in terms of a classification function c_i as was done in Section 5.8.2. Application of Equation 5.114 to the mill contents gives

$$m_i = \frac{f'_i + \zeta_{i-1}m_{i-1} + \sum_{j=1}^{i-1} \Delta_{ij}m_j}{1 + \zeta_i + \xi_i} \tag{5.119}$$

where

$$f'_i = \frac{p_i^F}{1 + C} + c_i m_i \tag{5.120}$$

and C is the fraction of the total stream that is returned to the mill by the classification action at the discharge end. The residence time to be used to calculate ζ_i, ξ_i, and Δ_{ij} in Equation 5.119 is

$$\tau' = \frac{M}{(1 + C)W} = \frac{\tau}{1 + C} \tag{5.121}$$

$$m_i = \frac{\dfrac{p_i^F}{1 + C} + \zeta_{i-1}m_{i-1} + \sum_{j=1}^{i-1} \Delta_{ij}m_j}{1 + \zeta_i + \xi_i - c_i} \tag{5.122}$$

Define a new variable

$$m_i^* = (1 + C)m_i \tag{5.123}$$

and this equation becomes

$$m_i^* = \frac{p_i^F + \zeta_{i-1}m_{i-1}^* + \sum_{j=1}^{i-1} \Delta_{ij}m_j^*}{1 + \zeta_i + \xi_i - c_i} \tag{5.124}$$

which can be solved recursively starting from $i = 1$ and $\zeta_0 m_0^* = 0$. The value of τ' must be established by iterative calculation using Equation 5.121 after C is recovered from

$$C = \sum_{i=1}^{N} c_i (1 + C) m_i = \sum_{i=1}^{N} c_i m_i^* \qquad (5.125)$$

The iterative calculation starts from an assumed value of τ' and is continued until the value of τ' stabilizes. The size distribution from the mill is recovered from

$$p_i = (1 - c_i) m_i^* \qquad (5.126)$$

The kinetic parameters k_i and κ_i and the breakage functions $A(D_i; D_j)$ and $B(D_i; D_j)$ must be estimated from experimental data.

5.12 Models for the specific rate of breakage in ball mills

The utility the kinetic model for breakage depends on the availability of robust models for the specific rate of breakage to describe specific milling conditions. Both functions are strong functions of the milling environment. Factors which affect the rate of breakage are the mill diameter, mill speed, media load and size and particle hold-up.

The most important functional dependence is between the specific rate of breakage and the particle size and methods for the description of this functional dependence are described below. The specific rate of breakage increases steadily with particle size which reflects the decreasing strength of the particles as size increases. This is attributed to the greater density of microflaws in the interior of larger particles and to the greater likelihood that a particular large particle will contain a flaw that will initiate fracture under the prevailing stress conditions in a mill. The decrease in particle strength does not lead to an indefinite increase in the specific rate of breakage. As the particle size becomes significant by comparison to the size of the smallest media particles, the prevailing stress levels in the mill are insufficient to cause fracture and the specific rate of breakage passes through a maximum and decreases with further increase in particle size. Some typical data are shown in Figure 5.35.

5.12.1 The Austin model for the specific rate of breakage

Austin represents the variation of the specific rate of breakage with particle size by the function

$$k(d_p) = \frac{S_1 d_p^{\alpha}}{1 + (d_p / \mu)^{\Lambda}} \qquad (5.127)$$

and it is usual to specify the particle size in mm and the specific rate of breakage in min^{-1}.

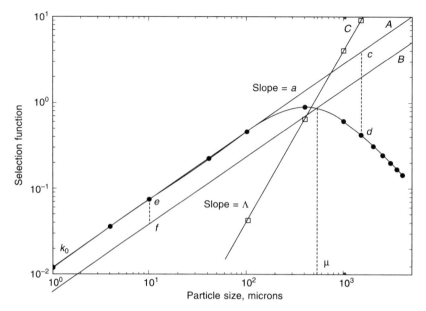

Figure 5.35 Graphical construction to calculate the parameters in Austin's model
for the selection function

It is useful to relate the individual parameters in this function to specific
features of the graph of $k(d_p)$ plotted against d_p. A typical plot of this function
is shown in Figure 5.35 and it has a maximum at a particle size somewhat
smaller than the parameter μ which essentially fixes the position of the
maximum. The maximum occurs in the specific rate of breakage because as
the particles get larger they are less likely to be broken during any typical
impact in the mill. The specific particle fracture energy decreases as size
increases according to Equation 5.9 but the rate of decrease reduces as the
particle size increases and eventually becomes approximately constant for
particles larger than a few millimeters for most ores. Consequently the particle
fracture energy increases at a rate approximately proportional to the particle
mass. Thus for a given impact energy larger particles have a smaller probability
of breaking since the energy that a particle absorbs from the impact must
exceed its fracture energy otherwise it will not break.
    The graphical procedure is implemented as follows:

   (i) Extend the initial straight line portion of the data curve as straight
       line A having equation

$$k = S_1 d_p^{\alpha} \tag{5.128}$$

   (ii) S_1 is evaluated at the intersection of the straight line with the ordinate
        $d_p = 1$ mm. α is equal to the slope of the line. S_1 is called the selection
        function at 1 mm.

(iii) Evaluate ratios such as c/d to form $S_1 d_p^\alpha / k(d_p)$ at a number of particle sizes as shown.

Plot $(S_1 d_p^\alpha / k(d_p)) - 1$ against the particle size as shown as line C in Figure 5.35. The slope of the resulting line is equal to Λ because, according to Equation 5.127,

$$\frac{S_1 d_p^\alpha}{k(d_p)} - 1 = (d_p/\mu)^\Lambda \tag{5.129}$$

Parameter μ can be evaluated in one of two ways.

Construct line B, parallel to line A and passing through a point f which has ordinates equal to $0.5e$.

This line intersects the data curve at abscissa value μ as shown.

Alternatively the size at which $k(d_p)$ is a maximum is given by

$$\frac{dk(d_p)}{dd_p} = 0 \tag{5.130}$$

which implies that

$$\frac{\mu}{d_{p\max}} = \left(\frac{\Lambda}{\alpha} - 1\right)^{\frac{1}{\Lambda}} \tag{5.131}$$

from which μ can be evaluated when Λ is known.

5.12.2 Scale-up of the Austin selection function

When developing scale-up rules for the specific rate of breakage, it is necessary to distinguish between parameters that are material specific and those that depend on the material that is to be milled and also on the geometrical scale of the mill that is to be used. The parameters α and Λ in the Austin model for specific rate of breakage are usually assumed to be material specific only while S_1 and μ depend on the geometrical scale.

The graphical construction that is outlined in the previous section reveals the role that each parameter in the Austin selection function plays in determining the specific rate of breakage in a ball mill. It is not difficult to determine the values of these parameters from data obtained in a batch milling experiment in the laboratory. They can also be estimated using standard parameter estimation techniques from the size distributions in samples taken from the feed and discharge streams of an operating mill.

In order to use this model for simulation of other mills it is necessary to use scale-up laws that describe how these parameters vary with variations of the mill size and the environment inside the mill. The dominant variables are the mill diameter D_m and the size of the balls that make up the media d_b. These variables together determine the average impact energy in the mill and both have a significant influence on the value of the constant S_1 in Equation

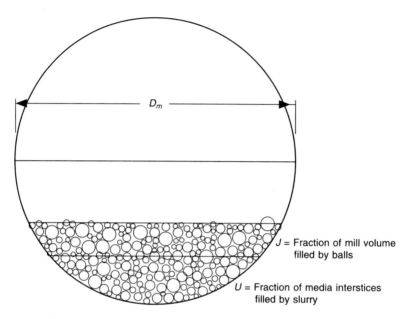

Figure 5.36 The geometrical constants for mill scale-up are defined by the load conditions when the mill is stationary

5.127. Because the specific rate of breakage is essentially a kinetic parameter, it obviously increases with the number of impacts that occur per second per unit volume in the mill. Geometrically similar mills having the same fractional filling by media and rotating at the same fraction of critical speed φ_c will produce nearly identical impact frequencies per unit volume. The impact frequency per unit volume should vary at a rate proportional to the speed of rotation. The variation of impact frequency with mill filling is rather more complex and purely empirical scale-up rules must be applied. The effect of interstitial filling is also modeled empirically to reflect the fact that not all of the slurry remains in the region of the tumbling media where energetic impacts occur. A pool of slurry can accumulate at the toe of the charge for example and this is largely devoid of impacts that cause breakage. As the interstitial fill fraction approaches 1.0 the impacts are increasingly cushioned by excess slurry between media particles. The scale-up law for parameter S_1 is

$$\frac{S_1}{S_{1T}} = \left(\frac{D_m}{D_{mT}}\right)^{N_1} \left(\frac{1 + 6.6J_T^{2.3}}{1 + 6.6J^{2.3}}\right) \left(\frac{\varphi_c - 0.1}{\varphi_{cT} - 0.1}\right) \left(\frac{1 + \exp[15.7(\varphi_{cT} - 0.94)]}{1 + \exp[15.7(\varphi_c - 0.94)]}\right)$$

$$\times \exp\left[-c(U - U_T)\right] \tag{5.132}$$

The subscript T in this equation refers to the variable determined under the test conditions for which the parameters are estimated and the corresponding variable without this subscript refers to the large scale mill that must be simulated. The variables J and U are defined by the load when the mill is stationary as shown in Figure 5.36.

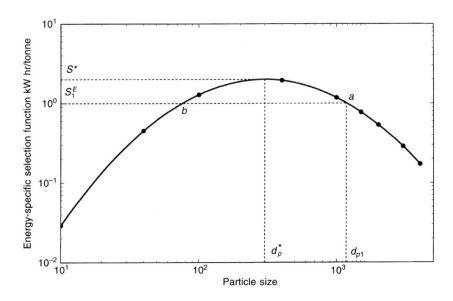

Figure 5.37 Graphical method for the determination of the parameters in the Herbst-Fuerstenau energy-specific rate of breakage

The media ball size also influences the specific rate of breakage. Smaller ball sizes produce less energetic impacts and each impact influences fewer particles in the immediate vicinity of the impact point between any two balls. The active zone in the slurry between balls is obviously smaller with smaller balls. Smaller balls are also less efficient at nipping larger particles. Offsetting these effects that tend to decrease the specific rate of breakage as ball size decreases is the increased frequency of impact that results from the increased number of smaller balls in the mill. The number of balls per unit volume varies as $1/d^3$. The net result of these competing effects is revealed by experiments which show that the specific rate of breakage scales as $1/d^n$ where n is approximately equal to 1 while the particle size at maximum specific rate of breakage increases in direct proportion to the ball size. The parameter μ, which defines the size at which the specific rate of breakage is a maximum, increases with ball size raised to a power close to 1 and also shows a power dependence on the mill diameter. Operating mills always contain a distribution of ball sizes and to accommodate this the scale-up is weighted in proportion to the mass of balls in each size.

$$k(d_p) = S_1 d_p^\alpha \sum_k \frac{m_k \left(\dfrac{d_T}{d_k} \right)^{N_0}}{1 + \left(\dfrac{d_p}{\mu_k} \right)^\Lambda} \tag{5.133}$$

where d_k is the representative diameter of the kth ball size class and m_k is the mass fraction of the ball size class in the mill charge. μ_k is given by

$$\frac{\mu_k}{\mu_{kT}} = \left(\frac{D_m}{D_{mT}}\right)^{N_2}\left(\frac{d_k}{d_T}\right)^{N_3}$$

(5.134)

Recommended values for the constants are $N_0 = 1.0$, $N_1 = 0.5$, $N_2 = 0.2$, $N_3 = 1.0$ and $c = 1.3$. Large size ball mills have been observed to operate inefficiently so that it is often recommended that S_1 should be scaled by the factor $(3.81/D_m)^{0.2}$ when the diameter of this type of mill exceeds 3.81 meters.

5.12.3 The Herbst-Fuerstenau model for the specific rate of breakage

The variation of the specific rate of breakage and of the breakage function with milling conditions can be substantially accounted for by describing the variation of these functions with the specific power input into the mill and examining how the specific power input varies with mill size and with design and operating conditions. This indirect method is particularly useful for scaling up the functions from batch scale and pilot scale test work. This is called the energy-specific scale-up procedure and was developed initially by Herbst and Fuerstenau (1980).

The specific rate of breakage out of size class i is proportional to the net specific power input to the mill charge.

$$k_i = S_i^E \frac{P}{M}$$

(5.135)

where P is the net power drawn by the mill exclusive of the no-load power that is required to overcome mechanical and frictional losses in the mill. M is the mass of the charge in the mill excluding the media. S_i^E is called the energy-specific selection function for particles in size class i. The energy-specific breakage rate is commonly reported in tonnes/kW hr. The essential feature of the Herbst-Fuerstenau model is that S_i^E is a function of the material only and does not vary with milling conditions nor with mill size. This assumption is equivalent to the postulate that the amount of breakage that occurs inside the mill is proportional to the amount of energy that has been absorbed by the material that is being milled and it is immaterial how the energy is actually imparted to the particles or at what rate. Thus the method can be used to scale up the breakage rates over broad operating ranges. In fact it has been found from experimental observations that S_i^E is a function of the ball media size distribution in the mill. It should be determined in a laboratory experiment using the same ball size distribution as the full-scale mill that is to be modeled or simulated.

It is usually convenient to use Equation 5.135 in the form

$$k_i\tau = S_i^E \frac{P}{W}$$

(5.136)

where W is the mass flowrate of solids through the mill. The product $k\tau$ can be used in Equations 5.84, 5.90 and 5.111 and neither the mean residence time τ of the solids in the mill nor the specific rate of breakage k_i need be known explicitly. This eliminates the need for complex empirical scale-up rules such as Equation 5.132. The ratio P/W, which is measured in kWhr/tonne, is the net specific power consumption in the mill.

The variation of energy-specific breakage rate with particle size is given by

$$\ln(S_i^E/S_1^E) = \zeta_1 \ln(d_{pi}/d_{p1}) + \zeta_2 [\ln(d_{pi}/d_{p1})]^2 + \ldots \quad (5.137)$$

which represents the logarithm of the energy-specific breakage rate as a power series of the logarithm of the particle size. Usually two terms in the series are sufficient to describe the variation in sufficient detail for most purposes so that the function becomes

$$\ln(S_i^E/S_1^E) = \zeta_1 \ln(d_{pi}/d_{p1}) + \zeta_2 [\ln(d_{pi}/d_{p1})]^2 \quad (5.138)$$

A typical plot of this function is shown in Figure 5.37.

The parameters in this functional form can be obtained from experimental data using the following simple graphical procedure.

(i) A suitable reference size d_{p1} is chosen to give the reference value S_1^E at point a on the curve. d_{p1} is usually taken as the top size for the problem on hand but that is not essential.

(ii) The turning point on the curve is located midway between points a and b and the coordinates of the turning point are (d_p^*, S^*).

(iii) The parameters in the selection function are related to the coordinates of the turning point by

$$\zeta_1 = \frac{2\ln(S^*/S_1^E)}{\ln(d_p^*/d_{p1})} \quad (5.139)$$

and

$$\zeta_2 = -\frac{\ln(S^*/S_1^E)}{[\ln(d_p^*/d_{p1})]^2} \quad (5.140)$$

These expressions are obtained by differentiating Equation 5.138 and setting the result to zero.

$$0 = \zeta_1 + 2\zeta_2 \ln(d_p^*/d_{p1}) \quad (5.141)$$

This must be solved simultaneously with

$$\ln(S^*/S_1^E) = \zeta_1 \ln(d_p^*/d_{p1}) + \zeta_2 [\ln(d_p^*/d_{p1})]^2 \quad (5.142)$$

which yields Equations 5.139 and 5.140.

This model for the variation of specific rate of breakage with particle size is useful to model the effect of ball size distribution in the mill. The parameter ζ_2 determines the sharpness of the maximum in the plot of S_i^E against the particle size as shown in Figure 5.37. ζ_2 must always be a negative number otherwise the graph will have a minimum rather than a maximum. Large numerical values of ζ_2 make the peak in the curve sharper and consequently the rate of breakage falls off rapidly at the smaller particle sizes. Conversely smaller values of ζ_2 make the peak flatter and breakage rates are maintained even at comparatively small sizes. Smaller media are associated with smaller values of ζ_2. ζ_1 determines the particle size at maximum specific breakage rate through the expression

$$\ln\left(\frac{d_p^*}{d_{p1}}\right) = -\frac{\zeta_1}{2\zeta_2} \tag{5.143}$$

5.12.4 The JKMRC model

The JKMRC group recommends that the ball mill should be modeled as a single perfect mixer with post classification. Equations 5.89 and 5.90 describe this situation and these equations can be written as

$$p_i^P = p_i^F - \frac{Mk_i p_i^P}{W(1+C)(1-c_i)} + \sum_{j=1}^{i-1} \frac{Mk_j b_{ij} p_j^P}{W(1+C)(1-c_j)} \tag{5.144}$$

The group $JK_i = Mk_j/W(1+C)(1-c_i)$ which includes both the specific rate of breakage and the classification or discharge function is considered to be an arbitrary function of the particle size which must be determined from operating data in the mill that is to be modeled or by comparison with an equivalent operating mill. The group JK_i is usually calculated from size distributions measured in the feed and discharge from a continuous mill at steady state using Equation 5.144 in the following form

$$JK_i = \frac{p_i^F - p_i^P}{p_i^P} + \sum_{j=1}^{i-1} JK_j b_{ij} \frac{p_j^P}{p_i^P} \tag{5.145}$$

Given a set of measured values for p_i^F and p_i^P and an assumed breakage function b_{ij}, the JK_i can be calculated starting with JK_1 and progressing to JK_2 and then to JK_3 in sequence until sufficient values have been estimated. A smooth cubic spline with four knots is then fitted to the experimental data and the smoothed values used in Equation 5.144 to calculate the mill performance.

5.12.5 Scale-up of the JKMRC model

The model can accommodate variations in the operating parameters of the ball mill through the scale-up relationship

$$JK_i \text{ proportional to } \frac{D_m^{2.5} L_e J(1-J)\varphi_c WI^{0.8}}{Q} \qquad (5.146)$$

In Equation 5.146 D_m is the mill diameter, L_e the effective length, Q is the volumetric flowrate of slurry to the mill, φ_c is the mill speed as a fraction of critical speed, J is the fractional mill filling by grinding media (see Figure 5.36) and WI is the Bond work index for the material that is to be ground in the mill. Using Equation 5.67 for the power drawn by the mill, JK_i can also be scaled according to the net mill power

$$JK_i \text{ proportional to } \frac{PWI^{0.8}}{Q} \qquad (5.147)$$

where P is the net power drawn by the mill.

Variations in media ball size affect JK_i in two ways. If a maximum occurs in the graph of JK_i against d_p, the particle size at the maximum, d_p^*, varies as d_b^2 where d_b is the ball diameter. Then for particle sizes smaller than d_p^*, JK_i is proportional to d_p^{-1} and for particle sizes larger than d_p^*, JK_i is proportional to d_b^2.

5.12.6 Specific rate of breakage from the impact energy spectrum

The selection function can be calculated from the fundamental breakage characteristics of the particles. The selection function is essentially a measure of the likelihood that a particle will be broken during a specific impact event. In order for a particle to be selected for breakage during the event it must be involved with the event (it must be in the impact zone between two media particles) and it must receive a sufficiently large fraction of the event impact energy so that its fracture energy is exceeded. Integrating over all the impacts in the mill gives

$$\text{Specific rate of breakage} = \int_0^\infty p(E)w(d_p, E) \int_0^1 P(eE, d_p)p(e)de\, dE \qquad (5.148)$$

where $p(E)$ is the distribution density for impact energies in the mill. e is the fraction of the impact energy that is captured by individual particles during the impact. These have been called the energy split factors (Liu and Schönert, 1996). $P(E, d_p)$ is the probability that a particle of size d_p will break when it captures energy E from the impact event.

The integral in Equation 5.148 has been evaluated with the following assumptions used for each of the terms in Equation 5.148.

Partition of energy among particles involved in the impact:

$$p(e) = \frac{0.3726}{(e+0.1)^{1.1}} \qquad (5.149)$$

Mass involved in the impact from single-impact measurements on beds of particles:

$$w(d_p, E) = 0.4 \, d_p^{1/2} E^{0.4} \text{ kg} \tag{5.150}$$

Distribution of impact energies in the mill based on DEM simulations:

$$p(E) = \beta \alpha_1^2 E \exp(-\alpha_1 E) + (1 - \beta)\alpha_2 \exp(-\alpha_2 E) \tag{5.151}$$

The probability of breakage $P(E, d_p)$ is log normal with

$$E_{50} = 56\left(1 + \frac{1}{d_p}\right)^2 \text{ J/kg} \tag{5.152}$$

These expressions were substituted into Equation 5.148 and the result is compared with the standard Austin function in Figure 5.38.

$$k = \frac{0.4 d_p^{0.5}}{1 + (d_p/10)^{2.51}} \tag{5.153}$$

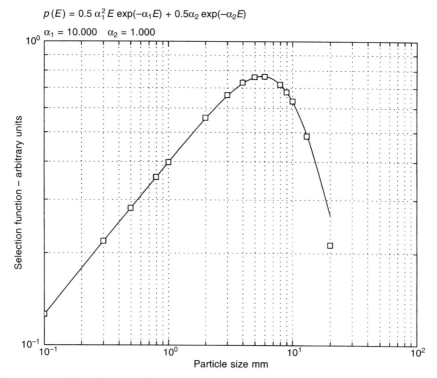

$p(E) = 0.5 \, \alpha_1^2 E \exp(-\alpha_1 E) + 0.5\alpha_2 \exp(-\alpha_2 E)$

$\alpha_1 = 10.000 \quad \alpha_2 = 1.000$

Figure 5.38 Specific rate of breakage calculated from Equation 5.148 (plotted points) compared to the Austin function Equation 5.153 (solid line)

5.13 Models for the specific rate of breakage in autogenous and semi-autogenous mills

The essential features of an autogenous or semi-autogenous mill are shown in Figure 5.39. The most significant difference between ball milling and autogenous milling is the presence of considerably larger particles of ore in the charge. These are added in the mill feed and act as grinding media. Consequently the average density of the media particles is considerably less than in the ball mill and this results in lower values for the specific rate of breakage when compared to ball or rod mills. The average density of the load is proportionately less also and as a result autogenous mills can be built with significantly larger diameters.

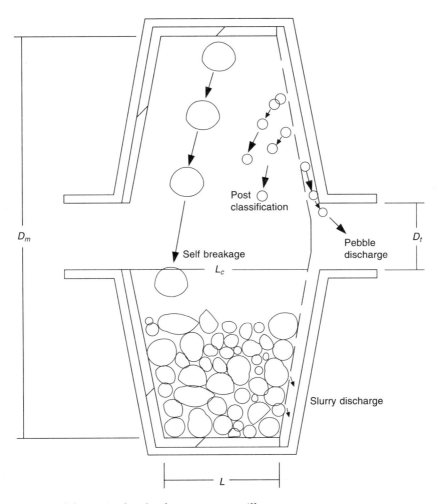

Figure 5.39 Schematic sketch of autogenous mill

Equation 5.124 defines the size distribution in the charge of an autogenous or semi-autogenous mill using the population balance model. The size distribution in the mill discharge can be calculated from the size distribution of the charge using Equation 5.126. In order to use these equations it is necessary to be able to calculate the selection functions and breakage functions for the separate breakage mechanisms that occur in the mill, namely attrition, chipping, impact fracture and self-breakage. Chipping and attrition occur on the surface of the particle and the inner core of the particle is not affected. Particles are subject to attrition through friction between particles and also between particles and the walls of the mill. Particles are chipped through impact against other particles or against the walls of the mills. The size distribution of the progeny from attrition is somewhat finer than that from chipping but these two subprocesses are normally lumped together. Goldman and Barbery (1988), Goldman *et al.* (1991) and Austin *et al.* (1987) have measured wear rates in comparatively small test mills in batch mode and all the data can be described by a wear model of the type

$$\kappa_i = \kappa d_{pi}^{\Delta} \qquad (5.154)$$

Austin *et al.* favor $\Delta = 0$ while Goldman *et al.* (1991) found $\Delta = 0.37$ in a 1.75 m diameter pilot plant mill and $\Delta = 1$ in a 0.75 m laboratory mill (Goldman *et al.* 1988). The surface specific wear rate depends on the milling environment decreasing as the proportion of fines increases in the mill charge and increasing with mill load and mill diameter. Careful experiments have shown that the surface specific wear rate on a particular particle decreases for several minutes after it has been introduced into the mill environment because its initial rough surface is subject to chipping and attrition that decreases as the particle becomes rounded. Because of this phenomenon the population balance equation should also allow for a distribution of sojourn times of the particles in the mill. This level of fine detail is not justified at the level of modeling that is described here.

The value of the specific attrition rate κ is ore specific and should be measured experimentally for the ore. An attrition method has been developed at Julius Kruttschnitt Mineral Research Center and is described in their 25th anniversary volume (Napier-Munn *et al.*, 1996). A sample of the material in the 38 mm to 50 mm size range is tumbled for 10 minutes. The size distribution of the charge after this time is determined by screening and is typically bimodal as shown in Figure 5.40. Estimates of the specific attrition rate and the attrition breakage function can be easily obtained from this graph. From Equation 2.109:

$$\frac{dm}{dt} = -\kappa \frac{\pi \rho_s x^{2+\Delta}}{2} = -\kappa \frac{3m}{x^{1-\Delta}} \qquad (5.155)$$

Since the particles do not change much in size during the test, x is a constant equal to d_f for the duration of the test.

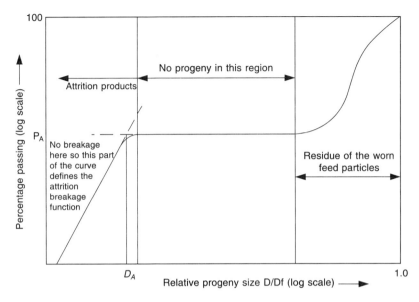

Figure 5.40 Schematic of a typical progeny size distribution from a 10 minute tumbling test

$$\frac{dm}{dt} = -3\kappa \, \frac{m}{d_f^{1-\Delta}} \tag{5.156}$$

where d_f is the geometric mean size of the original feed particles and simple integration gives

$$m(t) = m(0) \exp\left(-\frac{3\kappa}{d_f^{1-\Delta}} t\right) \tag{5.157}$$

and

$$\frac{P_A}{100} = \frac{m(0) - m(t)}{m(0)} = 1 - \exp\left(-\frac{3\kappa \Delta t}{dp^{1-\Delta}}\right) \approx \frac{3\kappa \Delta t}{d_f^{1-\Delta}} \tag{5.158}$$

Since the plateau in Figure 5.40 usually includes the point $D/D_f = 0.1$, P_A is closely related to the t_a parameter that is quoted in the JKMRC attrition test result. The fraction of the original particles that has degraded by attrition during the test is equal to P_A, the height of the plateau on the cumulative size distribution curve. Thus

$$P_A = 10 t_A$$

$$\kappa = \frac{d_f^{1-\Delta} P_A}{300 \Delta t} = \frac{d_p^{1-\Delta} t_A}{300 \times 60} \; \mathrm{m}^{1-\Delta} \tag{5.159}$$

The specific rate of breakage due to impact follows the pattern observed in ball mills and the Austin method can be used to model the specific rate of breakage due to impact fracture provided that due allowance is made for the lower density of the medium particles. In semi-autogenous mills due allowance must also be made for the presence of steel balls as well as the autogenous media. These effects are modeled by the Austin scale-up procedure by including the autogenous medium in the media size classes in Equation 5.133. However the lower density of the media particles must be allowed for and in Equation 5.133 both m_k and μ_k must be scaled by the ratio ρ_k/ρ_s where ρ_k is the density of media particles in media size class k and ρ_s is the density of the balls in the test mill.

$$m_k^{\text{pebbles}} = m_k^{\text{steel}} \frac{\rho^{\text{pebbles}}}{\rho^{\text{steel}}} \qquad (5.160)$$

and

$$\mu_k^{\text{pebbles}} = \mu_k^{\text{steel}} \frac{\rho^{\text{pebbles}}}{\rho^{\text{steel}}} \qquad (5.161)$$

The sharp decrease in the specific rate of breakage that is evident for particles that are too large to be properly nipped during an impact event is especially important in autogenous and semi-autogenous mills because the coarse feed supplies many particles in this size range. An intermediate size range exists in the autogenous mill in which the particles are too large to suffer impact breakage but are too small to suffer self-breakage. Particles in this size range can accumulate in the mill because they neither break nor are they discharged unless appropriate ports are provided in the discharge grate. This is the phenomenon of critical size build up.

The phenomenon of self-breakage is completely absent in ball mills but it plays an important role in autogenous milling. The larger the particle and the greater its height of fall in the mill, the larger its probability of self-breakage on impact. Particles smaller than 10 mm or so have negligible breakage probabilities and consequently very low values of the specific rate of self-breakage. Specific rates of breakage by impact fracture and self-breakage are additive and the variation of k_i with particle size over the entire size range is shown schematically in Figure 5.41. The overall model is obtained from Equations 5.127 and 5.7.

$$k_i = \frac{S_1 d_p^\alpha}{1 + (d_p/\mu)^\Lambda} + \text{drop frequency} \times G\left[\frac{\ln(E/E_{50})}{\sigma_E}\right] \qquad (5.162)$$

In Equation 5.162, E is the average kinetic energy per unit mass of a lump of size d_p when it impacts the liner or the charge after being released at the top of the mill during tumbling. This is calculated as the potential energy of the particle at a fraction of the inside diameter of the mill. E_{50} is the median particle fracture energy of a lump of size d_p. For larger lumps this is independent

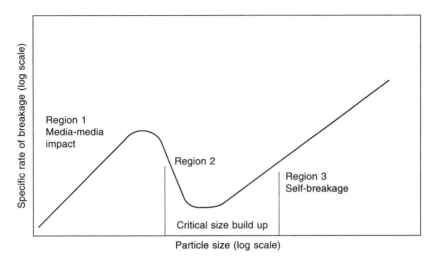

Figure 5.41 Schematic representation of the specific rate of breakage in SAG and FAG mills

of size as given by Equation 5.9 but is material specific. The drop frequency is calculated from the assumption that each lump will be dropped once per revolution of the mill.

$$E = fgD_m \quad J \tag{5.163}$$

and drop frequency $f = 0.705 \times \varphi_c / D_m^{1/2}$ from Equation 5.61.

5.14 Models for the breakage function in autogenous and semi-autogenous mills

The breakage function for impact and self-breakage is determined primarily by the impact energy level and the t_{10} method is used as the model as described in Section 5.4.2. For self-breakage, the impact energy is a function of the lump size and the height of the drop and this is calculated as the potential energy of the lump at half the inside diameter of the mill. For impact breakage the average energy is calculated as the net specific power input to the mill charge.

The breakage function for the products of attrition and chipping can be obtained from the size distribution of the products of the single batch attrition test because there is no significant rebreakage of the attrition products during the test. This breakage function is modeled using a simple logarithmic distribution:

$$A(x, D_A) = \left(\frac{x}{D_A} \right)^{\lambda_A} \tag{5.164}$$

D_A is the largest fragment formed by attrition and can be obtained from the measured attrition progeny size distribution as shown in Figure 5.40. λ_A is the slope of the straight line portion of the curve in the attrition product region.

The breakage function for impact fracture is modeled using the same models as for ball mills. The breakage function for self-breakage can be determined by dropping individual lumps of ore and determining the size distribution of the products as a function of the kinetic energy of impact.

5.15 Mill power and mill selection

5.15.1 The Bond method

The correlation between material toughness and power required in the comminution machine is expressed by the empirical Bond equation. The work done in reducing a mass of material from representative size d_{80}^F to representative size d_{80}^P is given by the Bond equation (5.33).

$$P_o = K\left(\frac{1}{(d_{80}^P)^{1/2}} - \frac{1}{(d_{80}^F)^{1/2}}\right) \text{ kWhr/ton} \tag{5.165}$$

A reference condition is the hypothetical reduction of 1 tonne of material from a very large size to a representative size of 100 microns. This reference energy is called the work index of the material *WI*.

$$WI = K\left(\frac{1}{(100)^{1/2}} - 0\right) = \frac{K}{10} \Rightarrow K = 10WI \tag{5.166}$$

$$P_o = 10WI\left(\frac{1}{(d_{80}^P)^{1/2}} - \frac{1}{(d_{80}^F)^{1/2}}\right) \tag{5.167}$$

in these equations d_{80} must be specified in microns.

The representative size is conventionally taken as the 80% passing size. *WI* can be determined from a standard laboratory test procedure. The Bond equation can be used for crushers, rod and ball mills. *WI* is usually different for these three operations and must be measured separately. The standard laboratory test for the measurement of the Bond work index was designed to produce an index that would correctly predict the power required by a wet overflow discharge ball mill of 2.44 m diameter that operates in closed circuit with a classifier at 250% circulating load. d_{80}^F and d_{80}^P in Equation 5.167 refer to the feed to the circuit as a whole and the product from the classifier.

The work index for the ore as measured using the standard laboratory method must be adjusted to account for various operating conditions before applying it to calculate the energy requirements of an industrial mill that differ from this standard. This is done by multiplying the measured work index by a series of efficiency factors to account for differences between the actual milling operation and the standard conditions against which the work index was originally calibrated.

The efficiency factors are:

EF_1:   Factor to apply for dry grinding in closed circuit in ball mills = 1.3.

EF_2:   Open circuit factor to account for the smaller size reduction that is observed across the mill itself (open circuit condition) as opposed to the size reduction obtained across the closed circuit. If two ball mill circuits, one open and the other closed, as shown in Figure 5.42, produce the same d_{80}^P the power required for the closed circuit is given directly by the Bond formula:

$$P_{CC} = 10\,\mathrm{WI}\left(\frac{1}{(d_{80}^P)^{1/2}} - \frac{1}{(d_{80}^F)^{1/2}}\right) \tag{5.168}$$

and the power required by the open circuit is given by

$$P_{OC} = 10 \times EF_2 \times \mathrm{WI}\left(\frac{1}{(d_{80}^P)^{1/2}} - \frac{1}{(d_{80}^F)^{1/2}}\right) \tag{5.169}$$

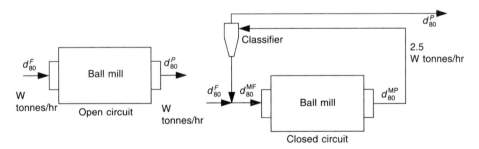

Figure 5.42 Application of Bond work index for calculating power requred for open and closed milling circuits

with $EF_2 = 1.2$.

If the two circuits are required to match their product passing size at a different % passing say d_{90}^P, the factor EF_2 will have the different value as given in Table 5.8.

EF_3:   Factor for variation in mill diameter. Larger mills are assumed to utilize power somewhat more efficiently than smaller mills. This factor is calculated from

$$EF_3 = \left(\frac{2.44}{D_m}\right)^{0.2} \quad \text{for} \quad D_m < 3.81$$
$$= 0.914 \quad \text{for} \quad D_m \geq 3.81 \tag{5.170}$$

Table 5.8 Bond work index efficiency factor for wet open circuit milling

Reference % passing	Open circuit efficiency factor EF_2
50%	1.035
60%	1.05
70%	1.10
80%	1.20
90%	1.40
92%	1.46
95%	1.57
98%	1.70

EF_4:   Oversize feed factor. The optimal d_{80}^F size for a ball mill that grinds material having a work index of WI kW hr/tonne is given by

$$F_O = 4 \times \left(\frac{14.3}{WI}\right)^{1/2} \quad \text{mm} \tag{5.171}$$

When the feed has a size distribution coarser than the optimum, the mill must draw more power to achieve the desired product size. The appropriate efficiency factor is given by

$$EF_4 = 1 + \frac{(WI - 7)(d_{80}^F - F_o)}{R_r F_o} \tag{5.172}$$

where R_r is the reduction ratio $\dfrac{d_{80}^F}{d_{80}^P}$

EF_5:   The work index in a ball mill increases when the reduction ratio decreases below 3 and the efficiency factor is given by

$$EF_5 = 1 + \frac{0.013}{R_r - 1.35} \tag{5.173}$$

To size a ball mill or rod mill that must process W tonnes/hr it is necessary to calculate the mill power required from

$$P = P_o \times W \tag{5.174}$$

and find a mill from the manufacturer's catalogue that can accept that power. If manufacturer's data is not available, Equations 5.65 or 5.67 can be used to select a suitable mill. This will not produce a unique design for the mill because many combinations of D_m and L will satisfy these equations and the aspect ratio of the mill must be chosen to ensure that the mill will provide sufficient residence time or specific power input to produce the required product size distribution as calculated from Equation 5.84 or 5.126. It is important to note that the geometry of the mill will determine the power input *not* the tonnage through the mill. The power consumed will give a certain amount of size reduction and the output size will be a function of the tonnage. By contrast a crusher will produce a fixed size reduction ratio and the power will vary to match the tonnage subject to the maximum power available from the motor that is installed on the crusher.

5.16 The batch mill

5.16.1 Batch grinding of homogeneous solids

The comminution properties of materials are often determined in the laboratory by following the size distribution during milling in a batch mill. In such a test, the mill is charged with a given mass of feed material and the mill is operated without continuous discharge and feed.

The size distribution in the charge in a batch mill changes continuously with time. The mass balance must be written for each size class.

For the top size:

$$M\frac{dm_1}{dt} = -k_1 M m_1 \tag{5.175}$$

It is usually adequate to assume that k_1 does not vary with time and this equation can be easily integrated to give

$$m_1 = m_1(0)e^{-k_1 t} \tag{5.176}$$

This solution plots as a straight line on log-linear coordinates.

For the next size down:

$$M\frac{dm_2}{dt} = -k_2 M m_2 + k_1 M m_1 b_{21} \tag{5.177}$$

$$\frac{dm_2}{dt} + k_2 m_2 = k_1 b_{21} m_1 = k_1 b_{21} m_1(0)e^{-k_1 t} \tag{5.178}$$

This is a first order linear differential equation which has a solution of the form:

$$m_2 = Ae^{-k_2 t} + Be^{-k_1 t} \tag{5.179}$$

where A and B are constants that must be determined from the differential equation and from the initial conditions. First the proposed solution is differentiated

$$\frac{dm_2}{dt} = -Ak_2 e^{-k_2 t} - Bk_1 e^{-k_1 t} \tag{5.180}$$

These expressions for w_2 and its derivative are substituted into the original differential equation

$$-Ak_2e^{k_2t} - Bk_1e^{-k_1t} + k_2Ae^{-k_2t} + k_2Be^{-k_1t} = k_1b_{21}m_1(0)e^{-k_1t} \quad (5.181)$$

which simplifies to

$$(-k_1B + k_2B)e^{-k_1t} = k_1b_{21}m_1(0)e^{-k_1t} \quad (5.182)$$

so that B is given by

$$B = \frac{k_1b_{21}m_1(0)}{k_2 - k_1} \quad (5.183)$$

A must be evaluated from the initial condition.

$$m_2(0) = A + B \quad (5.184)$$

which gives

$$A = m_2(0) - B \quad (5.185)$$

and

$$\begin{aligned} m_2 &= m_2(0)e^{-k_2t} - B(e^{-k_2t} - e^{-k_1t}) \\ &= m_2(0)e^{-k_2t} - \frac{k_1b_{21}m_1(0)}{k_2 - k_1}(e^{-k_2t} - e^{-k_1t}) \end{aligned} \quad (5.186)$$

The solution can be continued in this way to develop the solution from size to size. The solution is tedious but not impossible. It is better to develop a solution that works automatically for all sizes and which is especially suited to computer methods. Note firstly that a general solution can take two forms:

$$\begin{aligned} m_i &= \sum_{j=1}^{i} \alpha_{ij}e^{-k_jt} \\ m_i &= \sum_{j=1}^{i} \beta_{ij}m_j(0) \end{aligned} \quad (5.187)$$

The coefficients α_{ij} are not functions of time but are functions of the initial conditions and the coefficients β_{ij} are not functions of the initial conditions but they vary with the time. The coefficients α_{ij} can be developed through some recursion relationships as follows:

The differential equation that describes the variation of each of the size classes is

$$\frac{dm_i}{dt} = -k_im_i + \sum_{j=1}^{i-1} b_{ij}k_jm_j \quad (5.188)$$

The general solution is now substituted into this equation

$$\sum_{j=1}^{i} -\alpha_{ij}k_je^{k_jt} = -k_i\sum_{j=1}^{i}\alpha_{ij}e^{-k_jt} + \sum_{j=1}^{i-1}b_{ij}k_j\sum_{l=1}^{j}\alpha_{jl}e^{-k_lt} \quad (5.189)$$

Re-arranging and collecting terms

$$\sum_{j=1}^{i} \alpha_{ij}(k_i - k_j)e^{-k_jt} = \sum_{j=1}^{i-1} \sum_{l=1}^{j} b_{ij}k_j\alpha_{jl}e^{-k_lt} \tag{5.190}$$

The order of the double summation must now be reversed

$$\sum_{l=1}^{i-1} \alpha_{il}(k_i - k_l)e^{-k_lt} = \sum_{l=1}^{i-1} \sum_{j=l}^{i-1} b_{ij}k_j\alpha_{jl}e^{-k_lt} \tag{5.191}$$

The change in the limits on the double summation should be noted particularly. The region over which the double summation operates must not change as the order of summation is switched.

Now the terms are collected

$$\sum_{l=1}^{i-1} \left(\alpha_{il}(k_i - k_l) - \sum_{j=l}^{i-1} b_{ij}k_j\alpha_{jl} \right) e^{-k_lt} = 0 \tag{5.192}$$

The coefficient of each exponential must be zero if the summation is to be zero for every value of t,

$$\alpha_{il}(k_i - k_l) = \sum_{j=l}^{i-1} b_{ij}k_j\alpha_{jl} \tag{5.193}$$

which provides the value of each value of α except for α_{ii}.

$$\alpha_{il} = \frac{1}{k_i - k_l} \sum_{j=l}^{i-1} b_{ij}k_j\alpha_{jl} \quad \text{for} \quad i > l \tag{5.194}$$

α_{ii} can be obtained from the initial condition

$$m_i(0) = \sum_{j=1}^{i} \alpha_{ij} = \alpha_{ii} + \sum_{j=1}^{i-1} \alpha_{ij} \tag{5.195}$$

$$\alpha_{ii} = m_i(0) - \sum_{j=l}^{i-1} \alpha_{ij} \tag{5.196}$$

All the coefficients can be solved by recursion starting from $i = 1$.

$$\alpha_{11} = m_1(0)$$
$$\alpha_{21} = \frac{1}{k_2 - k_1}b_{21}k_1\alpha_{11} \tag{5.197}$$

etc.

This recursion is most useful in the form

$$\alpha_{ij} = c_{ij}a_j \quad \text{with} \quad j < i \quad \text{and} \quad c_{ii} = 1 \tag{5.198}$$

$$\alpha_{il} = c_{il}a_l = \frac{1}{k_i - k_l} \sum_{j=l}^{i-1} b_{ij}k_jc_{jl}a_l \tag{5.199}$$

$$c_{il} = \frac{1}{k_i - k_l} \sum_{j=l}^{i-1} b_{ij}k_jc_{jl}$$

$$\alpha_{ii} = a_i = m_i(0) - \sum_{j=1}^{i-1} c_{ij} a_j \qquad (5.200)$$

Then the c_{il}'s are independent of both the time and the initial conditions and they can be calculated once and for all from a knowledge of the specific breakage rate constants and the breakage function.

5.16.2 Batch grinding of heterogeneous solids

The batch comminution equation for heterogeneous solids is

$$\frac{dp_{ij}}{dt} = -k_{ij}p_{ij} + \sum_{l=1}^{j-1} \sum_{k=K_l^*}^{K_l^{**}} k_{kl} b_{ijkl} p_{kl} \qquad (5.201)$$

In Equation 5.201 i, j, k and l index the parent particle composition, the parent size, the progeny composition and the progeny size respectively. K_l^* and K_l^{**} are the left and right hand boundaries of region R' in the Andrews-Mika diagram for parent particles in size class l. b_{ijkl} is the discretized version of the function $b(g, d_p | g', d_p')$.

In practice it is convenient to decouple the size reduction process from the liberation process. This can be done by using the conditional breakage functions:

$$b_{ijkl} = b_{j,kl} b_{i,jkl} \qquad (5.202)$$

where $b_{j,kl}$ is the fraction of material breaking from class k, l that reports to size class j. $b_{i,jkl}$ is the conditional transfer coefficient from grade class k to grade class i given that the particle transfers from size class l to size class j. $b_{i,jkl}$ is usually represented as an Andrews-Mika diagram as described in Chapter 3. $b_{j,kl}$ is referred to as the 'size breakage function' and $b_{i,jkl}$ as the Andrews-Mika coefficients. Equation 5.202 is completely general and does not depend on the assumption of random fracture. $b_{j,kl}$ and $b_{i,jkl}$ are conditional distributions and must satisfy the conditions

$$\sum_{j=l+1}^{N} b_{j,kl} = 1 \qquad (5.203)$$

and

$$\sum_{i=1}^{12} b_{i,jkl} = 1 \qquad (5.204)$$

Appropriate models for the Andrews-Mika diagram are discussed in Chapter 3.

A solution to the heterogeneous batch comminution equation can be generated by exploiting the linearity of the differential equations to generate the general solution

$$p_{ij} = \sum_{l=1}^{j} \sum_{k=1}^{12} \alpha_{ijkl} e^{-k_{kl}t} \qquad (5.205)$$

The coefficients in Equation 5.205 are related to the selection and breakage functions and to the initial conditions using the following recursion relationships

$$\alpha_{ijmj} = 0 \quad \text{if } i \neq m \tag{5.206}$$

$$\alpha_{ijij} = p_{ij}(0) - \sum_{l=1}^{j-1} \sum_{k=1}^{12} \alpha_{ijkl} \tag{5.207}$$

$$\alpha_{ijmn} = \frac{\sum_{l=n}^{j-1} \sum_{k=1}^{12} k_{kl} b_{ijkl} \alpha_{klmn}}{k_{ij} - k_{mn}} \tag{5.208}$$

Note that the summations in Equation 5.208 run over the feeder regions and not over the attainable regions.

This solution to Equation 5.201 is based on the usual convention that breakage implies that all progeny leave the size class of the parent particle. The pathological case $k_{ij} = k_{mn}$ is occasionally encountered in practice. When it occurs, it is usually handled by making a slight adjustment to the parameters that define the relationship between the specific rate of breakage and the particle size to assure that no two values of k_{ij} are exactly equal.

Equation 5.205 represents a complete and convenient solution to the discrete version of the batch comminution equation with liberation and this solution produces the size distribution as well as the liberation distribution as a function of the time of grinding.

Bibliography

The bibliography dealing with comminution is large. The classic reference for the application of population balance methods is Austin *et al.* (1984). This book discusses the application of the population balance method and describes the careful experimental work that is required to measure the breakage and selection functions, which are central to the method, in the laboratory. The considerable work done at the Julius Kruttschnitt Mineral Research Centre on the application of these models to simulation and plant evaluation are recorded in two important texts by Lynch (1977) and by Napier-Munn *et al.* (1996). Useful research material on comminution can be found in two special issues of the *International Journal of Mineral Processing*, Volume 22, 1988 and combined Volumes 44 and 45, 1996.

Many good texts on fracture mechanics are available. A modern text that covers the material that is used here is Anderson (1995). Lawn (1993) gives an excellent account of the fracture of brittle solids.

Bearman *et al.* (1989), Middlemiss (1989) and Middlemiss and Tait (1990) have measured the fracture toughness of some minerals and ores.

Elementary breakage events have been studied in the laboratory using the drop weight method on beds of particles (Höffler, 1990; Bourgeois, 1993) and by slow compression of constrained particle beds in a piston and die configuration (Liu and Schönert, 1996; Fandrich *et al.*, 1997).

The concept of particle fracture energy was introduced by Baumgardt *et al.* in 1975. A method for the accurate measurement of this particle property using the fast load cell is described by Tavares and King (1998).

The t_{10} method for reconstructing the breakage function was developed by Narayanan and Whiten (1983) using data originally obtained using the pendulum impact test.

The models for the gyratory and jaw crushers is based on data from the *Nordberg Reference Manual*, Nordberg Inc., 4th edition 1993. Use of the logarithmic distribution for plant simulation and optimization was reported by Csoke *et al.* (1996).

The model used for cone crushers is based on Whiten *et al.* (1979), Whiten (1973) and Karra (1982).

The use of the energy-specific selection function to scale up ball mill operations is described in Herbst and Fuerstenau (1980), Herbst *et al.* (1986) and Lo and Herbst (1986, 1988).

Residence-time distributions for the solids in the mill have been measured by Kinneberg and Herbst (1984) and by Schneider (1995).

The population balance model for the autogenous and semi-autogenous mill is based on the analysis of Hoyer and Austin (1985). Details of the breakage mechanisms in these mills were investigated by Stanley (1974), Austin *et al.* (1987), Goldman and Barbery (1988), Goldman *et al.* (1991) and Leung *et al.* (1988).

Application of the classical Bond method is described by Rowland (1998).

The analysis of the batch mill with liberation is described in King and Schneider (1998).

References

Anderson, T.L. (1995) *Fracture Mechanics*. 2nd edition, CRC Press, p. 47.

Austin, L.G. (1990) A mill power equation for SAG mills. *Minerals and Metallurgical Processing*, pp. 57–62.

Austin, L.G., Klimpel, R.R. and Luckie, P.T. (1984) *Process Engineering of Size Reduction: Ball Milling*. Soc. Mining Engineers, New York.

Austin, L.G., Menacho, J.M. and Pearcy, F. (1987) *Proc. 20th Int. Symposium on the Application of Computers and Mathematics in the Mineral Industries*. S. Afr. Inst. Min. Metall., Johannesburg, pp. 107–126.

Baumgardt, S., Buss, B., May, P. and Schubert, H. (1975) On the comparison of results in single grain crushing under different kinds of load. *Proc. 11th Intl Mineral Processing Congress*, pp. 3–32.

Bond, F.C. (1952) The third theory of grinding. *Trans. AIME*, Vol. 193, pp. 484–494.

Bourgeois, F. (1993) Microscale Modeling of Comminution Processes. Ph.D Thesis, University of Utah.

Bearman, R.A., Pine, R.J. and Wills, B.A. (1989) Use of fracture toughness testing in characterising the comminution potential of rock. *Proc. Joint MMIJ/IMM Symposium*, Kyoto.

Csoke, B., Petho, S., Foldesi, J. and Meszaros, L. (1996) Optimization of stone-quarry technologies. *International Journal of Mineral Processing*, Vol. 44–45, pp. 447–459.

Dan, C.C. and Schubert, H. (1990) Breakage probability, progeny size distribution and energy utilization of comminution by impact. *Proc. 7th European Symposium on Comminution*, Conkarjev Dom, Ljubljana 1:160–179.

Fandrich, R.G., Bearman, R.A., Boland, J. and Lim, W. (1997) Mineral liberation by particle bed breakage. *Minerals Engineering*, Vol. 10, pp. 175–187.

Goldman, M. and Barbery, G. (1988) Wear and chipping of coarse particles in autogenous grinding: experimental investigation and modeling. *Minerals Engineering*, Vol. 1, pp. 67–76.

Goldman, M., Barbery, G. and Flament, F. (1991) Modelling load and product distribution in autogenous and semi-autogenous mills: pilot plant tests. *CIM Bulletin*, Vol. 8 (Feb.), pp. 80–86.

Herbst, J.A. and Fuerstenau, D.W. (1980) Scale-up procedure for continuous grinding mill design using population balance models. *Int. Jnl Mineral Processing*, Vol. 7, pp. 1–31.

Herbst, J.A., Lo, Y.C. and Rajamani, K. (1986) Population balance model predictions of the performance of large-diameter mills. *Minerals and Metallurgical Engineering*, pp. 114–120.

Höffler, A. (1990) Fundamental Breakage Study of Mineral Particles with an Ultrafast Load Cell Device. Ph.D Thesis, University of Utah.

Hoyer, D.I. and Austin, L.G. (1985) A simulation model for autogenous pebble mills. Preprint number 85-430, SME-AIME Fall Meeting, Albuquerque, Oct.

Karra, V.K. (1982) A process performance model for ore crushers. *Proc. 4th Int. Min. Proc. Congress Toronto*, Vol. III, pp. 6.1–6.14.

Kick, F. (1883) *Dinglers Polytechnisches Journal*, Vol. 247, pp. 1–5.

King, R.P. and Schneider, C.L. (1998) Mineral liberation and the batch comminution equation. *Minerals Engineering*, Vol. 11, pp. 1143–1160.

Kinneberg, D.J. and Herbst, J.A. (1984) A comparison of linear and nonlinear models for open-circuit ball mill grinding. *Intl. Jnl Mineral Processing*, Vol. 13, pp. 143–165.

Lawn, B. (1993) *Fracture of Brittle Solids*. 2nd edition, Cambridge University Press, Cambridge.

Leung, K., Morrison, R.D. and Whiten, W.J. (1988) An energy based ore-specific model for autogenous and semi-autogenous grinding. *Coper '87*, Universidadde Chile, pp. 71–85.

Liu, J. and Schönert, K. (1996) Modelling of interparticle breakage. *Int. Jnl Miner. Processing*, Vol. 44–45, pp. 101–115.

Lo Y.C. and Herbst, J.A. (1986) Consideration of ball size effects in the population balance approach to mill scale-up. In *Advances in Mineral Processing* (P. Somasudaran, ed.), Society of Mining Engineers Inc. Littleton, pp. 33–47.

Lo, Y.C. and Herbst, J.A. (1988) Analysis of the performance of large-diameter mills at Bougainville using the population balance approach. *Minerals and Metallurgical Processing*, pp. 221–226.

Lynch, A.J. (1977) *Mineral Crushing and Grinding Circuits. Their simulation, Optimisation, Design and Control*. Elsevier Scientific Publishing Company, Amsterdam.

Menacho, J.M. (1986) Some solutions for the kinetics of combined fracture and abrasion breakage. *Powder Technology*, Vol. 49, pp. 87–96.

Middlemiss, S.N. (1989) Effects of notch size and geometry and test techniques on fracture toughness of quartzite. Dept of Metallurgy and Materials Engineering, Univ. of the Witwatersrand. Report FRG/89/5.

Middlemiss, S.N. and Tait, R.B. (1990) Fracture toughness of quartzite as a function of various parameters including triaxial confining stress, temperature and under conditions of shear. Univ. of the Witwatersrand, Dept of Metallurgy and Materials Engineering Report FRG 90/1.

Morrell, S. (1996) Power draw of wet tumbling mills and its relationship to charge dynamics, Part 2: An empirical approach to modeling of mill power draw. *Trans. Instn Mining Metall. (Sect. C: Mineral Processing Extr Metall.)*, **105**, January–April, pp. C54–C62.

Napier-Munn, T.J., Morrell, S., Morrison, R.D. and Kojovic, T. (1996) Mineral Comminution Circuits. Their Operation and Optimisation. Julius Kruttschnitt Mineral Research Centre, Brisbane.

Narayanan, S.S. and Whiten, W.J. (1983) Breakage characteristics for ores for ball mill modelling. *Proc. Australas. Instn Min. Metall.*, No. 286, pp. 31–39.

Pauw, O.G. and Mare, M.S. (1988) The determination of optimum impact-breakage routes for an ore. *Powder Technology*, Vol. 54, pp. 3–13.

Rajamani, R.K., Mishra, B.K., Songfack, P. and Venugopal, R. (1999) Millsoft – simulation software for tumbling-mill design and trouble shooting. *Mining Engineering*, December, pp. 41–47.

Rittinger, R.P. von, (1857) Lehrbuch der Aufbereitungskunde, Ernst u. Korn, Berlin.

Rowland, C.A. (1998) Using the Bond work index to measure the operating comminution efficiency. *Minerals & Metallurgical Processing*, Vol. 15, pp. 32–36.

Rowland, C.A. and Kjos, D.M. (1980) Rod and Ball Mills. In *Mineral Processing Plant Design*, 2nd edition (A.L. Mular and R.B. Bhappu, eds), SME Littleton, CO, p. 239.

Schneider, C.L. (1995) The Measurement and Calculation of Liberation in Continuous grinding Circuits, Ph.D Thesis, University of Utah.

Stanley, G.G. (1974) Mechanisms in the autogenous mill and their mathematical representation. *S. Afr. Inst. Mining Metall.*, Vol. 75, pp. 77–98.

Tavares, L.M. and King, R.P. (1998) Single-particle fracture under impact loading. *Intl Journal Mineral Processing*, Vol. 54, pp. 1–28.

Whiten, W.J., Walter, G.W. and White, M.E. (1979) A breakage function suitable for crusher models. *4th Tewkesbury Symposium*, Melbourne, pp. 19.1–19.3.

Whiten, W.J. (1973) The simulation of crushing plants. *Application of computer methods in the mineral industry. Apcom 10*, S. Afr. Inst. of Mining and Metall., Johannesburg, pp. 317–323.

6
Solid–liquid separation

It is often necessary to separate the particulate solid material from the water which acts as a carrier medium for the solids. The type of solid–liquid separation procedure that must be used will depend on the nature of the slurry that must be dewatered and on the final products that are to be produced. Two kinds of solid–liquid separation processes are commonly used in mineral processing operations: sedimentation and filtration. The sedimentation processes are further subdivided into clarification and thickening.

6.1 Thickening

Thickening is an important process for the partial dewatering of comparatively dense slurries. The slurry is allowed to settle under gravity but the particles are close enough together to hinder each other during settling and they tend to settle as a mass rather than individually. The rate of settling is a fundamental characteristic of the slurry which must be determined experimentally for each slurry under appropriate conditions of flocculation in order to design and size an appropriate thickener. The rate of settling depends on the nature of the particles that make up the slurry and on the degree of flocculation that is achieved. For a particular flocculated slurry the rate of settling is determined chiefly by the local solid content of the slurry and will vary from point to point in the slurry as the local solid content varies.

The rate of settling of a flocculated slurry can be measured in a simple batch settling test. The method used to extract the information from the simple batch settling curve is known as the Kynch construction. It is based on a fundamental proposition by Kynch that the settling velocity is a function only of the local concentration of solid in the pulp. The Kynch construction is applied to the experimental batch settling curve and this establishes the relationship between the rate of settling of a slurry and the local solid content. The details of the Kynch construction and its theoretical derivation from the basic Kynch hypothesis are not given here since we concentrate on the application of the principle to the operation of continuous thickeners.

The settling flux, ψ, is the mass of solid crossing a horizontal plane in the slurry per unit time and specified per unit area of the plane.

Kynch postulated the fundamental principle that the settling velocity is a function only of the local concentration of solids in the slurry. The settling velocity is the velocity relative to fixed laboratory coordinates that will be recorded if a single settling particle is observed. Thus in an ideal slurry the settling velocity is not a function of particle size (true hindered settling) and not an explicit function of time or history of the settling slurry.

The flux is related to the settling velocity by

$$\psi = CV(C) \tag{6.1}$$

The relationship between the settling velocity and the concentration as shown in Figure 6.1 applies at every point in the slurry. C_s represents the concentration of an ideal fully settled incompressible sediment. In practice sediments are rarely incompressible because they continue to settle over time as water is expressed due to the compression of the flocs under the weight of the settled solids higher in the bed. Consequently it is not possible to determine C_s precisely but nevertheless it is convenient to assume a single well-defined value for C_s in the simple theory. It is more useful to represent the data shown in Figure 6.1 as a relationship between the settling flux, obtained as the product $\psi(C) = CV(C)$, and the concentration. This is shown in Figure 6.2.

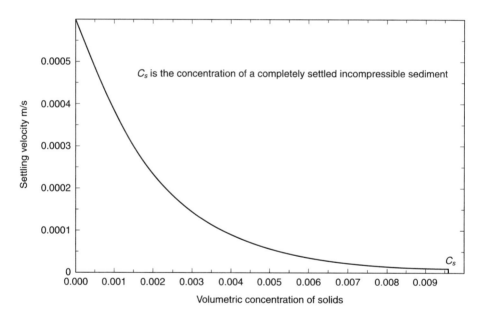

Figure 6.1 Typical relationship between the settling velocity and the local solids concentration

The point I on Figure 6.2 at which the flux curve has a point of inflection has particular relevance.

The concentration discontinuities that develop in sedimentation slurries dominate the behavior of ideal slurries in a thickener. In the batch thickener a discontinuity is not stationary but moves up or down depending on the concentration just above and just below the discontinuity but in a continuous thickener these discontinuities must be stationary for proper steady operation.

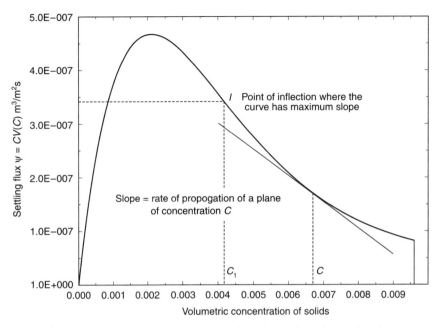

Figure 6.2 The data from Figure 6.1 converted to show the relationship between the settling flux ψ and the concentration C

6.1.1 *Continuous cylindrical thickener*

A cylindrical ideal Kynch thickener that operates at steady state is shown schematically in Figure 6.3. The feed slurry is introduced below the surface and a sharp interface develops at the feed level between the clear supernatant fluid and a slurry of concentration C_L. The feed slurry is assumed to spread instantly across the cross-section of the thickener and to dilute to concentration C_L. Obviously this is an idealization in a real thickener. Nevertheless it provides a useful simulation model. Lower down in the thickener, a layer of concentration C_M develops and at the bottom of the thickener the mechanical action of the rake releases some water from the fully settled pulp and the fully thickened slurry is discharged through the discharge pipe at concentration C_D. In practice these layers are not so well defined and the concentration in any one layer will vary with increasing depth. Nevertheless this ideal picture of the thickener is useful and leads to a useful design and simulation model. In particular the ideal Kynch thickener model assumes that the fully settled slurry is incompressible. In reality, the settled pulp is compressed toward the bottom of the compression zone as water is expelled from between the flocs in the pulp due to the weight of settled solid that must be supported.

The operation of the thickener is dominated by the behavior of these layers and the relationships between them. The concentration of solids in each of the layers is constrained by the condition that the thickener must operate at steady state over the long term. If the slurry is behaving as an ideal Kynch

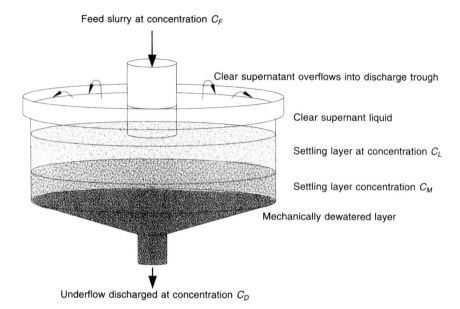

Figure 6.3 Schematic representation of the ideal Kynch thickener operating at steady state

slurry, well-defined sharp interfaces will develop in the thickener and the analysis below shows how these concentrations can be calculated.

The total settling flux relative to fixed coordinates at any level where the concentration is C must include the effect of the net downward volumetric flow that is due to the removal of pulp at the discharge in addition to the settling flux of the solid relative to the slurry itself. If the total flux is represented by $f(c)$ and the volumetric flux of slurry below the feed by q then

$$f(C) = qC + \psi(C) \tag{6.2}$$

In batch settling $q = 0$ so that $f(C)$ and $\psi(C)$ are identical. $f(C)$ is plotted for different values of q in Figure 6.4 using the data of Figure 6.1.

Consider any discontinuity as shown in Figure 6.3.
Flux of solid into discontinuity from above:

$$f(C^+) = C^+q + C^+V(C^+) \tag{6.3}$$

Flux of solid leaving from below:

$$f(C^-) = C^-q + C^-V(C^-) \tag{6.4}$$

In general

$$f(C^+) \neq f(C^-) \tag{6.5}$$

so that solid accumulates at the discontinuity. The accumulation is positive if

$$f(C^+) > f(C^-) \tag{6.6}$$

and vice versa. If the accumulation is positive the discontinuity moves upward a distance Δx during a time interval Δt and these are related by

$$A\Delta x(C^- - C^+) = [f(C^+) - f(C^-)]A\Delta t \tag{6.7}$$

the rate of movement of the discontinuity. The analysis leading to Equation 6.7 for the batch settler gives an expression for the rate at which a discontinuity will move in a continuous thickener.

$$\sigma(C^+, C^-) = \lim_{\Delta t \to 0} \frac{\Delta x}{\Delta t} = -\frac{f(C^+) - f(c^-)}{C^+ - C^+} \quad m/s \tag{6.8}$$

If the thickener is operating at steady state $\sigma(C^+, C^-)$ must be zero across every discontinuity. The right hand side of Equation 6.8 is the negative of the slope of the chord connecting two points on the flux curve and these chords must be horizontal as shown in Figure 6.4 to satisfy the steady state requirement. The concentrations in the layers on each side of a discontinuity make up a conjugate pair. These conjugate concentrations are further limited by the requirement that all concentration discontinuities must be stable as well as stationary. The stability of the interface requires that the higher conjugate concentration can exist only at concentration C_S or at concentration C_M at which the total settling flux has a local minimum. Thus as soon as the underflow volumetric flux q is fixed the conjugate concentrations can be determined by drawing the horizontal tangent to the total flux curve as shown by line A–B in Figure 6.4. If the minimum does not occur at a concentration less than C_S, the conjugate concentrations are found by drawing the horizontal line from the point G as shown by line G–F in Figure 6.4.

The flux curves shown in Figure 6.4 can be used to develop a simple ideal model of the continuously operating cylindrical thickener. The model is based on the requirement that at the steady state all the solid must pass through every horizontal plane in the thickener. In other words, the solid must not get held up anywhere in the thickener. If that were to happen, solid will inevitably accumulate in the thickener, which will eventually start to discharge solids in its overflow.

The flux through **any** horizontal plane in a steady state thickener must equal the feed flux

$$f_F = \frac{Q_F C_F}{A} \tag{6.9}$$

and the underflow flux

$$f_D = \frac{Q_D C_D}{A} = q C_D \tag{6.10}$$

where $q = Q_D/A$ is the total downward volumetric flux at any horizontal layer below the feed well.
Thus

$$f_F = qC + \psi(C) \tag{6.11}$$

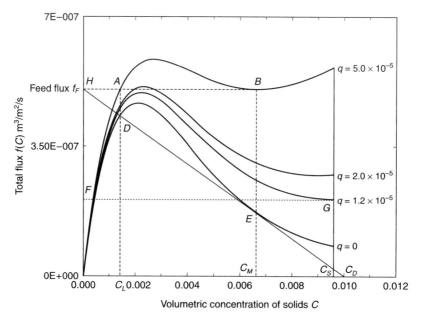

Figure 6.4 Graphical procedure to describe the steady state operation of an ideal Kynch thickener

where C is the concentration at any level in the thickener.

Equation 6.11 can be plotted on the $\psi(C)$ vs C axes as a straight line of slope $-q$ as shown as line HDE in Figure 6.4 which is plotted for the case $q = 5.0 \times 10^{-5}$ m/s.

Re-arranging Equation 6.11

$$\psi(C) = f_F - qC \tag{6.12}$$

which shows that the line intersects the $\psi(C)$ axis at $\psi = f_F$ and since

$$f_D = qC_D = f_F \tag{6.13}$$

it intersects the C axis at $C = C_D$.

The quantity $qC + \psi(C)$ is referred to as the demand flux because this flux must be transmitted through every horizontal level in the thickener otherwise the thickness could not operate at steady state.

It is not difficult to show that the straight line representing Equation 6.11 is tangential to the $\psi(C)$ function at point E, which has the same abscissa value C_M as the conjugate operating point B. The point E is directly below B in Figure 6.4. Likewise the intersection D between the straight line and the $\psi(C)$ curve has the same concentration as the lower conjugate point A.

Point A is defined by

$$f_F = qC_A + \psi C_A \tag{6.14}$$

and the point D is defined by

$$f_F - qC_D = \psi C_D \tag{6.15}$$

These two equations are true simultaneously only if $C_D = C_A = C_L$.

6.2 Useful models for the sedimentation velocity

A number of models have been proposed to relate the sedimentation velocity to the solids concentration and these have been used with varying success to describe the sedimentation of a variety of slurries that are important in industrial practice. The most commonly used model is based on the work of Richardson and Zaki (1954):

$$V = v_{TF}(1 - r_F C)^n \tag{6.16}$$

Here v_{TF} is the terminal settling velocity of an isolated floc and r_F is the floc dilution which is the volume of a floc per unit mass of contained solids. The recommended value for n is 4.65. Scott (1977) has applied this equation to the sedimentation behavior of flocculated Witwatersrand gold ore slurries and some typical data is compared to this model in Figure 6.5 with $v_{TF} = 0.778$ mm/s and $r_F = 0.973$ m^3 floc/tonne solid. The following values have been reported for Kaolin slurries at pH 6.0: $v_{TF} = 0.36$ mm/s and $r_F = 40$ m^3/m^3. Shirato *et al.* (1970) reported the following values for zinc oxide slurries $v_{TF} = 0.715$ mm/s and $r_F = 200$ and for ferric oxide slurries $v_{TF} = 6.42$ mm/s and $r_F = 17$.

An alternative model has been used by Shirato *et al.*

$$V = \alpha \frac{(1 - C)^3}{C} \exp(-\beta C) \tag{6.17}$$

Zinc oxide slurries in the range $0.02 \le C \le 0.26$ have been described using $\alpha = 5.7 \times 10^{-5}$ mm/s and $\beta = 6.59$.

Careful experimental work over wide concentration ranges by Wilhelm and Naide (1981) has shown that the settling velocity of flocculated slurries over restricted concentration ranges follows the power law:

$$\frac{1}{V} = aC^b \tag{6.18}$$

Therefore the complete settling velocity concentration relationship follows a simple equation:

$$\frac{v_{TF}}{V} = 1 + \alpha_1 C^{\beta_1} + \alpha_2 C^{\beta_2} + \dots \tag{6.19}$$

with $0 < \beta_1 < \beta_2 < \dots$ which describes a variety of slurries over wide ranges of C. Value for b range from 1.0 to as high as 10 or more. Usually no more than two power-function terms are required in this equation to describe the settling velocity over two or more orders of magnitude variation in the concentration.

In Equation 6.19, v_{TF} represents the terminal settling velocity of the individual flocs when they are widely separated from each other and settle as single entities. v_{TF} is quite difficult to measure but its value can be estimated using the methods of Section 4.5 if the size and effective density of the individual flocs can be measured.

Experimental data on the settling of coal refuse sludge measured by Wilhelm and Naide and of flocculated quartzite slurries typical of mineral processing operations are shown in Figure 6.5. Equation 6.19 describes the settling behavior of these flocculated slurries well and for these data is significantly better than the Richardson-Zakai model.

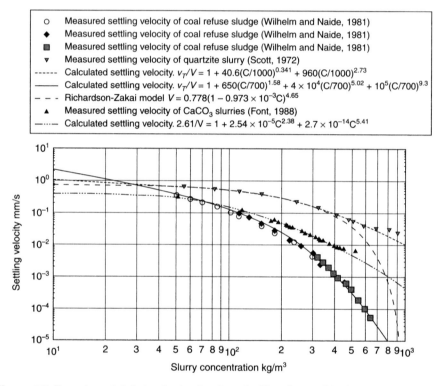

Figure 6.5 Experimental data obtained using the Kynch graphical construction and fitted by the extended Wilhelm-Naide equation

6.3 Simulation of continuous thickener operation

A simple simulation model can be constructed for the continuous cylindrical thickener using the ideal Kynch model. If the area of the thickener is given and the conditions in the feed pulp are known then

$$f_F = Q_F C_F \tag{6.20}$$

Thus the underflow concentration and the underflow pumping rate is fixed by the slope and abscissa intercept of the line HDE in Figure 6.4. This defines the flowrate and the composition of the pulp that is passed from the thickener on to the next unit in the flowsheet.

The maximum possible feed flux is fixed by the slope of the flux curve at the point of inflexion ψ'_I.

$$f_{M\,max} = \psi(C_I) + C_I \psi'_I \tag{6.21}$$

The method requires that a suitable model be available for the settling flux. This is calculated using settling velocities given by Equations 6.16, 6.17 or 6.19.

The Richardson-Zaki model for the sedimentation velocity can be used to build a simple but self-consistent simulation model for the ideal Kynch thickener. The maximum feed rate of solid that can be sent to a thickener of given diameter is fixed by the slope of the sedimentation flux curve at the point of inflexion. The sedimentation flux is given by Equation 6.1

$$\psi = v_{TF} C (1 - r_F C)^n \tag{6.22}$$

The point of inflexion is at

$$C_I = \frac{2}{r_F (n + 1)} \tag{6.23}$$

and the critical slope at the point of inflexion is given by

$$\psi'_I = \frac{v_{TF}}{(n + 1)^n} \left((n - 1)^n - 2n(n - 1)^{n-1} \right) \tag{6.24}$$

The maximum possible feed flux occurs when the operating line in Figure 6.4 is tangential to the flux curve at the critical point of inflexion. Thus

$$f_{F max} = \psi(C_I) - \psi'_I C_I$$

$$= \frac{4 v_{TF}}{r_F} \times \frac{n(n - 1)^{n-1}}{(n + 1)^{n+1}} \tag{6.25}$$

The maximum possible feed rate of solids to the thickener is

$$W_{F max} = \rho_s f_{F max} A \qquad \text{kg/s} \tag{6.26}$$

When the thickener is fed at a rate less than the maximum, the concentration of the underflow can be calculated from the intersection with the horizontal axis of the operating line that passes through the given feed flux on the vertical axis and which is tangential to the flux curve as shown in Figure 6.4. This requires the solution of a non-linear equation

$$C_M = \frac{\psi(C_M) - f_F}{\psi'(C_M)}$$

$$= \frac{C_M (1 - r_F C_M)^n - \dfrac{f_F}{v_{TF}}}{(1 - r_F C_M)^n - n r_F C_M (1 - r_F C_M)^{n-1}} \qquad (6.27)$$

for the intermediate concentration C_M after which the concentration of the pulp in the discharge is calculated from

$$\frac{C_D}{C_M} = \frac{f_F - \psi(C_M)}{f_F} \qquad (6.28)$$

When the more useful and widely applicable extended Wilhelm-Naide model is used for the sedimentation velocity, the analytical method used above does not produce nice closed-form solutions and numerical methods are required.

The construction illustrated in Figure 6.4 provides a rapid and simple design procedure for an ideal thickener based on the ideal Kynch theory. Either the underflow concentrations or the feed concentration can be specified and the other is fixed by the line drawn tangent to the settling flux curve. This also fixes the total volumetric flux q from which the required area of the thickener can be determined.

$$A = \frac{Q_D}{q} = \frac{Q_F C_F}{C_D q} \qquad (6.29)$$

Under the assumption that the settled pulp is incompressible, the maximum discharge pulp concentration is C_S and when the thickener discharges at this concentration, only one feed flux and one volumetric flux q is possible for the thickener as shown in Figure 6.4. In practice the thickener discharge concentration is always greater than the concentration in the lower settling zone although the simple Kynch theory provides no mechanism to describe this. This change in concentration can be ascribed to the mechanical action of the rake at the bottom of the settler which is able to move sediment of concentration C_S which would otherwise be immobile under the ideal assumption of an incompressible sediment. The conditions at the base of the thickener are illustrated in Figure 6.6. As the sediment is mechanically dragged down the conical base of the thickener, water can be released so that the discharge concentration C_D is greater than the conjugate concentration C_M. In practice the sediment will always be compressible and then the simple model that has been described needs some significant modification.

Figure 6.7 shows measured pulp density profiles in an industrial thickener operating normally and in an overloaded condition. The measured profiles agree with those expected in an ideal Kynch thickener except for the gradual increase in the pulp concentration between the greater conjugate concentrate C_M and the discharge concentration C_D. This can be ascribed to the compressibility of the sediment. The data is from Cross (1963). The thickener was 22.9 m in diameter with a 3 m cylindrical section and a cone depth of 1.55 m.

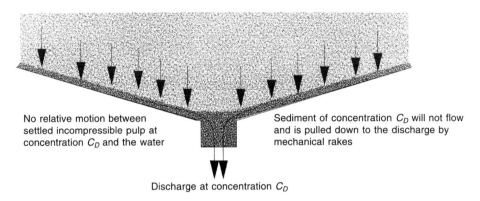

No relative motion between settled incompressible pulp at concentration C_D and the water

Sediment of concentration C_D will not flow and is pulled down to the discharge by mechanical rakes

Discharge at concentration C_D

Figure 6.6 Conditions at the floor of the thickener

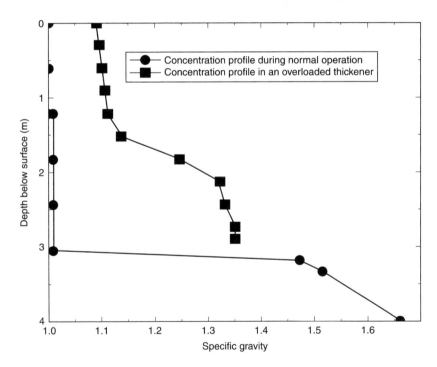

Figure 6.7 Measured density profiles in an industrial thickener. The specific gravity of the feed was 1.116 and the underflow discharged at a specific gravity of 1.660 under normal conditions

6.4 Mechanical dewatering of slurries

6.4.1 Dewatering screens

Slurries can be dewatered on dewatering screens although the dewatering is not as complete as can be achieved in a filter. A schematic of a dewatering screen is shown in Figure 6.8. The wet slurry is fed to one end of the screen. The solid material is retained on the screen while the water drains from the bed and flows through the screen. The vibration induces the bed of solids to move along the screen at a velocity that is a complex function of the mechanical action and of the properties of the bed itself. The vibration of the bed also induces an oscillating acceleration on the bed and the water it contains. This assists in overcoming the capillary retention forces in the pores and assists the dewatering process. The performance of any particular dewatering screen depends on a number of factors that include the distribution of particle size and shape, the nature of the screening aperture and its open area and the density of the solid component. No comprehensive model that accounts for all of these variables has been developed and the model presented here represents a first attempt to outline an approach that may prove to have some general applicability.

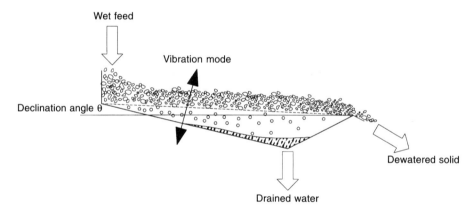

Figure 6.8 Schematic diagram of a dewatering screen

To build a useful model for the performance of the dewatering screen the bed transport mechanism and the dewatering mechanism must be modeled separately. The rate of transport of the bed along the screen is determined in terms of the gravitational parameter

$$G_y = \frac{g \cos\theta}{\omega^2 A_y} \tag{6.30}$$

where θ is the angle of declination of the screen, ω is the angular speed of vibration in radians/s and A_y is the amplitude of vibration in the direction normal to the screen surface. The dimensionless velocity of transport is defined as

$$\dot{X} = \frac{u}{\omega A_x} \tag{6.31}$$

where u is the velocity of travel along the screen and A_x is the amplitude of the vibration in the direction parallel to the screen surface.

A useful model for mineral slurries is

$$\dot{X} = a \exp(-bG_y) \tag{6.32}$$

The parameter b varies linearly with the slope of the screen θ

$$b = 1.44 - 2.32\theta \tag{6.33}$$

and parameter a is specific to the material. Ng *et al.* (1986) have measured $a = 1.87$ for a coal slurry having $d_{80} = 7.2$ mm and $d_{50} = 3.2$ mm with a feed water content of 11%. Unfortunately not much is known about the variation of this parameter with material type and values for other materials are not available. The amount of dewatering depends primarily on the time that the slurry spends on the screen, which is fixed by the velocity of travel and the length of the screen

$$t = \frac{L}{u} \tag{6.34}$$

The dewatering process is governed by the inertial effects that result from the rapid acceleration and deceleration of the material on the screen and by the requirement that the draining water must be able to leave the bottom of the screen by drip dislodgement. Complete dewatering is not possible and the minimum achievable moisture content is a function of several physical properties of the particle bed and also of the vibration mechanism. This minimum is referred to as the saturation moisture content. If S_∞ represents the saturation moisture content of the bed expressed as mass fraction of water in the material, the rate of dewatering can be described by the equation

$$\frac{dS}{dt} = -\alpha(S - S_\infty)^\beta \tag{6.35}$$

This equation can be integrated to give the saturation as a function of time

$$S - S_\infty = pt^{-q} \tag{6.36}$$

with

$$q = \frac{1}{\beta - 1} \tag{6.37}$$

and

$$p = (\alpha(\beta - 1))^{-q} \tag{6.38}$$

Combining Equations 6.34 and 6.36 gives the water content of the discharged slurry as

$$S = S_\infty + p\left(\frac{L}{u}\right)^{-q} \tag{6.39}$$

Ng (1990) has measured p and q for a coal slurry. p is a function of the solids to be dewatered but is apparently independent of the amplitude, frequency and mesh size of the screen. q increases with the mesh size of the screen reflecting the greater water discharge rate through the larger mesh. Their data give $p = 0.234$ and

$$q = 0.33 + 0.081(h - 1.0) \tag{6.40}$$

where h is the screen mesh size in mm.

Particles that are smaller than the mesh size can report to the underflow in proportion to the recovery of water.

The carrying capacity of the dewatering screen can be calculated approximately from

$$W = W_\infty(1 - e^{-4h}) \tag{6.41}$$

where W_∞/b is the capacity per unit width of screen at large mesh size. This limiting capacity is a function of the size consist of the feed slurry

$$W_\infty = 20 + 2.5d_{80}^F \text{ tonne/hr m} \quad \text{for } d_{80}^F < 12 \text{ mm}$$

$$= 50 + 0.3d_{80}^F \text{ tonne/hr m} \quad \text{for } d_{80}^F > 12 \text{ mm} \tag{6.42}$$

Illustrative example 6.1

Calculate the water content of a coal slurry after dewatering on a 0.6 mm mesh screen 3.66 m long by 0.914 m wide with angle of declination of 15°. Vibration mode has amplitude 2 cm at a frequency of 300 rpm and is inclined at 45° to the surface of the screen. The saturation water content of the coal is 7.35%.

$$\omega = \frac{300}{60} \times 2\pi = 31.42$$

$$G_y = \frac{9.81 \times \cos 15°}{31.42^2 \times 0.02 \times \sin 45°} = 0.679$$

$$b = 1.44 - 2.52 \times 0.262 = 0.780$$

$$\dot{X} = 1.87 \exp(-0.780 \times 0.679) = 1.101$$

$$A_x = 0.02 \times \cos 45° = 0.014 \text{ m}$$

$$u = 1.101 \times 31.42 \times 0.014 = 0.489 \text{ m/s}$$

$$t = \frac{3.66}{0.489} = 7.485$$

$$p = 0.234 \quad q = 0.33 + 0.081(0.6 - 1) = 0.298$$

The water content of the solid leaving the end of the dewatering screen can be calculated from Equation 6.39.

$$S = 0.0735 + 0.234 \times (7.485)^{-0.298} = 0.202 \qquad \text{kg water/kg slurry}$$

6.5 Filtration

Filtration is used for dewatering operations in mineral processing plants but it is often something of a bottleneck. As a result, alternative processing routes that avoid filtration are usually preferred. When filtration is used for dewatering, continuous vacuum filters are usually used. These filters are available as rotary drums, rotary disks and horizontal belt filters. Of these the rotary drum vacuum filter is the most widely used and a simple analysis of these filters is given here.

Generally the rate of flow of filtrate through a filter cake is sufficiently slow to make the Kozeny-Carman equation a valid descriptor of the process.

$$\frac{\Delta P}{\Delta x} = \frac{150(1 - \varepsilon)^2 \mu V_s}{\varepsilon^3 d_p^2} \tag{6.43}$$

which relates the flowrate of filtrate through the cake to the pressure gradient. In Equation 6.43:

V_s = superficial liquid velocity,
ε = filter cake porosity,
d_p = surface specific particle diameter which, for spherical particles, is related to the particle specific area per unit volume S_0 by $d_p = 6/S_0$.

If A_c is the interstitial cross-sectional area of the filter cake and L its thickness, Equation 6.43 can be written:

$$\frac{\Delta P A_c}{\mu L} = \frac{4.17(1 - \varepsilon)^2 S_0^2}{\varepsilon^2} Q \tag{6.44}$$

where Q is the volumetric flowrate of filtrate passing through the cake under the influence of the pressure gradient. The constant 4.17 in Equation 6.44 varies with the properties of the particles that make up the filter cake and a simplified form of the equation, called Darcy's law for porous media, is normally used

$$\frac{\Delta P A_C}{\mu L} = \frac{Q}{K} \tag{6.45}$$

where

$$K = \frac{\varepsilon^2}{4.17 (1 - \varepsilon)^2 S_0^2} \tag{6.46}$$

is the permeability of the filter cake.

It is convenient to work with the resistance of the bed $R = L/K$ and Darcy's law is written

$$\Delta P = \frac{\mu Q R}{A_c} \qquad (6.47)$$

The thickness of the cake can be related to the mass of solid in the cake per unit area M_c

$$M_c = \rho_s(1 - \varepsilon)\Delta x \qquad (6.48)$$

and the resistance of the cake is related to M_c by

$$R = \alpha M_c \qquad (6.49)$$

α is called the specific cake resistance.

Usually the resistance of the filter medium contributes a noticeable amount to the total resistance and it is necessary to add an equivalent resistance of the filter cloth or filter medium so that

$$R = R_{\text{cake}} + R_{\text{medium}} \qquad (6.50)$$

The medium resistance may not remain constant during the filtration because solids may penetrate into the medium and increase the resistance to the flow of filtrate. This effect is neglected here.

$$\Delta P = \Delta P_c + \Delta P_M = \mu Q \left(\frac{\alpha M_c}{A_c} + \frac{R_m}{A_c} \right) \qquad (6.51)$$

In a simple batch filtration experiment

$$Q = \frac{dV_f}{dt} \qquad (6.52)$$

where V_f is the volume of filtrate passed.

The mass of solid deposited in the filter cake is related to the amount of filtrate collected by

$$M_c = \frac{V_f C_f}{A_c} \qquad (6.53)$$

where C_s is the concentration of the feed slurry in kg solid/m^3 water. This makes no allowance for the solids that are in the liquid that is retained by the cake but this is usually negligible in practical applications.

$$\Delta P = \mu \frac{dV_f}{dt} \frac{1}{A_c^2} \alpha C_s (V_f + V_e) \qquad (6.54)$$

V_e = fictitious volume required to build up a layer of filter cake having the same resistance as the filter cloth.

This is a two-parameter equation describing batch filtration

$$A_c \frac{dt}{dV_f} = k_1 \frac{V_f}{A_c} + k_2 \tag{6.55}$$

which can be integrated at constant pressure to give

$$t = \frac{k_1}{2} \frac{V_f^2}{A_c^2} + k_2 \frac{V_f}{A_c} \tag{6.56}$$

The constants k_1 and k_2 can be evaluated from a simple batch filtration test by plotting t/V_f against the filtrate volume V_f.

$$\frac{t}{V_f} = \frac{k_1}{2A_c^2}V_f + \frac{k_2}{A_c} \tag{6.57}$$

which gives a straight line.

Equation 6.56 applies directly to rotary drum filtration since each element of the drum surface appears as a small batch filter during its sojourn in the slurry. The pressure drop across the cake is constant if the hydrostatic head variation is neglected. The submergence time is given by

$$t = \frac{\omega}{2\pi N} \tag{6.58}$$

where N is the rotation speed of the drum in revs/sec and ω is the submerged angle.

The filtration rate per unit area of drum surface is given by

$$q_f = \frac{V_f}{\pi DL}$$

where D is the diameter of the drum and L its length.

$$\frac{\omega}{\pi N} = k_1 q_f^2 + 2k_2 q_f$$

From which the capacity q_f can be easily calculated.

Illustrative example 6.2

The data given in the table were obtained in a batch filtration test on a single frame filter press.

The frame was 40 cm by 40 cm. The concentration of solid was 12 kg/m³ water. Viscosity of water = 0.001 Pa s. The pressure drop was held constant at 1.5×10^5 Pa.

Evaluate the specific cake resistance.

Time s	357	716	1107	1504	1966	2431	2924	3414	3922	4450	4994	5570
Filtrate volume collected m³	0.02	0.04	0.06	0.08	0.10	0.12	0.14	0.16	0.18	0.20	0.28	0.24

The data plots as a straight line as shown in Figure 6.9. The slope of the line is 2.54×10^4 and the intercept is 17160.

$$k_1 = 2.54 \times 10^4 \times 2A_c^2 = 8.13 \times 10^3$$

$$k_1 = \frac{\alpha \mu C_s}{\Delta P}$$

$$\alpha = \frac{8.13 \times 10^3 \times 1.5 \times 10^5}{0.001 \times 12} = 1.02 \times 10^{11} \quad m/kg$$

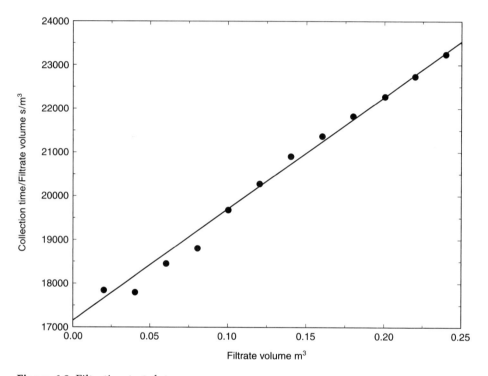

Figure 6.9 Filtration test data

Bibliography

The foundation of the modern theory of thickening was laid down by Kynch in 1952. The Kynch theory has been comprehensively studied by Concha and Barrientos (1993).

The analysis of the stability of the discontinuity is discussed in many places in the literature and Fitch (1983) gives a detailed argument.

The model for the dewatering screen is due to Ng *et al.* (1986) and Ng (1990) but no comprehensive study is available to predict how the parameters of the model vary with type of material and the nature of the equipment.

The capacity estimates for coal dewatering screens is based on data presented by Leonard (1979).

Practical implications of coal dewatering are discussed by Osborne (1988).

References

Concha, F. and Barrientos, A. (1993) A critical review of thickener design methods. *KONA*, No. 11, pp. 79–104.

Cross, H.E. (1963) A new approach to the design and operation of thickeners. *Jnl of South African Inst. Mining and Metallurgy*, Vol. 63, pp. 271–298.

Fitch, B. (1983) Kynch theory and compression zones. *AIChEJ*, Vol. 29, pp. 940–947.

Kynch, G.J. (1952) Theory of sedimentation. *Trans. Faraday Society*, Vol. 48, pp. 166–176.

Leonard, J.W. (1979) *Coal Preparation*. AIME, New York, pp. 12–11.

Ng, K.L. (1990) Dewatering performance of vibrating screen. *Proc. Instn Mech. Engrs*, Vol. 204, pp. 73–79.

Ng, K.L., Ang, L.A. and Chng, S.C. (1986) A computer model for vibrating conveyors. *Proc. Instn Mech. Engrs*, Vol. 200, pp. 123–130.

Osborne, D.G. (1988) *Coal Preparation Technology*. Graham and Trotman, Vol. 1, p. 522.

Richardson, J.F. and Zaki, W.N. (1954) Sedimentation and fluidization. *Trans Instn Chem. Engrs*, Vol. 32, pp. 35–53.

Shirato, M., Kata, H., Kobayashi, K. and Sakazaki, H. (1970) Analysis of thick slurries due to consolidation. *Jnl of Chemical Engineering of Japan*, Vol. 3, pp. 98–104.

Stolz, E.C. and Scott, K.J. (1972) Design, Operation and Instrumentation of thickeners. Combined Report Chamber of Mines of South Africa Project No. 11/504/64, October.

Wilhelm, J.H. and Naide, Y. (1981) Sizing and operating continuous thickeners. *Mining Engineering*, pp. 1710–1718.

7
Gravity separation

Differences in the density of individual particles can be exploited to effect a separation and concentrate the desired components. The driving force for separation can be the gravitational field of the earth or the vastly more intense fields that can be generated by centrifugal action.

Gravity separation operations can be classified into two broad groups: manufactured medium and autogenous medium devices. In the former class, the separation takes place in a liquid medium that is manufactured to have a density intermediate between the densities of the materials that are to be separated. Examples are the dense-medium drum and the dense-medium cyclone. In autogenous medium devices, the particulate material itself makes an environment that has an effective density that will induce separation of particles of different densities by stratification. Examples of autogenous separators are the jig, the sluice, the Reichert cone and the water-only cyclone. A third group of gravity separators that rely on more complex physical processes to effect a separation between particles of different density include the spiral concentrator and the shaking table.

7.1 Manufactured-medium gravity separation

The most common manufactured media used in industry are slurries of finely ground magnetite or ferrosilicon in water. The effective density of such a slurry can be controlled by adjusting the concentration of suspension. The size of the particles that make up the suspension must be much smaller than the particles that are to be separated. In order to be useful as a separating medium, the slurry must be stable over the required range of densities, which means that the suspension should not settle appreciably under gravity or in a centrifugal field if that is to be used to achieve separation. The suspension should also have low viscosity to promote rapid relative motion of particles that are to be separated. The medium must be easily separated from the two products after processing and for economic reasons the medium must be readily recoverable for re-use. Typical solids, with their respective specific gravities, that have been used for manufactured media are silica (2.7), barite (4.5), magnetite (5.18), ferrosilicon (6.8) and galena (7.8). The medium should be cheap, readily available and chemically stable. Suspensions of finely ground magnetite and ferrosilicon satisfy these criteria. Magnetite is generally cheaper than ferrosilicon and the latter is generally used only for separation and recovery of high density minerals such as diamonds. Dense media manufactured from magnetite are commonly used for the cleaning of coal and for the recovery and concentration of a wide variety of minerals such as iron ore, manganese ores, cassiterite, fluorite and others. Magnetite media typically have densities in the range 1250–2200 kg/m^3 and ferrosilicon suspension in the range 2900–3400 kg/m^3. Mixtures of magnetite and FeSi can be used for the manufacture of media of intermediate density.

The principle of dense medium separation is quite straightforward. The material to be separated is placed in the dense medium in which the lighter particles float and the heavier particles sink. The floats and sinks are then separated. Equipment to effect the separation varies widely in mechanical design since it is quite difficult to achieve efficient separation continuously. Centrifugal separators are considerably simpler in design than gravity separators.

7.2 Quantitative models for dense-media separators

Four identifiable physical factors define the separating performance of manufactured media separators: the separating density or cut point, the separating efficiency, the short circuit of feed to underflow and the short circuit of feed to overflow. These effects are best described by means of a partition curve.

Three typical partition curves for a dense-medium cyclone are shown in Figure 7.1. These curves show how material of different specific gravity will partition to the float fraction. Other dense-medium separators have similar partition curves. These partition curves reveal the four operating characteristics.

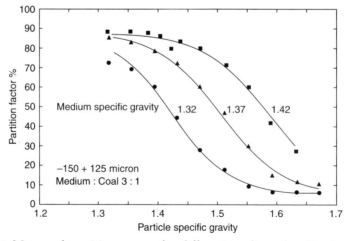

Figure 7.1 Measured partition curves for different medium densities in a 150 mm dense-medium cyclone

7.2.1 *The cut point*

The cut point is defined as the density ρ_{50} at which the partition function has the value 0.5. A particle having the density ρ_{50} has equal chance of reporting to the sink fraction as to the float fraction.

The cut point is an operating parameter that can be controlled fairly easily by variation of the medium density which obviously determines which particles tend to sink in the medium and which tend to float. In a static bath of dense medium the cut point is always equal to the density of the medium. However, in continuously operating equipment this is not always the case and the cut point can be greater or less than the medium density depending on whether the lighter or heavier fraction must move counter to the prevailing bulk flow of the medium in the equipment. For example in the dense-medium cyclone, the medium together with the material to be separated enters tangentially on the periphery of the cylindrical section and the bulk of the medium leaves through the vortex finder on the axis. There is therefore an overall flow of medium from outside inward and the inward velocity of the medium tends to drag particles with it towards the center. Any heavy particles that leave the cyclone in the underflow must move against this flow and therefore require a net negative buoyancy to overcome the viscous drag of the medium. Thus any particle that finds an equilibrium orbit in the cyclone must be denser than the medium. Since the particles on equilibrium orbits define the cut point:

$$\rho_{50} > \rho_m \tag{7.1}$$

where ρ_m is the density of the medium. The difference $\rho_{50} - \rho_m$ is called the cut point shift and this can be normalized with respect to the medium density to form the normalized cut point shift:

$$\text{NCPS} = \frac{\rho_{50} - \rho_m}{\rho_m} \tag{7.2}$$

The normalized cut point shift must obviously increase as particle size decreases and experimental data for coal in a dense-medium cyclone are correlated by the simple power law:

$$\text{NCPS} = 0.4 d_p^{-0.32} \tag{7.3}$$

over the range $0.015 < d_p < 1$ mm.

When the lighter fraction must move counter to a prevailing flow of medium, the NCPS must be negative. This is quite uncommon because dense medium drum separators are usually designed to provide a medium flow counter to the prevailing force field to ensure that the medium does not settle under gravity or the centrifugal field. This means that the sink material moves against a prevailing flow of medium and the cut point shift is positive. In general the NCPS can be neglected for dense-medium drum type separators for particles larger than a few millimeters.

An additional factor that influences the cut point shift is the settling of the dense-medium suspension particularly in centrifugal separators. This inevitably leads to a density differential between overflow and underflow media. The cut point shift increases with increased media density differential. Napier-Munn *et al.* (1992) have evaluated a number of modeling techniques that may be appropriate for the description of the underflow density in terms of the operating variables in the dense-medium cyclone. An empirical model that appears to correlate the data well for smaller diameter cyclones is

$$\frac{\rho_u + E}{\rho_m} = 1.872 Re_i^{0.0891} \left(\frac{D_u}{D_c}\right)^{-0.0872} \left(\frac{D_o}{D_c}\right)^{1.919} \tag{7.4}$$

where

$$Re_i = \frac{\rho_m v_i D_i}{\mu_a} \tag{7.5}$$

is the Reynolds number at the inlet. The symbols used in this correlation have the following meaning:

ρ_u = underflow density,
D_u = spigot diameter,
D_o = vortex finder diameter,
D_i = inlet diameter,
v_i = velocity in the cyclone inlet,
μ_a = apparent viscosity of the medium.

E is an error term that varies strongly with the cyclone diameter and must be estimated for the cyclone that is used.

The partition curves usually normalize well with respect to the cut point density so that partition curves determined at different medium densities will be superimposed when they are plotted using a normalized abscissa ρ/ρ_{50}. This effect is seen in Figure 7.2 where the experimental data of Figure 7.1 are plotted against the reduced or normalized density ρ/ρ_{50} and the data fall on a single curve.

7.2.2 Short circuit flows

The curve shown in Figure 7.2 is called the normalized partition curve and it is represented by the symbol $R(x)$ where x represents the normalized density ρ/ρ_{50}. The general characteristics of normalized partition functions are shown in Figure 7.3 and in practice the low- and high-density asymptotes are not at 100% and 0% respectively. A fraction of the feed is assumed to pass directly to the underflow and another fraction to the overflow directly without passing through the separation field of the separator. If α is the short circuit fraction to sinks and β the short circuit fraction to floats, the actual partition factor R can be related to an ideal or corrected partition factor R_c by

$$R(x) = \beta + (1 - \alpha - \beta)R_c(x) \tag{7.6}$$

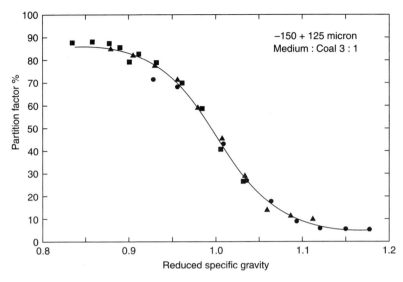

Figure 7.2 Normalization of the partition function for a dense-medium cyclone. The presence of short circuit flows is clearly evident in this data

Both short circuit flows are functions of particle size and the short circuit to sinks extrapolates at zero size to the recovery of water to sinks. An exponential dependence with size has been found in a 150 mm dense-medium cyclone.

$$\alpha = R_f e^{-bd_p} \tag{7.7}$$

with $b = 5.43$ mm^{-1} and d_p in mm. Short circuit to floats can usually be neglected except for material smaller than 0.5 mm.

7.2.3 Separation efficiency

The efficiency of separation is specified in terms of the corrected imperfection defined by

$$I_c = \frac{\rho_{25} - \rho_{75}}{2\rho_{50}} = \frac{\text{EPM}}{\rho_{50}} \tag{7.8}$$

where

$$R_c(\rho_{25}) = 0.25 \tag{7.9}$$

and

$$R_c(\rho_{75}) = 0.75 \tag{7.10}$$

The imperfection of the corrected partition curve is independent of the short circuit flows and it is better to base a predictive model on the corrected imperfection than on the imperfection of the actual partition curve.

The imperfection increases as particle size decreases and the partition curves become steadily flatter as particle size decreases as shown in Figure 7.3.

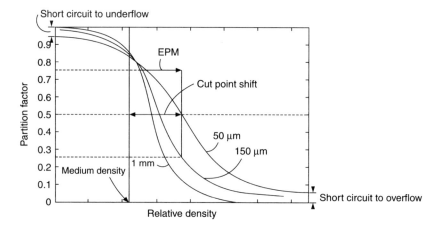

Figure 7.3 Generalized partition curves for dense-medium separators

The imperfection also increases as the size of the cyclone increases and for a small (150 mm diameter) cyclone a useful quantitative relationship for the imperfection of the corrected partition curve is

$$I_c - 0.013 = \frac{3.8}{d_p} \tag{7.11}$$

with d_p in µm.

7.2.4 *The corrected partition function*

A variety of empirical one-parameter expressions for the corrected partition curve have appeared in the literature. These are shown in Table 9.1 as functions of the normalized density $x = \rho/\rho_{50}$. Once a particular function has been chosen to model the corrected partition curve, the actual partition curve can be constructed from the four parameters α, β, ρ_{50} and I_c. It is then a simple matter to calculate the recovery of each particle type to floats and sinks in the separator.

The exponential sum function (Type 7 in Table 7.1) has been found to fit data for the dense-medium cyclone well, as shown in Figure 7.2.

In most large dense-medium vessels handling coarse material, I_c is usually assumed to be independent of particle size. Some representative values are given in Table 7.2.

The nine function types listed in Table 7.1 are shown in graphical form in Figures 7.4 and 7.5 to compare their shapes. Although similar to each other, they do exhibit some differences in shape so that one of the functions can fit

Table 7.1 Empirical equations for the corrected partition function for dense-medium separators. $x = \rho/\rho_{50}$

Type	Function	Relationship between I_c and λ
1	$R(x) = G[\lambda(1 - x)]$	$\lambda I_c = 0.674$
2	$R(x) = 1 - G(\lambda \ln x)$	$I_c = \sinh\left(\dfrac{0.674}{\lambda}\right)$
3	$R(x) = \dfrac{1}{1 + \exp(\lambda(x - 1))}$	$\lambda I_c = 1.099$
4	$R(x) = \dfrac{1}{1 + x^{\lambda}}$	$I_c = \sinh \dfrac{1.099}{\lambda}$
5	$R(x) = \exp(-0.693x^{\lambda})$	$2I_c = 2^{1/\lambda} - 0.415^{1/\lambda}$
6	$R(x) = 1 - \exp(-0.693x^{-\lambda})$	$2I_c = 2.411^{1/\lambda} - 0.5^{1/\lambda}$
7	$R(x) = \dfrac{e^{\lambda} - 1}{e^{\lambda} + e^{\lambda x} - 2}$	$2\lambda I_c = \ln\left(\dfrac{9e^{\lambda} - 6}{e^{\lambda} + 2}\right)$
8	$R(x) = \dfrac{1}{2} - \dfrac{1}{\pi}\tan^{-1}(\lambda(x - 1))$	$\lambda I_c = 1.0$
9	$R(x) = \dfrac{1}{1 + \exp(\lambda(x^n - 1))}$	$2I_c = \left(1 + \dfrac{1.099}{\lambda}\right)^{1/n} - \left(1 - \dfrac{1.099}{\lambda}\right)^{1/n}$ $2I_c < 2^{1/n}$

a particular set of measured experimental data better than the others. When choosing a suitable function to model a particular dense-medium separator the available measured data should be tested against all the functions and the one with the best fit can be chosen.

The functions are plotted with a relatively large imperfection of 0.4 in Figure 7.4. In Figure 7.5 the corresponding functions are shown at the lowest practicable imperfection that can be achieved in real equipment, namely 0.01. This would represent extremely good separation performance in a dense-medium separator. A probability scale is used for the ordinate in Figure 7.5 to emphasize the differences in shape among the different functions. In this coordinate system the Type 1 function in Table 7.1 will plot as a straight line and this is often considered to represent good normal operating behavior. Downward curvature below this straight line in the region $\rho < \rho_{50}$ is taken to indicate less than optimal operation of the dense-medium separator. Positive deviations above the straight line such as those produced by the Type 6 model are not usually observed in practice. Type 1 function is the only function in Table 7.1 that is symmetrical about the abscissa value ρ/ρ_{50}. All the others show varying degrees of asymmetry with Types 6, 8 and 9 showing the greatest asymmetry.

Table 7.2 Typical values of cut point shift and imperfection for dense-medium separators

Separator type	Cut point	Cut point shift	Sp. Gr. differential	Imperfection	Particle size	Reference
Chance cone	1.54			0.021		Coal
	1.37			0.01	+50 mm	Leonard (1979)
	1.37			0.03	12.5 × 6.5 mm	
		0.033		0.015		
Dense-medium vessel	1.41			0.020		Coal
	1.35			0.018	50 × 25 mm	Leonard (1979)
				0.033	12.5 × 6.3 mm	
Dense-medium cyclone	1.49			0.019	$1/_2″ × 3/_8″$	Coal
	1.50			0.017	$3/_8″ × 1/_4″$	Weiss (1985)
	1.51			0.021	$1/_4″ × 8\#$	
	1.53			0.025	8# × 14#	
	1.57			0.051	14# × 28#	
Dense-medium cyclone	1.325	0.045		0.014	1″ × 28#	Coal
	1.488	0.142		0.015	1″ × 28#	Weiss (1985)
Dense-medium cyclone	3.65	0.59	0.39	0.043	3/16 × 28#	Iron ore
	3.385	0.54	0.51	0.028	3/16 × 28#	Weiss (1985)
Dense-medium cyclone	3.065	0.455	0.41	0.036	3/16 × 65#	Chromium ore
	3.095	0.345	0.27	0.045	3/16 × 65#	Weiss (1985)
	2.907	0.217	0.28	0.018	3/16 × 35#	
	3.065	0.215	0.25	0.027	3/16 × 35#	
Dense-medium cyclone	2.82	0.42	0.41	0.028	3/16 × 28#	Magnesite
	2.4	0.55	0.58	0.037	3/16 × 28#	Copper ore
						Davies et al. (1963)

Vorsyl separator	1.38	0.14			
	1.5	0.45	0.025	2 × 0.5 mm	Vanangamudi et al. (1992)
Dynawhirl-pool	1.36	0.050	0.035	31.5 × 0.75 mm	Coal Leonard (1979)
	1.52	0.058	0.038	6.3 × 0.75 mm	
	1.37	0.025	0.018	6.3 × 0.42 mm	
Larcodems	1.658	0.14	0.032	100 × 0.5 mm	Cammack (1987)
	1.656	0.14	0.038	25 × 8 mm	
	1.818	0.18	0.176	2 × 0.5 mm	

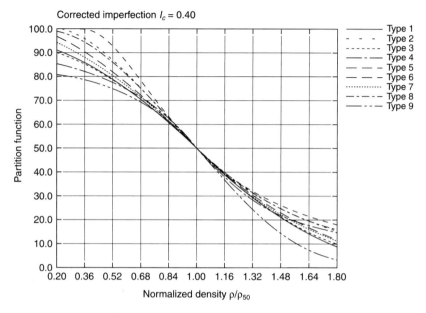

Figure 7.4 Nine models from Table 7.1 for the partition function. All models have the same value of the imperfection $I_c = 0.4$

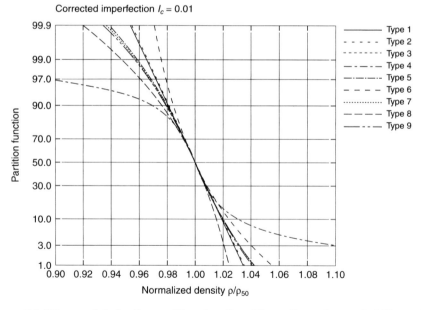

Figure 7.5 Nine models for the partition function. The ordinate is a probability scale. All models have the same value of the imperfection $I_c = 0.01$

7.3 Autogenous media separators

Manufactured dense media usually consist of a suspension of heavy particles that are sufficiently fine to remain in uniform suspension under the force field (gravity or centrifugal) that prevails in the separating unit. The particles to be separated can move more or less independently of each other and are free to float or sink depending on their density relative to the apparent density of the medium suspension. Although particle–particle interactions are significant in a manufactured medium separator, they are usually neither sufficiently frequent nor intense to significantly influence the cut point that is achieved in the separator. However, particle–particle interactions can have a significant influence on the separating efficiency that is actually achieved in the equipment. In an autogenous medium separator, particle–particle interactions are significant and dominant since it is the close interactions between the particles that generate the dense environment in which particles can separate.

The primary phenomenon that occurs in an autogenous medium separator is stratification in the dense bed of particles. Whenever a bed of particles is disturbed and is allowed to settle, the particles will exhibit some tendency to stratify. The heavier particles will tend to find their way to the bottom of the bed and the lighter particles will tend to migrate to the top of the bed. Autogenous media separators rely on stratification of the bed to effect a separation and they are designed to promote rapid and effective stratification of the particle bed.

The stratification behavior of a particle bed can be described in terms of the potential energy of all the particles in the bed. Stratification acts to minimize the total potential energy of the bed by rearrangement of the particles in the bed. This principle was used by Mayer to provide a quantitative description of the behavior of separation equipment such as the jig. However, a consideration of potential energy alone does not provide an adequate description of the actual stratification process as it occurs in practical operations. No matter how effective the stratifying action of the machine is, perfect stratification is never possible. A multitude of random processes influence the behavior of each particle during the time that it passes through the stratification zone of the equipment and these combine to destroy the ideal stratification pattern that is predicted by potential energy considerations alone. The ideal stratification pattern tends to create sharply defined layers in the bed. Sharp boundaries between layers cannot be maintained in the face of the variety of random perturbations that are felt by the particles. The sharp boundaries consequently become diffuse by a diffusion type mechanism which always acts to oppose the ideal stratification pattern.

Lovell *et al.* (1994) give a colorful graphic description of the relationship between the pulsation of the jig bed and the motion of the particles as can be sensed by inserting the hand into the bed. 'At the peak of the pulsion stroke, the particles must be free to move relative to each other and during the suction stroke they should be firmly held by the suction of the bed.' It is this combination of relative motion of the particles and the holding action of the bed during the suction stroke that produces a stable stratification profile that can be split at the end of the jig chamber to produce the desired products.

7.3.1 A quantitative model for stratification

A useful model for stratification that is successful in describing the behavior of operating industrial equipment can be developed by considering the interaction of the potential energy profile that drives the stratification process together with the random diffusion process that tends to break down the ideal stratification profile.

The variation of potential energy when two particles of different density interchange positions in a settled bed is the driving force for stratification. Consider an isolated particle of density ρ in a bed of particles all having density $\bar{\rho}$. The change in potential energy when the isolated particle changes position with a particle at the average density can be calculated as shown in Figure 7.6.

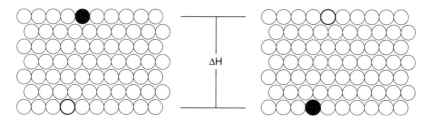

Figure 7.6 Change of potential energy when a heavy particle changes position in a bed of particles

$$\Delta E = E(\text{particle at } H + \Delta H) - E(\text{particle at } H)$$

$$= v_p\rho(H + \Delta H)g + v_p\bar{\rho}H - v_p\bar{\rho}(H + \Delta H)g - v_p\rho Hg$$

$$= v_p g(\rho - \bar{\rho})\Delta H \qquad (7.12)$$

The rate of change of potential energy as the particle of density ρ increases its height in a bed of average density $\bar{\rho}$ is therefore given by

$$\frac{dE}{dH} = v_p g(\rho - \bar{\rho}) \qquad (7.13)$$

This potential energy gradient causes the particle of density ρ to migrate upward or downward depending on the sign of $\rho - \bar{\rho}$. If $\rho > \bar{\rho}$ the particle will move down and *vice versa*. The rate at which particles move relative to the bed is proportional to the energy gradient and the migration velocity is given by $u(dE/dH)$. u is called the specific mobility of the particle and is defined as the velocity at which a particle penetrates the bed under a unit potential energy gradient. u is a strong function of particle size and shape and the bed expansion mechanism but is independent of the particle density. The flux of particles of density ρ in a bed of average density $\bar{\rho}$ caused by the potential energy gradient is given by

$$n_s = -C_\rho u \frac{dE}{dH} \quad \text{m}^3/\text{m}^2\text{s} \tag{7.14}$$

C_ρ is the concentration of particles of density ρ in the bed expressed as the solid volume fraction. The negative sign reflects the fact that each particle will tend to move down the potential energy gradient so as to ultimately minimize the total potential energy in accordance with Mayer's principle. n_s is called the stratification flux. Combination of Equations 7.13 and 7.14 gives

$$n_s = -C_\rho u v_p g (\rho - \bar{\rho}) \tag{7.15}$$

Opposing the stratification flux is the diffusive flux due to the random particle–particle and particle–fluid interactions within the bed. This flux is described by a Fickian equation of the type

$$n_D = -D \frac{dC_\rho}{dH} \tag{7.16}$$

The diffusion coefficient D is dependent on the particle size, shape and bed expansion mechanism.

A dynamic state of stratification equilibrium exists in the bed when the tendency of the particles to stratify under the influence of the potential energy gradient is exactly balanced by tendency to disperse under the influence of the concentration gradient that is created by the stratification action. This dynamic equilibrium is defined by

$$n_D = -n_s \tag{7.17}$$

so that

$$\frac{dC_\rho}{dH} = -\frac{u g v_p C_\rho}{D}(\rho - \bar{\rho}) \tag{7.18}$$

This can be written in terms of the relative height $h = \dfrac{H}{H_b}$, where H_b is the total bed depth and the specific stratification constant

$$\alpha = \frac{u g v_p H_b}{D} \quad \text{m}^3/\text{kg} \tag{7.19}$$

$$\frac{dC_\rho}{dh} = -\alpha C_\rho (\rho - \bar{\rho}(h)) \tag{7.20}$$

α is independent of particle density but will be a strong function of particle size and bed expansion mechanism. The average bed density is a function of h as indicated in Equation 7.20 and is given by

$$\bar{\rho}(h) = \int_0^\infty \rho C_\rho(h)\, d\rho \tag{7.21}$$

A solution to Equation 7.20 gives the vertical concentration profile of particles of density ρ. No boundary conditions can be specified *a priori* for Equation 7.20 because it is not possible to specify the concentration of any particle type at either the top or bottom or at any intermediate level in the bed.

However, the solution to Equation 7.20 must satisfy the conditions

$$\int_0^1 C_\rho\, dh = C_\rho^f \quad \text{for all } \rho \tag{7.22}$$

where C_ρ^f is the concentration of particles of density ρ in the feed to the bed and

$$\int_0^\infty C_\rho\, d\rho = 1 \quad \text{for } 0 \le h \le 1 \tag{7.23}$$

In practice the particle population is discretized into n grade classes, each of which has a unique density and Equations 7.20–7.23 are written in discrete form

$$\frac{dC_i(h)}{dh} = -\alpha C_i(h)(\rho_i - \bar{\rho}(h)) \quad \text{for } i = 1, 2, \ldots n \tag{7.24}$$

with

$$\bar{\rho}(h) = \sum_{i=1}^n C_i(h)\rho_i \tag{7.25}$$

$$\sum_{i=1}^n C_i(h) = 1 \quad \text{for all } h \tag{7.26}$$

$$C_i^f = \int_0^1 C_i(h)\, dh \quad \text{for all } i \tag{7.27}$$

7.3.2 Stratification of a two-component system

The set of differential Equations 7.24 can be solved easily in closed form when the bed consists of only two components.

Formal integral of Equation 7.24 gives

$$C_i(h) = C_i(0) \exp\left(-\alpha\rho_i h + \alpha \int_0^h \bar{\rho}(y)\, dy\right) \quad \text{for } i = 1 \text{ and } 2 \tag{7.28}$$

The average density of particles at level h in the bed for the two-component system is

$$\bar{\rho} = \frac{\rho_1 C_1(h) + \rho_2 C_2(h)}{C_1(h) + C_2(h)} \qquad (7.29)$$

The boundary condition $C_i(0)$ represents the concentration of particles of type i at the base of the bed. This is an unknown and must be calculated for each species but the solution must satisfy the two conditions 7.26 and 7.27. Equation 7.26 requires

$$C_1(0) + C_2(0) = 1 \qquad (7.30)$$

at every level in the bed.

The vertical density profile $\bar{\rho}(h)$ is also unknown so that Equation 7.28 cannot be used in isolation to calculate the vertical concentration profile of the different particle types. However, the ratio of the concentrations of the two species in a two-component system is independent of $\bar{\rho}(h)$.

$$\frac{C_1(h)}{C_2(h)} = \frac{C_1(h)}{1 - C_1(h)} = \frac{C_1(0)}{C_2(0)} \exp\left[-\alpha(\rho_1 - \rho_2)h\right] \qquad (7.31)$$

from which

$$C_1(h) = \frac{\dfrac{C_1(0)}{C_2(0)} \exp\left[-\alpha(\rho_1 - \rho_2)\,h\right]}{1 + \dfrac{C_1(0)}{C_2(0)} \exp\left[-\alpha(\rho_1 - \rho_2)h\right]} \qquad (7.32)$$

Equation 7.32 gives the vertical concentration profile of species 1 as a function of the ratio $C_1(0)/C_2(0)$ at the bottom of the bed. This ratio is fixed by the composition of the material that is fed to the jig bed and can be evaluated from Equation 7.27.

$$C_1^f = \int_0^1 C_1(h)\,dh \qquad (7.33)$$

Integration of Equation 7.33 using Equation 7.32 and solution for $C_1(0)/C_2(0)$ gives:

$$\frac{C_1(0)}{C_2(0)} = \frac{1 - \exp\left[-\alpha(\rho_1 - \rho_2)C_1^f\right]}{\exp\left[-\alpha(\rho_2 - \rho_1)C_2^f\right] - 1} \exp\left[\alpha(\rho_1 - \rho_2)\right] \qquad (7.34)$$

which is easy to evaluate when the composition of the feed is known. Substitution of this ratio into Equation 7.32 gives the vertical concentration profile. Typical concentration profiles are shown in Figure 7.7.

The grade and recovery in the two product streams can be calculated once the equilibrium stratification profiles have been calculated. The yield of total solids to the heavier fraction can be obtained by integrating the concentration profile.

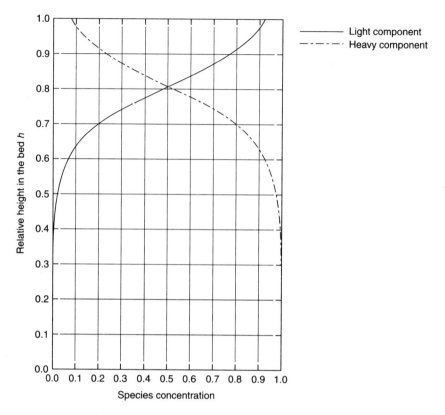

Figure 7.7 Calculated equilibrium stratification profiles for a binary mixture having components with densities 2670 and 3300 kg/m^3 and initial concentration 0.2 of the lighter component. Specific stratification constant is 0.02

Equipment that is designed to induce stratification of the particles and then to separate them always has a mechanism that enables the stratified bed to be split at some horizontal position. The top layers of the bed contain the lighter particles and the lower layers will include the heavier particles.

The mass yield of solids in the lighter product obtained by slicing the bed at a relative height h_s is given by

$$Y(h_s) = \frac{\int_{h_s}^{1} \bar{\rho}(h)\,dh}{\int_{0}^{1} \bar{\rho}(h)\,dh} \tag{7.35}$$

The recovery of the ith density component to the lighter product at a height h_s is given by

$$R_i(h_s) = \frac{\int_{h_s}^1 C_i(h)\,dh}{\int_0^1 C_i(h)\,dh} = \frac{\int_{h_s}^1 C_i(h)\,dh}{C_i^f} \tag{7.36}$$

The concentration of component i in the lighter product is given by

$$C_i^P = \frac{C_i^f R_i(h_s)}{\sum_i C_i^f R_i(h_s)} \tag{7.37}$$

Equations 7.35, 7.36 and 7.37 can be used to generate grade–recovery and recovery–yield curves using the relative splitting height h_s as the operating variable.

7.3.3 Stratification in multicomponent beds

When the material to be processed contains more than two particle types, the convenient analytical solution given in the previous section is no longer available and a numerical technique must be used to solve the system of differential Equations 7.24 subject to the conditions 7.25 to 7.27.

One obvious approach to the solution of the system of Equations 7.24 is to assume a set of $C_i(0)$, integrate Equations 7.24 numerically and check the resulting profiles against Equation 7.27. Using an optimization procedure, the initial guesses of $C_i(0)$ can be refined and the procedure repeated. This procedure, however, converges only slowly and exhibits weak stability characteristics.

A much more efficient procedure for integrating the equilibrium equations can be developed using the following method.

Integration of the stratification Equations 7.24 from the bottom of the bed with the unknown initial condition $C_i(0) = C_i^0$ gives

$$C_i(h) = C_i^0 \exp\left[-\alpha\rho_i h + \alpha \int_0^h \bar{\rho}(u)\,du\right] \quad i = 1, 2, 3 \ldots n \tag{7.38}$$

Substituting Equation 7.38 into Equation 7.27

$$\int_0^1 C_i^0 \exp\left[-\alpha\rho_i h + \alpha \int_0^h \bar{\rho}(u)\,du\right] dh = C_i^f \tag{7.39}$$

and rearranging gives an estimate of the concentration of each species at the bottom of the bed

$$C_i^0 = \frac{C_i^f}{\int_0^1 \exp\left[-\alpha\rho_i h + \alpha \int_0^h \bar{\rho}(u)\,du\right] dh} \tag{7.40}$$

An iterative method is started by guessing an initial average density profile $\bar{\rho}(h)$ and applying Equation 7.40 to calculate C_i^0 for each particle type.

Accumulation of numerical errors i prevented by normalizing the $C_i(0)$ to make

$$\sum_i C_i(0) = 1 \tag{7.41}$$

at this stage in the calculation.

These initial values $C_i(0)$ can be used in Equation 7.38 to get an estimate of the concentration profiles $C_i(h)$ for each particle type and then a new average density profile can be calculated using Equation 7.25. The procedure is repeated and continued until the calculated average density profile no longer changes from one iteration to the next.

This iterative procedure is efficient and robust and is capable of generating solutions rapidly for any number of components encountered in practice. The ideal Meyer profile has been found to provide a good starting point and this is easy to calculate.

The solution for a hypothetical four-component mixture is illustrated in Figure 7.9, which shows a typical equilibrium stratification profile and compares it to the ideal Mayer profile that would be achieved without dispersion.

Specific stratification constant values range from zero for a perfectly mixed bed with no stratification to infinity for perfect stratification. Typical practical values of alpha range from 0.001 m³/kg for a poor separation to 0.5 m³/kg for an exceedingly accurate separation. Although independent of the washability distribution of the feed, the specific stratification constant is dependent on the size and shape of the particles and on the type of equipment and its operating conditions. Experience has shown that this method is rapidly convergent over a wide range of conditions. As an example the calculated concentration profiles of nine washability fractions of coal in an industrial Baum jig are shown in Figure 7.8.

7.3.4 Performance of continuously operating single- and double-stage jigs

The mineral jig is designed specifically to stratify a bed of solid particles and a number of different designs have evolved over the years. These all have a common basic operating principle so that it is feasible to build a generic model that can be used to simulate the operation of any continuous jig.

Three distinct subprocesses can be identified in a continuous jig: bed stratification produced by the vertical pulsation of the bed, longitudinal transport of the bed due to water flow along the axis of the jig and splitting of the bed into concentrate and tailing layers.

The water velocity in the longitudinal direction is not uniform and vertical velocity profile exists in the continuous jig. In each compartment of the jig the various bed layers move at different velocities, with the upper layers generally moving faster than the lower ones. The velocities of particles at the

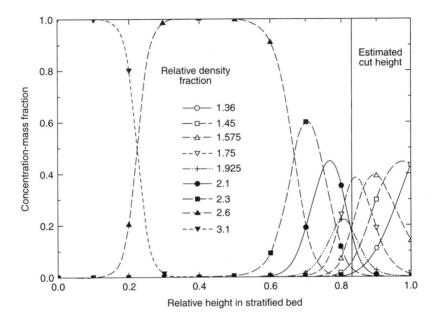

Figure 7.8 Calculated stratification profiles for nine washability fractions of coal in an industrial jig (Tavares and King, 1995)

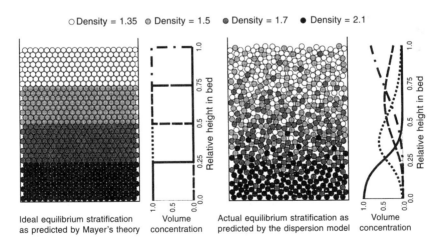

Figure 7.9 The effect of dispersion on the equilibrium stratification profile in a multicomponent mixture. The beds shown represent accurate simulations using the ideal Mayer theory and the dispersion model. The calculated vertical concentration profiles are also shown. A value of $\alpha = 0.03$ was used to simulate the dispersed bed

end of the bed depend primarily on the velocities of the layers to which they report. The velocity profile at the discharge end of the bed determines the discharge rate of each layer as it leaves the jig. In a continuous jig the overall material balance must be satisfied for each component and the integral condition Equation 7.27 must reflect the velocity profile and is replaced by

$$C_\rho^f = \frac{\int_0^1 C_\rho(h)V(h)\,dh}{\int_0^\infty \int_0^1 C_\rho(h)V(h)\,dh\,d\rho} \tag{7.42}$$

where $V(h)$ is the longitudinal velocity of the bed at height h.

The velocity profile must be known in order to integrate Equation 7.42. The actual longitudinal velocity at each bed height can only be determined by experiment for any particular equipment type and it will depend on the operating conditions in the unit. Defining a dimensionless velocity by $v(h) = (V(h)/V_{max})$ and discretizing Equation 7.42 we get

$$C_i^f = \frac{\int_0^1 C_i(h)v(h)\,dh}{\sum_{j=1}^n \int_0^1 C_j(h)v(h)\,dh} \tag{7.43}$$

The continuous jigging model for monosize feeds consists of Equations 7.24 subjected to the integral conditions given in Equations 7.25 and 7.43. Let the denominator of Equation 7.43 be equal to a constant, say $1/\beta$, then

$$C_i^f = \beta \int_0^1 C_i(h)v(h)\,dh \tag{7.44}$$

A similar iterative solution procedure to the one described in Section 7.3.2 is used to model equipment with a significant velocity profile. Substituting Equation 7.44 into Equation 7.38 and rearranging,

$$\beta C_i^o = \frac{C_i^f}{\int_0^1 \exp\left[-\alpha\rho_i h + \alpha \int_0^h \bar{\rho}(u)\,du\right]v(h)\,dh} \tag{7.45}$$

βC_i^o is determined iteratively as before and C_i^o is obtained from

$$C_i^o = \frac{\beta C_i^o}{\sum_{j=1}^n \beta C_j^o} \tag{7.46}$$

In the jigging process, particularly for well-operated jigs, the longitudinal velocity is expected to increase monotonically from the bottom to the top of the bed. The limited amount of direct measurements of velocity profiles found in the literature, however, does not allow the development of precise models for the velocity profile. A simple but effective empirical model for the velocity profile, which does not appear to conflict with the available experimental data is used here and is given by

$$v(h) = \kappa h + (1 - \kappa)h^2 \qquad (7.47)$$

where κ is a parameter of the model.

In a continuous jig, the separation of stratified layers is affected by refuse and middlings discharge mechanisms. Additional misplacement of clean coal and refuse is believed to happen as a result of the turbulence produced by the operation of these removal systems. The absence of any quantitative experimental work on this effect and the wide variety of refuse ejectors encountered in practice does not allow the precise modeling of this sub-process at present. A simple model which considers that the turbulence produced by splitting the stratified layers creates a perfectly mixed region of relative thickness 2δ is proposed tentatively here. The yield of solids to the lighter product can now be calculated by

$$Y(h_s, \delta) = \frac{\dfrac{1}{2} \displaystyle\int_{h_s-\delta}^{h_s+\delta} \overline{\rho}(h)v(h)\,dh + \int_{h_s+\delta}^{1} \overline{\rho}(h)v(h)\,dh}{\displaystyle\int_{0}^{1} \overline{\rho}(h)v(h)\,dh} \qquad (7.48)$$

The recovery of the ith density component to the lighter product is given by

$$R_i(h_s, \delta) = \frac{\dfrac{1}{2} \displaystyle\int_{h_s-\delta}^{h_s+\delta} C_i(h)v(h)\,dh + \int_{h_s+\delta}^{1} C_i(h)v(h)\,dh}{\displaystyle\int_{0}^{1} C_i(h)v(h)\,dh} \qquad (7.49)$$

The density distribution of the lighter product by volume is given by

$$C_i^p(h_s, \delta) = \frac{C_i^f R_i(h_s, \delta)}{\displaystyle\sum_{j=1}^{n} C_j^f R_j(h_s, \delta)} \qquad (7.50)$$

The effect of the velocity profile can be seen by comparing the simulated beds in Figures 7.9 and 7.10. In each case the same hypothetical four-component mixture was used to simulate the stratified beds. A thick layer of refuse accumulates in the bottom of the moving bed as a result of the longer residence times experienced by the heavier material in the jig. The concentration of the separate components in the lighter product stream is governed by Equation 7.50 and differs significantly from the equilibrium concentration in the bed itself.

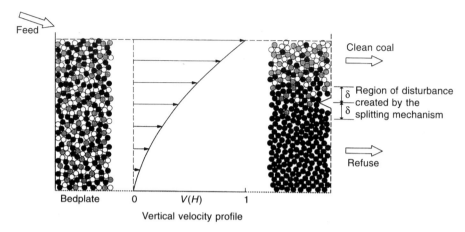

Figure 7.10 Development of the stratification profile in a continuous jig. The stratified bed is split at the right hand end as shown (Tavares and King, 1995)

This model can also be used for simulating separations in multiple compartment jigs. The first compartment is modeled as a single compartment jig and the second compartment is considered to be a single compartment with the feed equal to the light product from the first compartment. This is illustrated schematically in Figure 7.11. Thus Equation 7.43 becomes

$$C_i^p = \frac{\displaystyle\int_0^1 C_i^{(2)}(h) v_2(h)\, dh}{\displaystyle\sum_{j=1}^n \int_0^1 C_j^{(2)}(h) v_2(h)\, dh} \tag{7.51}$$

where C_i^p is the density distribution in the light product from the first

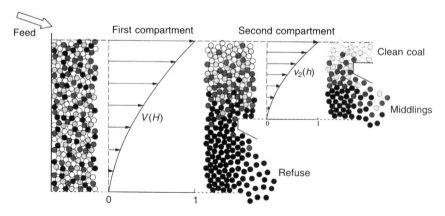

Figure 7.11 Stratification profiles in a continuous two-stage jig (Tavares and King, 1995)

compartment and $C_i^2(h)$ and $v_2(h)$ are the equilibrium concentration and velocity profiles in the second compartment, respectively.

The same computational procedure can be used for the second stage as was used for the first stage.

The data in Table 7.3 show that the value of α varies between the first and second stages of all the industrial jigs that have been analyzed. The most probable reason for this is the fact that the bed does not usually achieve equilibrium stratification profile in the first stage because of insufficient residence time. This shows up as an apparently low value of α. Since the solids entering the second stage are already partially stratified, it is easier to reach the equilibrium stratification profile in the limited residence time that is available.

Any complex jig can be modeled by an appropriate combination of the stratification model with models of the material transport and bed splitting phenomena that take place in the equipment. This versatile stratification model can be used to describe the behavior of other autogenous gravity separation devices such as the pinched sluice and the Reichert cone.

7.3.5 A model for the pinched sluice and the Reichert cone

The pinched sluice and the Reichert cone are two simple devices that rely on stratification to effect a separation of particles of different densities. Unlike the jig, these devices rely on a natural phenomenon to induce stratification rather than the purposeful jigging action of the jig.

The principles that govern the concentrating action of the pinched sluice and the Reichert cone are comparatively straightforward. The bed of particles is washed down the inside surface of the inverted cone as a sliding bed. The turbulence in the flowing stream induces stratification in the sliding bed of particles with the heavy particles tending to move towards the floor of the sliding bed. At the bottom of the cone, a slit in the floor of the cone allows the lower layers of particles in the bed to be discharged while the upper layers pass over the gap. The details of the cone arrangement are shown in Figure 7.12.

Because of the steadily decreasing area available for flow as the particle bed slides down the cone, the thickness of the bed increases significantly toward the apex of the cone. This thickening of the bed promotes the stratification of the particles because the lighter particles migrate upward through the particle layers more readily and more quickly than the heavier particles.

The fraction of the bed that is recovered through the slots is controlled by variation of the upstream relief at the slot as shown in Figure 7.13. The slot insert can be adjusted at nine vertical positions from the highest to the lowest. The higher the upstream relief, the greater the proportion of the particle bed that is recovered to the concentrate.

Table 7.3 Application of the multicomponent continuous model to the operation of industrial jigs

Jig	Size (mm)	Stage	Measured yield (%)	Estimated value of α	Calculated cut height h_s	Calculated bed density at the cutter position $\overline{\rho}(l_s)$
Baum 1	150 × 1.4	Primary	30.6	0.027	0.56	1.78
		Secondary	13.1	0.091	0.39	1.56
Baum 2	100 × 1.4	Primary	44.5	0.046	0.65	1.45
Baum 3	50 × 1	Primary	20.6	0.009	0.49	2.10
		Secondary	30.1	0.028	0.56	1.95
Baum 4	50 × 0.6	Primary	5.0	0.006	0.27	2.56
		Secondary	71.7	0.114	0.83	1.77
Batac 1	12 × 0.6	Primary	7.4	0.019	0.30	3.40
		Secondary	96.4	0.113	0.98	1.68
Batac 2	19 × 0.6	Primary	14.8	0.028	0.39	2.23
		Secondary	17.3	0.072	0.42	1.66
Batac 3	50 × 0.6	Primary	15.2	0.068	0.42	1.70

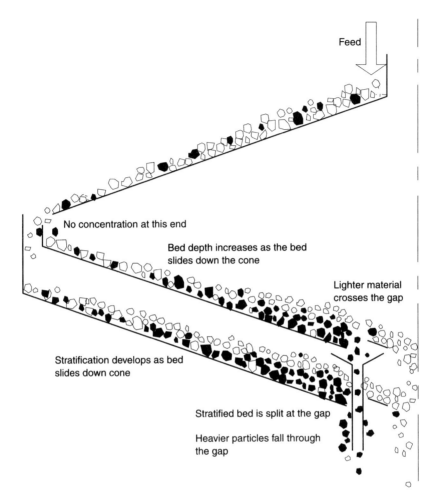

Figure 7.12 Details of the concentrating action of the Reichert cone. A double cone is illustrated

   Two well-documented experimental studies on the behavior of the Reichert provide some excellent data from which the essential nature of the operation of the Reichert cone can be established and a useful predictive model can be developed. Forssberg and Sandstrain (1979) and Holland-Batt (1978).
   Schematics of the four standard Reichert cone modules are shown in Figures 7.14 and 7.15. A basic cone module can consist of a single separating cone or the flow can be split on to two cone surfaces, as shown in Figures 7.14 and 7.15 which allows the cone to handle a greater feed. Two multiple cone modules are common: the DSV module which consists of a double cone followed by a single cone with a variable gap, and the DSVSV module which has an extra single cone with variable gap to clean the concentrate further. The two cleaning stages of the DSVSV module makes this configuration more suitable for cleaning

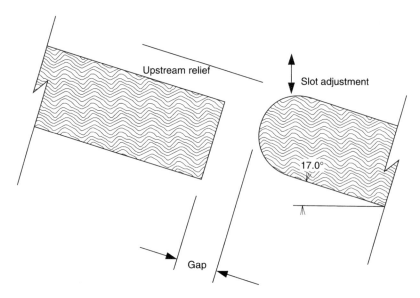

Figure 7.13 Details of the gap at the lower end of the cone. The vertical position of the bullnose is determined by the slot setting (Holland-Batt, 1978)

duties while the DSV module is better suited to roughing and scavenging. Actual cone units that are used in industrial practice are assembled as stacks of these standard modules. Two common unit stacks are shown in Figure 7.15. Nine stack configurations are commonly used in practice and these are SV, DSV, 2DSV, 3DSV, 4DSV, DSVSV, 2DSVSV, 3DSVSV and 2DSVSV.DSV with the symbols having the following meaning: D = double cone with fixed gap, SV = single cone with variable gap. The numerical prefix indicates the multiple of basic configuration that is repeated in the stack. The internal flow pattern for the heavies and the lights in the stack is fixed on installation and would not normally be altered afterwards. The normal internal flow arrangement of the 4DSV and 2DSVSV.DSV configurations is shown in Figure 7.15.

The variation of the concentrate flow with the feedrate to the cone surface for three different slot settings is shown in Figure 7.16. The data from the investigations cited above for a variety of ores are remarkably consistent and show clearly the effects of both feedrate and slot settings.

There are two phenomena that occur on the cone surface that must be modeled in order to simulate the operating performance of a simple cone: the stratification profile that develops in the sliding bed during its passage down the cone surface, and the splitting action of the bullnose splitter at the gap.

The split that is achieved is determined primarily by the vertical position of the bullnose splitter as shown in Figure 7.13. The standard cone provides nine pre-specified slot positions to control the bullnose. Slot 1 corresponds to maximum and slot 9 to minimum cut at the gap. Once the slot position has

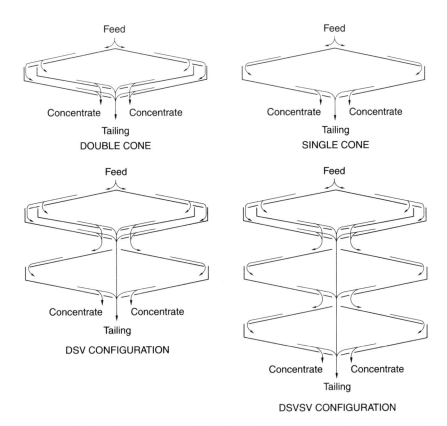

Figure 7.14 The four basic cone modules from which operating Reichert cone stacks are assembled

been chosen, the amount of solid taken off by the gap is a function of the total solids rate that flows down the cone surface.

Good experimental data are available to model the effects of both total solid flow and slot position on the underflow through the slot. The data for slot positions 1, 5 and 9 are shown in Figure 7.16. The data for other slot positions can easily be obtained by interpolation of the data in Figure 7.16. Obviously the amount flowing through the gap cannot exceed the total flow on the cone and the data have a distinct break at the point of intersection with the line *underflow = total flow* which is a straight line of slope 1 through the origin as shown in Figure 7.16. Each slot setting has a critical feedrate that must be exceeded to ensure that at least some material clears the concentrate gap and reports to tailings. When the feedrate falls below the critical flow, all the solid exits through the gap and no concentration is achieved.

The pattern of behavior shown in Figure 7.16 has been confirmed by extensive tests on other ore types (see Holland-Batt, 1978, Figure 13). The linear relationship between concentrate flow and the total flow on the cone is not surprising. The more material that is presented to the splitter the greater the

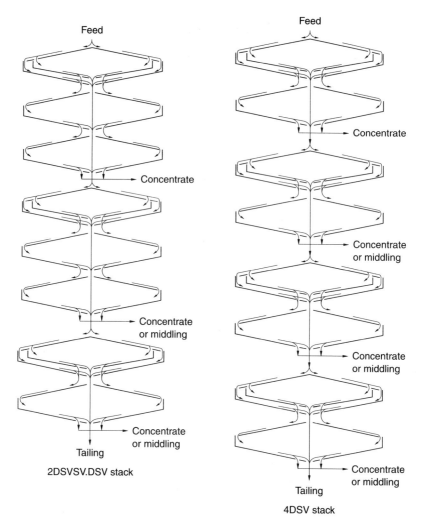

Figure 7.15 2DSVSV.DSV and 4DSV Reichert cone stacks that are commonly used for cleaning and roughing duties respectively

flow through the gap. The point of convergence, POC in Figure 7.16, and the slopes of the lines vary with ore type and Table 7.4 summarizes the data that are available. These variations are probably associated primarily with the shape of the particles but this effect has never been investigated.

In order to calculate the composition of the two discharge streams from the cone, it is necessary to model the stratification of the solid particles that occurs during the passage of the sliding bed down the cone surface. Stratification is a complex process, but a good realistic quantitative description of this phenomenon on the cone is provided by the equilibrium profile that balances the potential energy decrease that occurs with stratification against the random

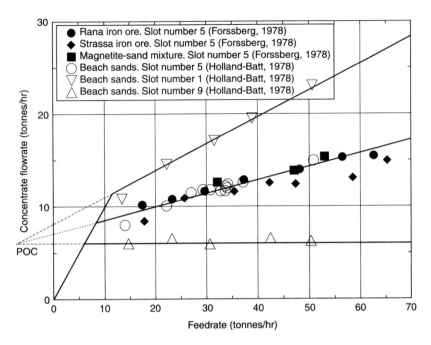

Figure 7.16 Concentrate take-off rate as a function of the feedrate to the cone and the variable slot setting. This data obtained on a single isolated cone

Table 7.4 Parameters that define the flow split at the gap of a single cone

Ore type	Coordinates of POC	Slope of the line	Reference
Beach sands	(–7.3, 6.0)	0.0359 × (9–slot #)	Holland-Batt, 1978
Iron ore and magnetite-sand mixtures	(–31.0, 2.5)	0.0359 × (9–slot #)	Forssberg, 1978, Forssberg and Sandström, 1981
Witwatersrand quartzite	(0.0, 3.5)	0.055 × (9–slot #)	Holland-Batt, 1978

particle motions in the sliding bed. This model of stratification is described in Section 7.3.1 and has been found to provide an excellent description of the operating behavior of Reichert cones. Figure 7.17 shows experimental data obtained with iron ore on a single Reichert cone. The stratification model is seen to describe the behavior of the cone well. The experimental data points were obtained at different slot settings, varying feed flowrates and solid content in the feed. The model predictions were made using the binary stratification model with a specific stratification constant of 0.0008 m^3/kg. This model is complex and highly nonlinear since it allows for the obvious fact that the stratification of the bed is governed by the make-up of the feed material. The model is computationally intensive since it requires the solution of a set of coupled differential equations – one for each type of particle – with integral boundary conditions. Once the behavior of a single cone can be modeled, the behavior of an entire cone stack can be synthesized by combining the models for the single cones appropriately. The double cone is synthesized as two single cones in parallel, each one receiving one half of the feed to the stage.

The calculation of the performance of a single module and a composite cone stack proceeds as follows:

1. Calculate the composition (distribution of particle size and particle composition) of the feed to the stage. This may be the gap underflow from the previous stage or it may be the combined concentrates and/or rejects from several stages higher up in the cone stack or from other units in the plant or it may be the plant feed itself.
2. Use the stratification model to calculate the equilibrium stratification profile of the sliding bed at the gap using the feed from step 1 as the model input.
3. Calculate the gap take-off flowrate from the flow model described by Figure 7.16 and Table 7.4.
4. Split the calculated stratified bed at the horizontal level that will give the correct gap take-off established in step 3.
5. Integrate the stratified bed from the floor to the splitting level and from the splitting level to the top of the bed to calculate the composition of the concentrate and discard streams.
6. Use these streams to constitute the feeds to the lower cones on the stack and repeat steps 1–5 as necessary.
7. Combine all the concentrates and discards from the cone stack appropriately to constitute the concentrate, middlings, and tailings from the unit.

The composition of the concentrate stream is established by the stratification that occurs while the bed of solids is sliding down the inside surface of the separating cone. Provided that the residence time on the cone is sufficiently long to allow the equilibrium stratification pattern to be established, the composition of the concentrate stream can be calculated using the equilibrium stratification profile as described in Sections 7.3.2 and 7.3.3.

It is comparatively straightforward to apply the model developed above to each stage in a stack and then to combine the concentrate streams appropriately to model the entire cone unit. This is the approach taken in MODSIM, which allows any of the standard configurations of the cone to be simulated.

Although each individual cone in the stack produces only a concentrate and tailings stream, the cone stack as a whole can be configured to produce a concentrate, a tailing, and, if required, a middling. A middling stream can be constituted from the concentrate streams produced by the lower cones in

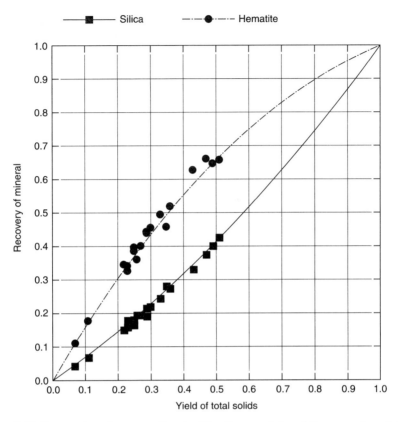

Figure 7.17 Comparison between the stratification model and the experimental performance of a single-stage Reichert cone

the stack as shown in Figure 7.15. If a middling stream is produced, it is usually returned to the feed of the stack.

The value of α that is achieved on a Reichert cone is significantly lower than that achieved in a jig. The stratification action of the cone is simply not as efficient as in the jig. The experimental data that is available indicates that the value of α should be about 0.002 on the cone.

7.4 Generalized partition function models for gravity separation units

The use of the partition curve is the most widely used method to describe the operation of any gravity separation unit and the general characteristics of the partition function are described in Section 7.2.3. It is always possible to describe the operation of any gravity separator by means of a partition function even if the partition function itself depends on the nature of the feed material, as is always the case for autogenous gravity separators. In fact the partition curve determined on an operating unit can be used to diagnose the health of the operation. Ideally, partition functions should produce steep symmetric curves that show no short circuiting and thus asymptote to the limits 1.0 and 0.0 at $\rho = 0$ and $\rho = \infty$ respectively. Deviations from the ideal can be attributed to various design and operational inadequacies and Leonard and Leonard (1983) provide a convenient tabulation of causes for poor partition functions in a variety for coal washing units. Although the partition curve is an excellent diagnostic tool it is not entirely satisfactory for simulation because of the difficulty of predicting the partition curve for any particular item of equipment.

A generalized procedure attributed to Gottfried and Jacobsen (1977) attempts to address this problem. The generalized partition curve is estimated in terms of a target specific gravity of separation for the proposed unit. The target specific gravity is the ρ_{50} point on the partition curve plotted as a composite for all sizes. This point is in fact fixed by the size-by-size behavior of the material in the separator and by the size-by-size composition of the feed. In any gravity separation operation, the partition function on a size-by-size basis varies in a systematic manner through the variations in ρ_{50} and the imperfection for each size fraction. This is illustrated for a dense-medium cyclone in Figure 7.18. Both the cut point and the imperfection increase as the size of the particles decrease. The data shown is fairly typical for industrial

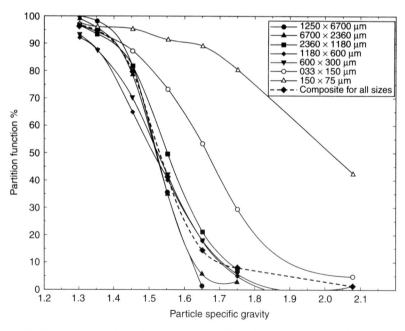

Figure 7.18 Measured partition functions for coal in a dense-medium cyclone. Individual size fractions are shown as well as the composite over all sizes. Data from Sokaski and Geer (1963)

gravity separators and the cut point for the individual sizes is modeled by

$$r = \frac{SG_s}{\overline{SG}_s} = f\left(\frac{d_p}{\overline{d}_p}\right) \tag{7.52}$$

where SG_s is the cut point for particles of size d_p, $\overline{SG}_s$ is the overall cut point for the composite feed and $\overline{d}_p$ is the average particle size in the feed. This normalization of the Gottfried-Jacobsen procedure with respect to average particle size in the feed is quite robust and correlates much of the data available in the literature and the function $f(\bullet)$ is not very sensitive to the feed composition and allows, at least approximately, for fairly wide variations in the size distribution in the feed. $\overline{SG}_s$ in Equation 7.52 is the target cut point and this target fixes the operating point of the particular unit. The function $f(\bullet)$ has been evaluated for a number of gravity separation units and the following exponential form fits the data reasonably well

$$r = a + b\exp\left(-c\frac{d_{pi}}{\overline{d}_p}\right) \tag{7.53}$$

Values of a, b and c for some different units are given in Table 7.5.

Table 7.5 Parameters to relate cut points for narrow size fractions to the composite cut point

Unit type	a	b	c
Water-only cyclone	0.8	0.6	1.26
Dense-medium cyclone	0.98	0.1	1.61
Shaking table	0.97	0.3	7.7
Coal washing drum	0.98	0.1	1.61

Once the target specific gravity and the average particle size in the feed are known, the cut point for each size of particle can be obtained from Equation 7.53. The cut point can be used with any appropriate generalized partition function such as those given in Table 7.1. An appropriate amount of short circuiting to either floats or sinks can also be applied if it is anticipated that this kind of inefficiency will be present in the equipment.

Osborne (1988) recommends that the variation of EPM with particle size, equipment size and separation density should be computed using a series of factors

$$\text{EPM} = f_1 f_2 f_3 E_s \tag{7.54}$$

f_1 is a factor accounting for the variation of EPM with particle size, f_2 a factor accounting for variation of EPM with equipment size and f_3 a manufacturer's guarantee factor usually in the range 1.1 to 1.2. E_s is a standard function representing the variation of EPM with separation density for each type of equipment. E_s for various types of coal washing equipment are:

Dense-medium cyclone: $E_s = 0.027\rho_{50} - 0.01$
Dynawhirlpool: $E_s = 0.15\rho_{50} - 0.16$
Dense-medium bath: $E_s = 0.047\rho_{50} - 0.05$
Baum jig: $E_s = 0.78(\rho_{50}(\rho_{50} - 1) + 0.01)$
Water-only cyclone: $E_s = 0.33\rho_{50} - 0.31$
Shaking table and spiral concentrator: $E_s = \rho_{50} - 1$

For dense-medium cyclones f_1 varies from 2 to 0.75 as particle size varies from 0.5 mm to 10 mm.

For dense-medium vessels f_1 varies from 0.5 for coarse coal to 1.4 for small coal.

Because both the Baum jig and the water-only cyclone are autogenous gravity separators, f_1 is dependent on the size distribution and washability of the feed material. Values of f_1 for jigs range from 3 to 0.5 as the average size of the feed material varies from 1 mm to 100 mm.

Bibliography

The model for the Reichert cone is based on data presented by Holland-Batt (1978) and Forssberg and Sandström (1979; 1981).

References

Cammack, P. (1987) The LARCODEMS – a new dense medium separator for 100 mm– 0.5 mm raw coal. *Aufbereitungs-Technik*, Vol. 28, pp. 427–434.

Davies, D.S., Dreissen, H.H. and Oliver, R.H. (1965) Advances in hydrocylone heavy media separation technology for fine ores. *Proc. 6th Intl Mineral Processing Congress* (A. Roberts, ed.), Pergamon Press, pp. 303–321.

Forssberg, E. (1978) Data from Reichert Cone tests. Private communication to author.

Forssberg, E. and Sandström, E. (1979) Utilization of the Reichert cone concentrator in ore processing. *Industrie Mineral-Mineralurgie*, pp. 223–243.

Forssberg, E. and Sandström, E. (1981) Operational characteristics of the Reichert cone in ore processing. *Proc. 13th International Mineral Processing Congress* (J. Laskowski, ed.), Warsaw 1979. Elsevier Scientific Publishing Co. Vol. II, pp. 1424–1481.

Gottfried, B.S. and Jacobsen, P.S. (1977) Generalized distribution curve for characterizing the performance of coal-cleaning equipment. USBM, Washington, RI8238.

Holland-Batt, A.B. (1978) Design of gravity concentration circuits by use of empirical mathematical models. *Proc. 11th Commonwealth Mining and Metallurgical Congress*, Institution of Mining and Metallurgy.

Leonard, J.W. (ed.) (1979) *Coal Preparation*, 4th edition. SME, Lyttleton, CO.

Leonard, I.V. and Leonard, J.W. (1983) Using Tromp Curves to Diagnose Performance Problems in Coal Cleaning. In *Basic Mathematics and Computer Techniques for Coal Preparation and Mining* (K.K. Humphreys and J.W. Leonard, eds), Marcel Dekker Inc., pp. 71–79.

Lovell, H.L., Moorehead, R.G., Luckie, P.T. and Kindig, J.K. (1994) Hydraulic Separation. Chapter 7 in *Coal Preparation* (J.W. Leonard, ed.), 5th edition, SME, Lyttleton, CO, p. 307.

Napier-Munn, T.J., Scott, I.A., Tuteja, R., Davis, J.J. and Kojovic, T. (1992) Prediction of underflow medium density in dense medium cyclones. In *Hydrocyclones: Analysis and Applications* (L. Svarovsky and M.T. Thew, eds), Kluwer Academic Publishers, London, pp. 191–214.

Osborne, D.G. (1988) *Coal Preparation Technology*, Vol. 1. Graham and Trotman Ltd, Chapter 8.

Sokaski, M. and Geer, M.R. (1963) Cleaning unsized fine coal in a dense-medium cyclone pilot plant. USBM, Washington, RI 6274.

Tavares, L.M. and King, R.P. (1995) A useful model for the calculation of the performance of batch and continuous jigs. *Coal Preparation,* Vol. 15, pp. 99–128.

Vanagamudi, M., Mitra, J. and Rao, T.C. (1992) Analysis of performance of a 76 mm Vorsyl separator. *Minerals Engineering*, Vol. 5, pp. 93–101.

Weiss, N.L. (1985) *SME Mineral Processing Handbook*, Vol. 1. Society of Mining Engineers, New York.

8
Magnetic separation

The differences in magnetic properties among minerals can be utilized under favorable conditions to effect a separation. Magnetic separators can be very effective and recent developments in their design have extended the range of minerals that are potentially suitable for concentration in magnetic separators.

8.1 Behavior of particles in magnetic fields

Permanent magnets are familiar objects and the fact that the opposite poles of two magnets attract each other strongly is a manifestation of the magnetic field that is generated in the space surrounding a magnetic pole. All solid substances react with a magnetic field and display definite properties when placed in a magnetic field. For our purposes solids may be classified as ferromagnetic, paramagnetic or diamagnetic. Ferromagnetic substances such as iron and magnetite react very strongly with the magnetic field and display a significant ability to be attracted towards a magnetic pole. Paramagnetic materials are affected much less by a magnetic field but, like ferromagnetic materials, are attracted towards a magnetic pole. Diamagnetic materials are repelled from a magnetic pole but this tendency is always very weak.

The magnetic field that surrounds a magnet or that is generated by an electric current is described quantitatively by the magnetic field strength H. The dimensions of H in the SI system are amps/m which reflect the strength of field created by an electric current that flows through unit length of conductor. H is vectorial in character and will have direction as well as magnitude. Thus the field strength at every point in space is described by the three components H_x, H_y and H_z of the field. The force exerted on an isolated magnetic pole of strength m Wb is given by

$$F = mH \tag{8.1}$$

A commonly used unit for H is the Oersted. This is the unit of the so-called emu system and although this system is being superseded by the SI system almost all data on magnetic properties are recorded in the literature in the emu system. The conversion factors from emu to SI are given in Table 8.1.

Note in the SI system:

$T = Wb/m^2 = kg/A \, s^2 = N/A \, m$

$Wb = N \, m/A$

$H = Wb/A = m^2 \, kg/A^2 \, s^2$

Table 8.1 Conversion factors for converting from the emu system to the SI system

Variable	Emu unit	SI unit	Conversion factor
Magnetic field strength, H	Oersted	Ampere/meter	$10^3/4\pi$
Magnetic induction, B	Gauss	Weber/m^2 (Tesla)	10^{-4}
Magnetic polarization, J	Emu	Weber/m^2 (Tesla)	$4\pi10^{-4}$
Magnetic pole strength, m	Emu	Weber	$4\pi10^{-8}$
Magnetic dipole moment, M	Emu	Weber meter	$4\pi10^{-10}$
Magnetic flux, Φ	Maxwell	Weber	10^{-8}
Magnetic permeability, μ	Dimensionless	Henry/meter	$4\pi10^{-7}$
Volume susceptibility, κ	Emu/cm^3	Dimensionless	4π
Mass susceptibility, χ	Emu/g	m^3/kg	$4\pi10^{-3}$

The magnetic permeability of free space $\mu_0 = 1$ emu $= 4\pi10^{-7}$ H/m.

The magnetic induction B can be regarded as the magnetic flux per unit area.

The magnetic polarization J can be regarded as the magnetic dipole per unit volume Wb m/m^3.

When a solid particle is placed in a magnetic field the particle disturbs the field and the field affects the particle in such a way that the particle becomes magnetized. This magnetization of the particle is important from the point of view of the magnetic separation of particles because a particle can be made to move in the magnetic field by virtue of the magnetic moment that is induced in the particle.

In any region of space the magnetic induction B is related to the magnetic field strength by

$$B = \mu H \tag{8.2}$$

where μ is the magnetic permeability of the medium.

The magnetic induction has two components: one due to the magnetic field and the second that results from the field produced by the magnetization of the region itself. The first component would be present even if the region were completely evacuated. Inside a particle, for example, the field is given by Equation 8.2.

$$B' = \mu H'$$

$$= \mu_0 H' + J' \tag{8.3}$$

J' is the magnetic induction that results from the magnetization of the particle and is measured in units of magnetic dipole moment per unit volume (Wb m/m^3). J' is a measure of the magnetic dipole that is induced in the particle material. Consider unit volume of material. A magnetic dipole is induced which can be represented by two poles of strength m separated by a distance δ inclined at θ to the x-coordinate direction. The geometry and orientation of the particle in relation to the magnetic field determines the angle θ and the dipole does not normally lie parallel to the vector J'.

The effective moment parallel to the x-coordinate direction is $m\delta \cos \theta$ and is related to the x-component of J' by

$$m\delta \cos \theta = J'_x \tag{8.4}$$

for unit volume of medium. Similar expressions can be written for the other coordinate directions.

The magnetic polarization J' is proportional to the field strength inside the particle, H', and

$$J' = \mu_0 \kappa H' \tag{8.5}$$

where κ is the volume susceptibility of the material. The permeability and volume susceptibility are not independent because

$$B' = \mu_0 H' + \mu_0 \kappa H' = \mu_0(1 + \kappa)H' \tag{8.6}$$

and by comparison with Equation 8.2

$$\mu = \mu_0(1 + \kappa) \tag{8.7}$$

The relative permeability, μ_r, of the material is given by

$$\mu_r = \mu/\mu_0 = 1 + \kappa \tag{8.8}$$

Diamagnetic and paramagnetic materials have small negative and positive values of κ respectively and these are independent of the field strength H. Ferromagnetic materials have larger positive values of κ and in addition the susceptibility, and hence the permeability, is a function of H. In fact in ferromagnetic materials J' does not continue to increase indefinitely as H increases but approaches a limiting value which is the familiar phenomenon of saturation.

Before discussing the magnetization of the particle, the disturbance of the field by the particle must be considered. The magnetic permeability of the particle will in general be different to that in the surrounding medium and the magnetic field is distorted in the neighborhood of the particle. The extent of the distortion depends on the shape of the particle and the value of its permeability. When the exterior field is uniform far from the particle, the field inside the particle is uniform also and is directed parallel to the exterior field and in the same direction as shown in Figure 8.1. For a number of different particle shapes

$$H' = \frac{H}{1 + \kappa N} \tag{8.9}$$

where
$\quad H$ is the external magnetic field strength,
$\quad H'$ is the magnetic field strength inside the particle,
$\quad \mu_r$ is the relative permeability of the particle, and

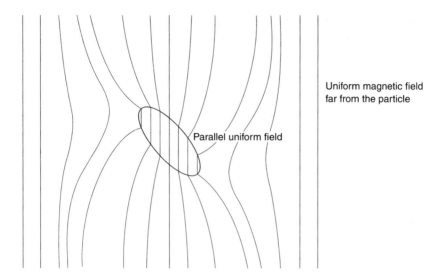

Figure 8.1 Distribution of the magnetic field caused by a paramagnetic particle. The magnetic field inside the particle is parallel to the inducing field

N is a geometric factor that depends on the shape of the particle and is called the demagnetization factor.
N is 0.333 for spherical particles, 0.27 for a cylinder of length equal to its diameter and 0.0172 for cylinders of length equal to 10 times the diameter. These values are taken from Anderson (1968).

Equation 8.9 can be derived simply by noting that the field inside the body is decreased because the induced magnetic poles in the particle create a field that opposes the applied field

$$H' = H - NJ'/\mu_o = H - N\kappa H' \tag{8.10}$$

This phenomenon is known as demagnetization.

Actually N is also vectorial and we should write

$$H'_x = \frac{H_x}{1 + \kappa N_x} \tag{8.11}$$

Thus H'_x is not perfectly parallel to H unless the particle is free to orient itself parallel to the field.

8.2 Forces experienced by a particle in a magnetic field

A small magnetic dipole placed in a nonuniform magnetic field experiences a translational force which is determined by the field strength and the strength, orientation and dimensions of the dipole. Consider a small particle of paramagnetic material placed in a magnetic field as shown in Figure 8.2.

Consider a small dipole of pole strength m Wb situated in the x–y plane as shown in Figure 8.2. The magnetic field is represented by the vector H as shown. The force on the north seeking pole is represented by Fn and the force on the south seeking pole is represented by Fs. These two vectors are parallel to the magnetic field H. The south seeking pole is situated at (x, y) and the north seeking pole at $(x + \Delta x, y + \Delta x)$. These forces create a couple that tends to orientate the dipole to lie parallel to the magnetic field, but, if the field is not uniform, they generate also a net translational force that can be related to the magnetic field by the following simple analysis.

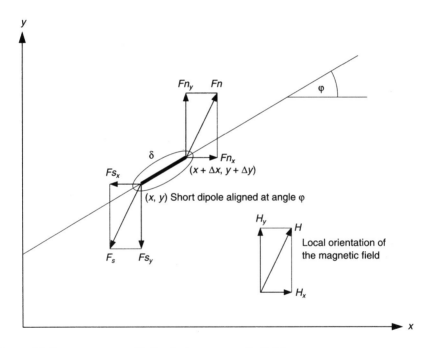

Figure 8.2 Forces on a small dipole in a magnetic field

The translational force in the x-direction is

$$F_x = Fn_x - Fs_x$$

$$= mH_x(x + \Delta x, y + \Delta y) - mH_x(x, y)$$

$$= m\Delta H_x \tag{8.12}$$

$$= m\left(\frac{\partial H_x}{\partial x}\Delta x + \frac{\partial H_x}{\partial y}\Delta y\right)$$

If the dipole length is δ, $\Delta x = \delta \cos \theta$ and $\Delta y = \delta \sin \theta$

$$F_x = m\delta\left(\frac{\partial H_x}{\partial x} \cos \theta + \frac{\partial H_x}{\partial y} \sin \theta\right) \tag{8.13}$$

Likewise the y-component of the translational force is given by:

$$F_y = m\delta\left(\frac{\partial H_y}{\partial x} \cos \theta + \frac{\partial H_y}{\partial y} \sin \theta\right) \tag{8.14}$$

Using Equation 8.4 these become:

$$F_x = v_p J_x' \frac{\partial H_x}{\partial x} + v_p J_y' \frac{\partial H_x}{\partial y} \tag{8.15}$$

and

$$F_y = v_p J_x' \frac{\partial H_y}{\partial x} + v_p J_y' \frac{\partial H_y}{\partial y} \tag{8.16}$$

where v_p is the volume of the particle.

This entire analysis can be extended to three dimensions in a straightforward manner.

When a small particle of magnetic susceptibility κ is placed in the magnetic field, the induced magnetism in the particle creates the effective dipole moment. The magnetic moment per unit volume J' in the particle is related to the field, H, by Equations 8.5 and 8.9.

$$J' = \frac{\mu_0 \kappa H}{1 + \kappa N} \tag{8.17}$$

Both J' and H are vectors and, as Equation 8.17 implies, the induced magnetic moment J' is parallel to the field H.

Substitution in Equation 8.15 gives:

$$F_x = \frac{v_p \mu_0 \kappa}{1 + \kappa N}\left(H_x \frac{\partial H_x}{\partial x} + H_y \frac{\partial H_x}{\partial y}\right) \tag{8.18}$$

The components of the magnetic field strength can be written in terms of the gradient of a magnetic potential φ as follows:

$$H_x = -\frac{\partial \varphi}{\partial x}$$
$$H_y = -\frac{\partial \varphi}{\partial y} \tag{8.19}$$

which yields the symmetry relationships:

$$\frac{\partial H_y}{\partial x} = -\frac{\partial^2 \varphi}{\partial x \partial y} = -\frac{\partial^2 \varphi}{\partial y \partial x} = \frac{\partial H_x}{\partial y} \tag{8.20}$$

and

$$\frac{\partial H_x}{\partial y} = -\frac{\partial^2 \varphi}{\partial y \partial x} = -\frac{\partial^2 \varphi}{\partial x \partial y} = \frac{\partial H_y}{\partial x} \tag{8.21}$$

Thus

$$F_x = \frac{v_p \mu_0 \kappa}{1 + (\mu_r - 1)N} \left(H_x \frac{\partial H_x}{\partial x} + H_y \frac{\partial H_y}{\partial x} \right) \tag{8.22}$$

This can be written:

$$F_x = \frac{0.5 v_p \mu_0 \kappa}{1 + \kappa N} \frac{\partial H^2}{\partial x} \tag{8.23}$$

where $H^2 = H_x^2 + H_y^2 + H_z^2$ is the square of the magnetic field strength. Equivalent expressions can be derived for the y and z components of the force experienced by the particle in the magnetic field.

If χ is the mass susceptibility of the particle and m_p the mass of the particle

$$v_p \kappa = m_p \chi \tag{8.24}$$

$$\kappa = \rho_p \chi \tag{8.25}$$

where ρ_p is the density of the particle.

Equation 8.22 shows clearly that the strength of translational force is proportional to product of the magnetic field strength and its gradient. Consequently magnetic separators must be constructed so that high fields containing strong field gradients are present. Methods for achieving this will be discussed later.

The force can be written as a product of two factors:

$$V_m = \frac{0.5 v_p \kappa}{1 + \kappa N} \tag{8.26}$$

and

$$f_m = \mu_0 \nabla H^2 \tag{8.27}$$

V_m has the dimensions of m³ and is called the magnetic volume of the particle. f_m has dimensions N/m³ and is called the magnetic force density. The value of V_m is determined by the characteristics of the particles and is particle specific. The magnetic force density is a function primarily of the machine design and magnetic field generated by the machine. It is therefore a machine-specific parameter. However, if the particles are subject to magnetic saturation in the working field, the magnetic force density will also be a function of the saturation magnetization of the particle. The force acting a particle in a magnetic field is the product of V_m and f_m.

$$F = V_m f_m \tag{8.28}$$

The concept of magnetic volume of a particle is especially convenient for practical calculations because the magnetic volume of a composite mineral particle is the sum of the magnetic volumes of the separate components. The magnetic volume of component i is calculated from

$$V_{mi} = \frac{0.5 v_i v_p \kappa_i}{1 + N\kappa_p} \tag{8.29}$$

where v_i is the volume fraction of component i and v_p and κ_p are the volume and magnetic susceptibility of the particle.

Illustrative example 8.1

Calculate the radial component of the force on a small spherical paramagnetic particle of diameter 100 µm in the neighborhood of a magnetically saturated ferromagnetic wire of radius a. The magnetic field has the following components in cylindrical coordinates.

$$H_r = \left(H_s \frac{a^2}{r^2} + H_0 \right) \cos \phi$$

$$H_\theta = \left(H_s \frac{a^2}{r^2} - H_0 \right) \sin \phi$$

$$H_z = 0$$

H_s is the field inside the magnetically saturated wire and H_0 is the strength of the magnetizing field.

Use the following data:

$$H_s = 2 \times 10^3 \text{ A/m}$$

$$H_0 = 3 \times 10^3 \text{ A/m}$$

$$a = 1 \text{ mm}$$

Magnetic susceptibility of the particle $= 1.36 \times 10^{-3}$

The particle is situated at coordinate location $r = 1.05$ mm and $\phi = 0°$

Solution

The force on the particle is given by

$$F = \frac{0.5 \mu_0 \kappa v_p}{1 + N\kappa} \nabla H^2$$

where vector notation has been used to represent F and H.

The square of the field strength is obtained from the components of the field strength vector

$$H^2 = H_r^2 + H_\phi^2 + H_z^2$$

$$H^2 = \left(H_s \frac{a^2}{r^2} + H_0 \right)^2 \cos^2\phi + \left(H_s \frac{a^2}{r^2} - H_0 \right)^2 \sin^2\phi$$

The radial component of the force is given by

$$F_r = \frac{0.5\mu_0 \kappa v_p}{1 + N\kappa} \frac{\partial H^2}{\partial r}$$

$$\frac{\partial H^2}{\partial r} = -4\left(H_s \frac{a^2}{r^2} + H_0 \right) H_s \frac{a^2}{r^3} \cos^2\phi - 4\left(H_s \frac{a^2}{r^2} - H_0 \right) H_s \frac{a^2}{r^3} \sin^2\phi$$

$$\frac{\partial H^2}{\partial r} = -4\left(\frac{2 \times 10^3}{1.05^2} + 3 \times 10^3 \right) \frac{2 \times 10^3}{1.05^3 \times 10^{-3}}$$

$$\frac{\partial H^2}{\partial r} = 3.327 \times 10^{10}$$

The magnetic volume of the particle is:

$$V_m = \frac{0.51 \times 36 \times 10^{-3} \, \pi/6(100 \times 10^{-6})^3}{1 + 0.3331 \times 36 \times 10^{-3}}$$

$$= 3.559 \times 10^{-16} \text{ m}^3$$

The magnetic force density is:

$$f_m = \mu_0 \frac{\partial H^2}{\partial r}$$

$$= 4\pi \times 10^{-7} \, 3.327 \times 10^{10} = 4.18 \times 10^4 \text{ N/m}^3$$

The radial component of the force experienced by the particle is:

$$F_r = 3.559 \times 10^{-16} \times 4.18 \times 10^4 = 1.49 \times 10^{-11} \text{ N}$$

Notice that this is a small force!

8.3 Magnetic properties of minerals

Differences in magnetic susceptibilities between minerals may be exploited for their separation in a machine that will permit the particles to move under the influence of the magnetic field according to Equation 8.22 and under the influence of some other field such as gravity or a shearing field in a flowing liquid. Particles with high susceptibilities will be influenced strongly by the magnetic field. Particles of different susceptibilities will follow different paths under the influence of the two interacting fields and can thus be physically separated.

Magnetic susceptibilities of minerals are difficult to measure and the values are strongly influenced by impurities. The susceptibilities of most minerals are low (typically less than 10^{-7} kg^{-1}) and very high fields and field gradients are required to effect a useful separation. A comprehensive tabulation of susceptibilities of minerals is not available and in any case the same mineral from various sources can exhibit significant variations. Minute inclusions of magnetite for example will drastically affect the apparent susceptibility.

A brief tabulation is given in Table 8.2.

Table 8.2 Magnetic susceptibilities of minerals at 20°C (Lawver and Hopstock, 1974)

Mineral	Chemical composition	χ (m^3/kg)
Quartz	SiO_2	-6.19×10^{-9}
Calcite	$CaCO_3$	-4.80×10^{-9}
Sphalerite	β-ZnS	-3.23×10^{-9}
Pyrite	FeS_2	$\geq 3.57 \times 10^{-9}$
Hematite	α-F_2O_3	$\geq 259 \times 10^{-9}$
Goethite	α-FeOOH	$\geq 326 \times 10^{-9}$
Siderite	$FeCO_3$	123×10^{-8}
Fayalite	$FeSiO_4$	126×10^{-8}

Values for other minerals are given by Taggart (1945) and Gaudin (1939) but these values should be treated with caution. Svoboda (1987) gives a comprehensive summary of the available data on measured values of the magnetic susceptibility of minerals that can be separated by magnetic methods.

Effective values for magnetic susceptibilities may be obtained in the mineral dressing laboratory by use of the Franz isodynamic separator. The mineral must be available as liberated particles and requires calibration with crystals of known susceptibility.

8.4 Magnetic separating machines

Production machines must operate continuously and the particles must move through the field in a continuous stream. Separation is effected by inducing the particles of higher susceptibilities to take a path through the magnetic field, which will deflect them to the concentrate discharge ports. The magnetic forces on most particles are very small and the high intensity regions of the magnetic field are small in extent. Consequently the machines must be designed to achieve a separation based on a comparatively weak effect which can only influence the particle for a short time during its passage through the machine. The mechanical design of magnetic separators is crucial to the effectiveness of their operation but the difficulty in relating the fundamental properties of the particle mechanics in magnetic fields to the detailed behavior of the machines has hampered the development of machines capable of high selectivity and good recovery of minerals. That improvements can be effected by the use of high magnetic fields is obvious and the improvements in wet high intensity magnetic separators has led to a growing list of applications over recent years.

The basis of performance calculations for both wet and dry magnetic separators is given below. The use of wet magnetic separators is preferred in the plant environment. Dry separators are limited to treating ores that can be liberated at particle sizes greater than 75 μm. Very small particles are severely influenced by air currents and tend to agglomerate and adhere to the pole surfaces of the machine. Machines that can handle slurries do overcome these problems to a certain extent and also offer the advantage of elimination of the drying stage when wet grinding is used.

8.4.1 Production of the magnetic field

The quantitative analysis presented above demonstrates that high gradients in the magnetic field are required as well as high field strength if a significant force is to act on weakly paramagnetic particles in the machine. Magnetic fields are normally produced by electromagnets and the pole pieces designed to create the high field gradients. Designs vary considerably among the different manufacturers and only the general principles are described here.

The field is established between two pole pieces, one of which is considerably narrower than the other so that the field intensity is greatest in its neighborhood. This concentration of the field produces the necessary gradient in the field to impart the required force on the paramagnetic particles. Two configurations are shown schematically in Figure 8.3.

A matrix of ferromagnetic material is used in wet high intensity magnetic separators rather than fixed or moving poles of the type shown above. The matrix provides a high concentration of points of high field intensity and can be made from small spheres, expanded metal sheet, 'wedge-wire' bars or grooved metal plates.

The magnetic field is most conveniently produced by electromagnets rather than permanent magnets. The field strength of an electromagnet may be conveniently varied by varying the current. The calculation of the field strength generated between the poles of an electromagnet can be done simply only if the geometry of the coil, the core, the poles and the air gap is very simple. Hughes (1956) presents a simple example. This simple method is based on the concept of the magnetic circuit which is made up of the core and pole pieces of the electromagnet and the air gaps in which the separating field is generated. The various parts of the circuit are in series or parallel and the complete circuit forms a ring. A schematic representation is shown in Figure 8.4.

The magnetic circuit shown in Figure 8.4 can be regarded as consisting of six segments in series: the core, pole 1, air gap 1, the keeper, gap 2 and pole 2. In order to relate the magnetic field strength in the gaps to the current in

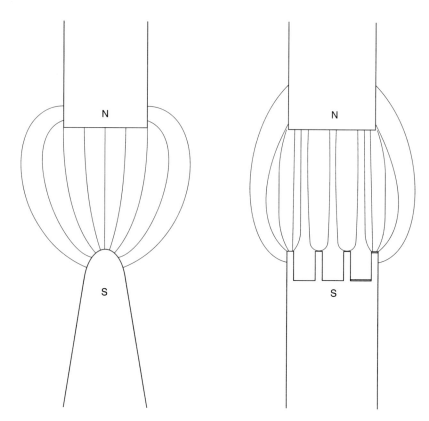

Figure 8.3 Simple poles for the generation of distorted magnetic fields

the coil the following assumptions are made. The average magnetic flux across any cross-section of the circuit is constant along the entire length of the circuit, fringing and leakage of magnetic field is negligible, the pole pieces and gaps can be represented by an effective cross-sectional area and length, and the magnetic fields are additive for each of the segments in series.

If Φ is the magnetic flux in the circuit, the field in segment i, H_i, is given by

$$H_i = \frac{B_i}{\mu_i} = \frac{\Phi}{a_i\mu_i} \qquad (8.30)$$

from Equation 8.2 μ_i is the permeability of the material making up segment i and a_i is the effective cross-sectional area of segment i.

The total magnetic field for the entire circuit which is to be generated by the coil is obtained by summing over each of the separate segments. In the SI system, H_i is obtained from the amperes/meter of path length to be provided by the coil. Therefore the current in a coil having only a single turn is given by

$$I_1 = \Sigma \, H_i\ell_i \qquad (8.31)$$

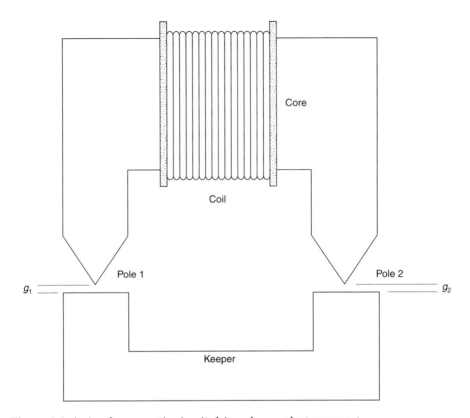

Figure 8.4 A simple magnetic circuit driven by an electromagnet

$$I_1 = \Phi \sum_i \frac{\ell_i}{a_i \mu_i} \tag{8.32}$$

where ℓ_i is the effective length of segment I. The term $\ell_i / a_i \mu_i$ is the magnetic reluctance of segment i of the magnetic circuit.

Electromagnets are normally constructed with coils having many turns. The current required is reduced in proportion to number of turns so that

$$I_n = \frac{I_1}{n} \tag{8.33}$$

is the current required in a coil having n turns.

The average field strength in the gaps in Figure 8.4 is given by

$$H = \frac{\Phi}{a_g \mu_g} \tag{8.34}$$

$$H = \frac{n I_n / a_g \mu_g}{\sum \ell_i / a_i \mu_i} \tag{8.35}$$

a_g is the effective cross-sectional area of the gap and μ_g the permeability of air ($4\pi \times 10^{-7}$ H/m). Equation 8.35 indicates that the average field strength is linearly related to the current in the coil but this is only apparent. The permeabilities of the core materials are strong functions of Φ and hence the current so the relationship is fairly strongly nonlinear. However, over moderate ranges of current the relationship between H and I_n is nearly linear and an empirical relationship of the type

$$H = \alpha I_n^m \tag{8.36}$$

with $m \simeq 1$ has been found to be satisfactory in many cases.

Equation 8.35 gives an indication as to how the field strength can be increased by good machine design. Attention can be given to the reduction of each of the terms $\ell_i/a_i\mu_i$ for the component parts of the circuit. The permeabilities of the core and keeper should be high and the gaps should be as small as possible.

Illustrative example 8.2
Calculate the current that is required to produce a flux density of 1.2 Tesla in the air gap of the magnetic circuit shown in Figure 8.5. The relative permeability of the iron core is 300 and that of the air core is 1.0. The coil has 800 turns.

If the magnetic field is assumed to be uniform everywhere in the gap, evaluate the three components of the field strength vector in the gap.

Solution
In the gap $\Phi = 1.2 \times 0.04 \times 0.05 = 2.40 \times 10^{-3}$ Wb and this is constant throughout

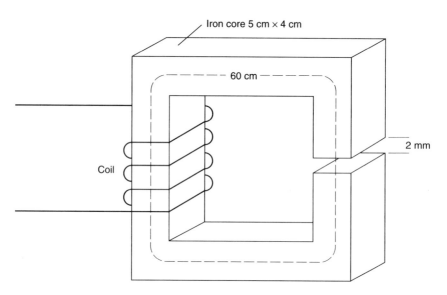

Figure 8.5 Magnetic circuit for Example 8.2

the circuit. Ampere-turns required can be calculated by adding the ampere-turns required for each segment of the magnetic path.

$$\text{Ampere-turns} = nI_n = \phi\Sigma \frac{\ell_i}{a_i\mu_i}$$

$$= \frac{2.4 \times 10^{-3}}{4\pi \times 10^{-7}0.04 \times 0.05}\left(\frac{0.002}{1} + \frac{0.598}{300}\right)$$

$$= 3813$$

The current required for this application is $3813/800 = 4.77$ A.

The group $\ell_i/a_i\mu_i$ is called the magnetic reluctance of segment I.

8.5 Dry magnetic separation

In dry magnetic separators the individual particles of mineral ore passed through the magnet field and the more highly magnetic particles are deflected into a path that carries them into the concentrate discharge. A diagram of a rotating drum separator is shown in Figure 8.6.

The roll of an induced roll separator is either laminated or serrated as shown so that many points of high field intensity are produced on the surface. The particles are fed on to the surface of the rotating roll and the magnetic particles adhere to the surface of the roll at the points of high field intensity. The field at the roll surface is greatest directly opposite the pole piece and, as the roll rotates, the field at the surface decreases until it is at a minimum at the lowest point. Somewhere during the quarter rotation the centrifugal force on an adhering particle will exceed the magnetic force of adhesion and the particle will part company with the roll. If this is sooner rather than later, the particle will not enter the concentrate stream.

8.5.1 *Hopstock model for the dry low intensity magnetic separator*

Figure 8.6 shows a simple arrangement for a rotating drum magnetic separator. The particulate material is fed to the top of the drum and is carried downward as the drum rotates. The paramagnetic particles are attracted by the magnetic field inside the drum. The nonmagnetic particles tend to be lifted from the surface of the drum by centrifugal force. The particles can be separated depending on the position of separation and a tailing, a middling and a magnetic concentrate are usually produced as shown.

A simple analysis can be used to develop a simple but workable model for this type of magnetic separator. This analysis is due to Hopstock (1975).

Near the surface of the drum the magnetic field strength can be described by two components in cylindrical coordinates with origin at the center of the drum:

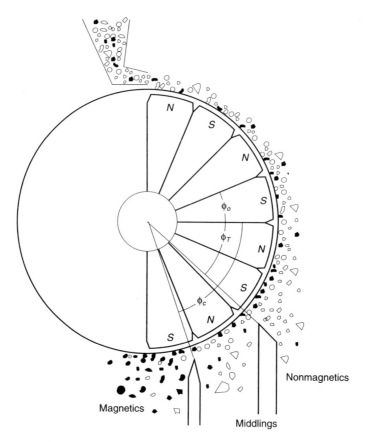

Figure 8.6 Dry drum magnetic separator

$$H_r = H_0 \cos\left(\pi\frac{\varphi}{\varphi_0}\right) \exp\left(\frac{-\pi(r-R)}{\varphi_0 R}\right) \qquad (8.37)$$

$$H_\varphi = H_0 \sin\left(\pi\frac{\varphi}{\varphi_0}\right) \exp\left(\frac{-\pi(r-R)}{\varphi_0 R}\right) \qquad (8.38)$$

H_0 is the field strength at the drum surface.

The model relies on a knowledge of the radial force that is experienced by a particle near the surface of the drum. This can be calculated as the product of the magnetic force density and the magnetic volume of the particle.

The square of the field strength close to the surface of the drum is obtained from Equations 8.37 and 8.38 as

$$H^2 = H_r^2 + H_\varphi^2 + H_z^2$$

$$= H_0^2 \exp\left(-\frac{2\pi(r-R)}{\varphi_0 R}\right) \qquad (8.39)$$

The radial component of force on a particle near the drum surface is given by:

$$F_r = V_m \mu_0 \frac{\partial H^2}{\partial r}$$

$$= -V_m \mu_0 H_0^2 \frac{2\pi}{\varphi_0 R} \exp\left(-\frac{2\pi(r-R)}{\varphi_0 R}\right) \qquad (8.40)$$

The negative sign indicates that the radial component of the force on a paramagnetic particle is directed inward.

At the surface of the drum, $r = R$ and

$$F_r = -V_m \mu_0 H_0^2 \frac{2\pi}{\varphi_0 R} \qquad (8.41)$$

The radial component of the magnetic force density at the surface of the drum is:

$$f_m = \frac{2\pi \mu_0 H_0^2}{\varphi_0 R} \qquad (8.42)$$

The behavior of a particle on the surface of the drum is governed by a force balance.

The centrifugal force is given by:

$$F_C = \rho_p v_p \omega^2 R \qquad (8.43)$$

and the radial component of the gravitational force is:

$$F_{Gr} = \rho_p v_p g \sin \varphi \qquad (8.44)$$

The particle will detach from the surface of the drum when

$$F_C + F_{Gr} > F_r \qquad (8.45)$$

The critical rotation speed for detachment of a particle of magnetic volume V_m is obtained by equating the magnetic and centrifugal forces:

$$\omega_{Crit} = \left(\frac{2\pi V_m \mu_0}{\varphi_0 \rho_p v_p}\right)^{1/2} \frac{H_0}{R} \qquad (8.46)$$

Because the ratio V_m/v_p is independent of particle size, this model predicts that the separating action of the rotating drum magnetic separator is independent of particle size and this has been confirmed by experiment over a size range from about 1 mm to 6 mm.

The theoretical point of detachment of a particle magnetic volume V_m is given by

$$\sin \varphi_d = \frac{V_m f_m}{\rho_p v_p g} - \frac{\omega^2 R}{g} \qquad (8.47)$$

In theory the drum separator should recover to tailing all particles that have $\varphi_T < \varphi_d$ and should recover to middling all particles that have $\varphi_T < \varphi_d < 90°$.

In practice the behavior of real materials in the magnetic separator deviate from the predicted pattern for a variety of reasons. Particle–particle interactions have been neglected in the analysis and these can be significant in an industrial machine because of the need to achieve economically high production rates and that in turn dictates that the bed of particles on the drum must be several mm thick. This will translate to a bed several layers thick if the particles are smaller than about 1 mm. Significant variations in the magnetic field strength on the surface of the drum caused by serrations and grooves or other design features aimed at intensification of the gradients in the magnetic field are another significant cause of imperfect behavior. In fact the value of the magnetic force density will be distributed over a range of values within the particle bed on the surface of the drum and the model should reflect this.

Imperfect separation of the desired mineral also results from the imperfect correlation between particle composition and the magnetic susceptibility or magnetic volume of the particle. The conditional distribution density of V_m for particles of mineral grade g can be quite wide and this should be reflected in the model. If $p(f_m)$ is the distribution density for the magnetic force density within the particle bed on the surface of the drum and $p(V_m \mid g)$ is the conditional distribution of particle magnetic volume for particles of grade g, the yield to tailing is given by:

$$Y_T = \int_0^{\varphi_T} \int_0^\infty \int_0^\infty \int_0^\infty \sin^{-1}\left(\frac{V_m f_m}{\rho_p v_p g} - \frac{\omega^2 R}{g}\right) p(f_m)p(V_m \mid g)p(g)df_m dV_m dg\, d\varphi$$

$$(8.48)$$

8.5.2 Capacity of the separator

The tonnage handled by the rotating drum separator is given by:

$$\text{Tonnage} = \omega R \rho_b b \qquad (8.49)$$

ρ_b = bulk density of layer.
b = depth of bed.

Generally a bed depth equal to particle diameter is desirable but this might lead to capacities that are too low. Then a depth of several particle layers will be used. To increase tonnage, the drum should rotate at a rate close to the critical value for separation of the lowest grade particle that should report to the tailing.

Then

$$\text{Tonnage} = \rho_b b \left(\frac{\pi \kappa \mu_0}{\varphi_0 \rho (1 + \kappa N)}\right)^{1/2} H_0 \qquad (8.50)$$

which is independent of R!

8.6 Wet high intensity magnetic separation

Dobby and Finch (1977) have developed an empirical model for the operation of the wet high intensity magnetic separator. The authors demonstrate that recovery of magnetic particles in a whims is dependent on the magnetic susceptibility and size of each particle. The probability of capture is proportional to the ratio of the magnetic capture force and the hydrodynamic force that tends to dislodge the particle.

F_m = Magnetic force is proportional to $d_p^3 \rho \chi HG/d_w$

$$G = H \text{ if } H < H_s$$

or

$$G = H_s \text{ if } H > H_s$$

F_u = Hydrodynamic force is proportional to $d_p \mu u_f$

d_w is the diameter of the matrix wire and H_s is the saturation magnetic field strength within the matrix material. U is the average velocity of the slurry as it flows through the matrix.

The recovery of a particle of size d_p and magnetic susceptibility χ should be a function of the group

$$\frac{F_M}{F_H} = \frac{d_p^2 \rho \chi H^2}{\mu u} \tag{8.51}$$

Their experimental data are correlated better by the group

$$M_L = \frac{HG(\rho\chi)^{1.2} d_p^{2.5}}{u^{1.8} L_m^{0.8}} \tag{8.52}$$

where

$$L_M \text{ is the matrix loading } = \frac{\text{Mass fraction of magnetics in feed}}{\text{Mass of matrix}}$$

The recovery of particles of size d_p and volumetric magnetic susceptibility κ is modeled by

$$R_M(\chi, d_p) = 0.5 + B \log_{10}\left(\frac{M_L}{M_{50}}\right) \text{ for } M_{50}10^{-0.5/B} \leq M_L \leq M_{50}10^{0.5/B}$$

$$= 0 \quad \text{for } M_L < M_{50}10^{-0.5/B} \tag{8.53}$$

$$= 1.0 \quad \text{for } M_L > M_{50}10^{0.5/B}$$

M_{50} is the magnetic cut point at which the recovery is 0.5. M_{50} and B are parameters that are specific to a particular machine but independent of the material that is treated. Dobby and Finch found $B = 0.348$ and $M_{50} = 1.269 \times 10^{-2}$ for the machine that they tested.

The particle population in the feed to the magnetic separator will be distributed over a range of values of χ and d_p so that the total yield of solids to magnetics is given by

$$Y = \sum_i \sum_j R_m\left(\chi(g), d_p\right) p_{ij}\left(g \mid d_p\right) p_j(d_p) + \sum_j R_p(d_p) p_j(d_p) \qquad (8.54)$$

where $R(d_p)$ is the fractional retention of particles of size d_p by physical entrapment. In practice it is easier to measure the distribution of magnetic susceptibilities in a sample of the feed material directly in a laboratory experiment using the isodynamic separator or rotating drum separator and then to determine the average value of the grade g for each of the separated fractions. The distribution of particle grades in each magnetic fraction can be determined by image analysis techniques. In that case the yield to magnetics is given by

$$Y = \sum_k \sum_i \sum_j R_M\left(\chi, d_p\right) p_{kij}\left(\chi \mid g, d_p\right) p_{ij}\left(g \mid d_p\right) p_j(d_p) \qquad (8.55)$$

with

$$p_{kij}(\chi \mid g, d_p)\, p_{ij}(g \mid d_p) = p_{ikj}(g \mid \chi, d_p) p_{kj}(\chi \mid d_p) \qquad (8.56)$$

Bibliography

Magnetic separation methods are comprehensively surveyed in books by Svoboda (1987) and Gerber and Birss (1983).

References

Anderson, J.C. (1968) *Magnetism and Magnetic Materials.* Chapman and Hall Ltd, London, p. 24.

Dobby, G. and Finch, J.A. (1977) An empirical model of capture in a high gradient magnetic separator and its use in performance prediction. XIIth Int. Min. Proc. Congr., Sao Paulo, Brazil, Special Publication, Vol. I, pp. 128–152.

Gaudin, A.M. (1939) *Principles of Mineral Dressing.* McGraw-Hill, New York, p. 436 and p. 438.

Gerber, R. and Birss, R.R. (1983) *High Gradient Magnetic Separation.* Research Studies Press, Chichester.

Hopstock, D.M. (1975) Fundamental aspects of design and performance of low-intensity dry magnetic separators. *Trans SME* 258, pp. 221–227.

Hughes, E. (1956) *Fundamentals of Electrical Engineering.* Longmans Green and Co., London, p. 47.

Lawver, J.E. and Hopstock, D.M. (1974) Wet magnetic separation of weakly magnetic materials. *Minerals Sci. Engng,* Vol. 6, pp. 154–172.

Svoboda, J. (1987) *Magnetic Methods for the Treatment of Minerals.* Elsevier, Amsterdam.

Taggart, A.F. (1945) *Handbook of Mineral Dressing.* John Wiley and Son, pp. 13–17.

9
Flotation

Flotation is the most widely used mineral separation method. It is the preferred method of mineral recovery for many of the most important minerals that are recovered and large tonnages of ore are processed by flotation annually.

The underlying principles of the flotation process are well established but it has proved to be unusually difficult to build quantitative predictive models that can be used to simulate the operation of flotation cells in typical industrial circuits. The reason for the difficulty lies in the complexity of the many micro processes that combine to produce the overall result, which is the separation of different mineral species by virtue of the differential surface conditions that can be induced on the various minerals. In the flotation cell, an agitated slurry is aerated by introducing a cloud of air bubbles that are typically about a millimeter in size. The agitation of the slurry is sufficient to keep the solid particles in suspension although the suspension is usually not uniform, with the larger heavier particles tending to remain in the lower parts of the cell.

The rising bubbles can and do collide with the suspended solid particles and those particles that have appropriate surface characteristics can attach to a rising bubble and can therefore be carried upward, eventually reaching the surface of the slurry. Each bubble will have many encounters with particles during its rise through the slurry and a bubble can carry several particles to the top of the slurry.

A more or less stable layer of froth is maintained on the surface of the slurry. Particles that are attached to bubbles will tend to remain attached at air–water interfaces when the bubble enters the froth layer. Particles that are retained in the froth are recovered at the lip of the froth weir at the edge of the flotation cell. The recovery of the froth is accomplished by the natural mobility of the froth, which causes it to flow over the weir, and the recovery is sometimes assisted by mechanical paddles.

Within this general macroscopic view of the flotation process as a whole, several distinct micro processes can be identified. Each of these plays a role in determining how individual solid particles will respond while they are in a flotation cell. Prior to any flotation taking place, a number of detailed chemical processes occur on and exceedingly close to the surfaces of the particles. This is a chemical conditioning step and it is necessary in order to ensure that differential hydrophobicity is achieved with respect to the different mineral species that are present. Minerals that have strongly hydrophobic surfaces have a greater chance of recovery into the froth phase than particles that are weakly hydrophobic or hydrophilic. The chemical conditioning of mineral surfaces for flotation has been the subject of sustained research efforts for the best part of a century and a great deal is known about the surface chemistry and the role the chemical factors play in the aqueous phase and on the particle surfaces. However most of this understanding is qualitative with comparatively little information emerging that allows quantitative prediction of such important variables as contact angle and other measures of hydrophobicity. As a result, quantitative models of flotation cell performance do not at present make any significant use of quantitative chemical parameters such as for example the pH of the slurry and the concentration of chemical collectors or frothers to define overall process behavior. Hopefully the situation will change in the future and an understanding of the surface chemistry will play its appropriate role in the development of truly predictive quantitative models for flotation.

Much the same can be said for the froth phase. Although a great deal is known about the structure of froths and foams, it is not possible to make quantitative predictions about froth stability and froth mobility and their effect on the ability to hold and ultimately recover mineral particles that enter the froth attached to bubbles.

When formulating a quantitative model for the flotation process, it is necessary to start from the premise that, by virtue of appropriate chemical conditioning of the slurry, minerals will exhibit varying levels of surface hydrophobicity, and, consequently, a separation of particles is possible, based on the proportion of different mineral exposure on the surface. Likewise it is presumed that a more or less stable and mobile layer of froth will persist on the surface of the slurry that will gather and recover a proportion of the adhering mineral particles.

9.1 A kinetic approach to flotation modeling

Almost all successful models of the flotation process have been based on the premise that flotation is a kinetic process. In this way a model can be formulated in terms of a rate of flotation, which can be quantified in terms of some of the many chemical and physical factors that define the environment inside a flotation cell. The formulation of a suitable rate model is not straightforward and it must be based on an analysis of the individual subprocesses that affect an individual particle in the flotation environment. These subprocesses can be identified by noting that a particle must successfully complete the following steps in order to be recovered in the froth phase of a flotation cell:

1. The particle must achieve a level of hydrophobicity that will permit it to attach to a rising bubble.
2. The particle must be suspended in the pulp phase of the cell.
3. The particle must collide with a rising bubble.
4. The particle must adhere to the bubble.
5. The particle must not detach from the bubble during passage through the pulp phase.

6. The particle must not detach from the bubble as the bubble leaves the pulp phase and enters the froth phase.
7. The particle must not detach and drain from the froth during the passage of the froth to the weir.

In order to incorporate these subprocesses into a useful kinetic model it is necessary to consider the pulp phase and the froth phase separately because the kinetic processes that occur are different in each. The particles are considered to be in one of four possible states in the flotation cell. These states are illustrated in Figure 9.1. The four states are: suspension in the pulp phase, attached to the bubble phase, attached to the air–water interface in the froth phase and suspended in the Plateau borders of the froth phase.

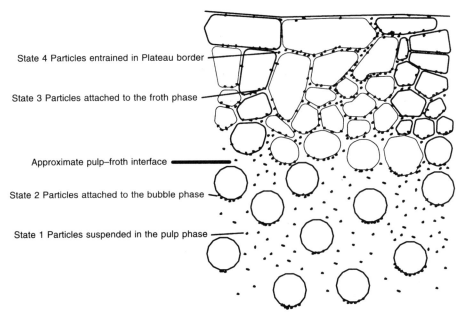

Figure 9.1 Four states in which particles exist in a flotation cell

9.1.1 *Pulp phase*

The pulp phase is aerated so that bubbles are formed continuously and rise through the pulp as an upward moving cloud. The solid particles are suspended throughout the pulp by the circulation that is induced by the agitator. At any point in the pulp the bubbles are moving upward relative to the pulp at their local rise velocity and the particles are moving downward at their local settling velocity. This does not mean that the bubbles are traveling vertically upward and the particles vertically downward because there is considerable lateral and vertical motion of the liquid, which is necessary to keep the particles suspended in the pulp. Nevertheless it is this relative velocity between particles and bubbles that is primarily responsible for collisions between particles and bubbles.

Particles can leave the pulp phase through one of two routes: by collision with and attachment to a bubble or direct to the froth phase by entrainment at the pulp–froth interface. The collision and attachment subprocesses are discussed in some detail later in this chapter and they govern, to a large extent, the kinetic behavior of the particles in a flotation cell.

Particles can re-enter the pulp phase by detachment from a bubble or by draining from the Plateau borders in the froth.

9.1.2 Bubble phase

The bubble phase consists of a cloud of bubbles that rise through the pulp phase. A particle transfers from the pulp phase to the bubble phase by a process of bubble–particle collision and particle attachment. Once a particle is attached to a bubble it moves with the bubble and eventually it will reach the froth phase unless it is detached from the bubble, in which case it returns to the pulp phase. Bubbles become steadily more and more heavily loaded with attached particles as the bubble passes through the pulp.

9.1.3 Froth phase

Particles enter the froth phase when a bubble carrying the particles crosses the pulp–froth interface. The froth phase floats on top of the pulp phase and is formed by the bubbles that break through the surface of the pulp phase. The water in the flotation cell is conditioned to promote the formation of a more or less stable froth. The bubbles do not burst when they cross the interface and each one carries a skin of water into the froth phase. The bubbles move close together, with a single film of liquid separating the individual bubbles. When the bubbles are small, these boundary films are curved. The liquid drains from the films, which become thinner and eventually break causing adjacent bubbles to coalesce. The bubbles become larger and larger and polyhedral in shape. At the junction of the polyhedral edges of the bubbles, Plateau borders form which act as a network of more or less tubular conduits through which the draining liquid can move downward. The particles that remain attached to the surface of bubbles in the froth move with the froth until they are eventually recovered in the concentrate froth when it passes over the froth lip of the flotation cell. Particles can become detached from the liquid film and transfer from the froth phase to the Plateau borders in the froth from where they can drain back into the pulp phase.

9.1.4 Entrained phase

Particles are entrained in the Plateau borders of the froth and will tend to settle down toward the pulp–froth interface under the influence of gravity. This settling is assisted further by the draining of the water as the films between bubbles in the froth become thinner and possibly break. This water flows through the network of Plateau borders and eventually re-enters the pulp phase. Particles enter the entrained phase by detachment from the air-water surfaces in the froth and also by direct entrainment from the pulp phase immediately below the pulp–froth interface on the top of the pulp.

9.2 A kinetic model for flotation

The kinetic model for flotation is based on the following simple principle. A single bubble rising upward through the pulp phase will collide with suspended particles and some of these particles will attach to the bubble surface and will travel upward with the bubble. The rate at which bubbles collide with particles depends on the size of the bubbles and the size of the particles, their relative velocities, and the concentration of particles in the pulp. The number of particles that can collide with the bubble can be calculated in the following way.

The volume of pulp that is swept by a bubble per second is $\frac{\pi}{4}D_{bh}^2 U_b$ where U_b is the local rise velocity of the bubble and D_{bh} the bubble diameter projected on to the horizontal plane. This is generally different to the volume equivalent bubble diameter D_{be} because a rising bubble distorts into an ellipsoidal shape unless it is very small.

A population balance approach is taken so that the population of particles in the cell is conceptually divided into classes so that particles in any one class are all similar in size and composition. The number of potential collisions with the particles of type ij is proportional to the concentration C_{ij} of these particles in the pulp phase. The diameter of the swept volume from which a particle of size d_{pi} may be captured is $D_{bh} + d_{pi}$. The number of potential collisions with particles of size d_{pi} is given by:

Number of potential collisions with particles of type

$$ij = \frac{\pi}{4}(D_{bh} + d_{pi})^2 (U_b + v_{Tij})C_{ij} \qquad (9.1)$$

where U_b is the velocity of rise of the bubble and v_{Ti} is the free fall velocity of the particle. These are taken to be the terminal velocities in the gravitational field in both cases, although in the turbulent circulating flow inside a flotation cell, terminal velocities may not be achieved by all particles and bubbles.

The rate of transfer of particles from the pulp phase to the bubble phase (transitions from state 1 to state 2 in Figure 9.1) is given by

Ideal rate of transfer = number of potential collisions

$$\times \text{ number of bubbles/unit volume}$$

$$= \frac{\pi}{4}(D_{bh} + d_{pi})^2 (U_b + v_{Tij})C_{ij} \times \frac{G_v \tau_b}{\frac{\pi}{6}D_{be}^3} \quad \text{kg/m}^3 \text{ of cell volume} \qquad (9.2)$$

where G_v is the specific aeration rate in m^3 air/s m^3 cell volume and τ_b is the average bubble residence time in the cell. D_{be} is the effective spherical diameter of the bubble.

In a real flotation cell, this ideal rate of transfer is never achieved because not all of the potential collisions actually occur and, of those that do occur, not all are successful in achieving adhesion between the bubble and the particle. The ideal rate is accordingly modified to reflect these inefficiencies.

Rate of transfer

$$= \frac{3}{2}(D_{bh} + d_{pi})^2 (U_b + v_{Tij}) C_{ij} \times \frac{G_v \tau_b}{D_{be}^3} \times E_{Cij} E_{Aij} (1 - E_{Dij}) \text{ kg/m}^3 \text{ of cell}$$

(9.3)

where E_{Cij} is the fraction of particles that are in the path of the bubble which actually collide with it. E_{Aij} is the fraction of bubble–particle collisions that lead to successful attachment.

E_{Dij} is the fraction of particles of type ij that are detached from the bubble during the time that it takes the bubble to rise through the pulp phase. The development of a quantitative kinetic model for the flotation process starts with models for E_{Cij}, E_{Aij} and E_{Dij}.

The subprocesses of collision, attachment and detachment can be considered to be independent since they are governed by essentially different forces. The collision process is dominated by the local hydrodynamic conditions around the bubble. The attachment process is dominated by the short-range surface forces and by the drainage and rupture characteristics of the thin liquid film between the bubbles and contacting particles. The detachment process is governed by stability considerations of the multi-particle aggregates that are formed on the bubble surface following successful attachment of the particles. These processes are largely independent of each other and, as will be shown below, require different models for their quantitative description.

The efficiency of bubble capture is reduced by the presence of other adhering particles which cover a portion of the bubble surface area. A necessary result of this kinetic model is that any bubble will become progressively covered with attached particles during its passage through the pulp phase. This behavior has been confirmed by direct experimental observation and is clearly shown in the photographic sequence in Figure 9.2. These photographs of a single bubble were taken to record the increasing load of particles during the lifetime of the bubble in the pulp. The essential elements of the collection of particles by bubbles are clearly illustrated in these photographs. The sweeping action of the fluid motion, which makes the particles slide over the bubble and accumulate on the lower surface, is strikingly evident. This behavior influences both the collision efficiency E_c and the adhesion efficiency E_A.

9.2.1 Particle–bubble collisions

Not every particle in the path of a rising bubble will collide with the bubble because, as the bubble advances through the water, it forces the water aside

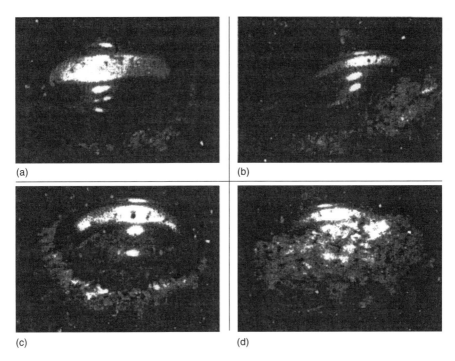

(a)   (b)

(c)   (d)

Figure 9.2 Loading of a single bubble during flotation. (a) After 3 seconds. (b) After 10 seconds. (c) After 30 seconds. (d) After 125 seconds (King *et al.*, 1974)

and this tends to carry the particles out of the path of and around the bubble. The processes that are at work can be seen in Figure 9.3, which shows the streamlines that are formed around a bubble as it advances through the pulp. The streamlines shown in Figure 9.3 were calculated for two extreme flow regimes: Stokes flow, which applies when the bubble Reynolds number is very much less than unity, and potential flow, which applies when the bubble Reynolds number is very much larger than unity. The streamlines show the trajectories that a small neutrally buoyant particle will take during the encounter with the bubbles. Such a particle will collide with the bubble only if it is on a streamline that has its closest approach to the bubble less than or equal to the radius of the particle. The streamlines shown in Figure 9.3 were calculated so that they are uniformly distributed a long way in front of the bubble. This means the two sets of streamlines are directly comparable and it is easy to see that the choice of flow model will have a significant effect on the value of E_c. Potential flow leads to much higher collision efficiencies than Stokes flow.

It can be seen from Figure 9.3 that a neutrally buoyant particle of diameter $d_p = 0.1\ D_b$ must be no more than about 17% of a bubble radius off the center line of the collision path if it is to make contact with the bubble surface during the flypast under Stokes flow. On the other hand a neutrally buoyant particle of the same size would touch the bubble surface if it were as far out as 56% of the bubble radius under potential flow conditions. Obviously the

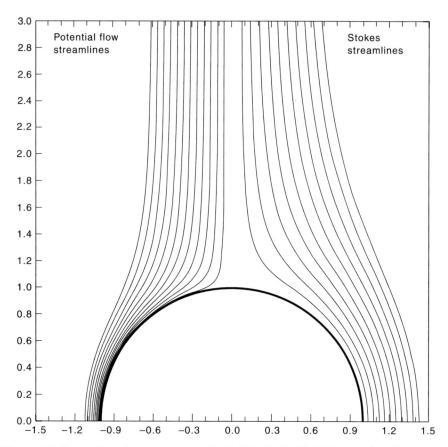

Figure 9.3 Streamlines for water around a rigid spherical bubble calculated assuming potential flow (left hand side) and Stokes flow (right hand side)

collision efficiency E_c for particles of this kind is quite small in either case but is significantly lower for Stokes flow conditions. This agrees with intuition since a faster moving bubble will sweep up particles in its path more effectively than a slow moving bubble.

The streamlines plotted in Figure 9.3 result from the analytical solutions of the continuity equations around solid spheres moving through Newtonian fluids. These solutions can be applied to the motion of a bubble through water provided that the surface of the bubble can be assumed to be immobile. This is usually assumed when analyzing flotation processes because of the relatively large quantities of surfactant that are available to concentrate at the bubble surface. These mathematical solutions under Stokes and potential flow conditions can be found in standard texts on fluid mechanics.

The streamlines are governed by the following equations.

$$\Psi = \frac{\psi}{U_b R_b^2} = \sin^2\theta \left(\frac{1}{2}\left(\frac{r}{R_b}\right)^2 - \frac{3}{4}\frac{r}{R_b} + \frac{R_b}{4r} \right) = \sin^2\theta \, \mathscr{F}_{St}(r) \qquad (9.4)$$

for Stokes flow and

$$\Psi = \frac{\psi}{U_b R_b^2} = \sin^2\theta\left(\frac{1}{2}\left(\frac{r}{R_b}\right)^2 - \frac{R_b}{2r}\right) = \sin^2\theta \mathscr{F}_{Pot}(r) \tag{9.5}$$

for potential flow. r and θ are the spherical spatial coordinates with the bubble center as origin. U_b is the velocity of rise of the bubble and R_b is the radius of the bubble.

Neither of the two extreme flow regimes, Stokes or potential flow, will be applicable under all conditions in operating flotation cells and a model is required that will hold for all values of the bubble Reynolds from low (Stokes regime) to high (potential flow). Using published streamline patterns for rigid spheres moving in fluids over a range of bubble Reynolds numbers as a guide, Yoon and Luttrell (1989) used a weighted sum of the Stokes and potential flow solutions to describe the streamlines around a sphere at Reynolds numbers in the range from 0 to 100.

$$\Psi = \sin^2\theta(\alpha \mathscr{F}_{st}(r) + (1 - \alpha)\mathscr{F}_{Pot}(r)) = \sin^2\theta \mathscr{F}(r) \tag{9.6}$$

with $\mathscr{F}(r)$ given by

$$\mathscr{F}(r) = \frac{1}{2}\left(\frac{r}{R_b}\right)^2 - \frac{3}{4}\alpha\frac{r}{R_b} + \left(\frac{3}{4}\alpha - \frac{1}{2}\right)\frac{R_b}{r} \tag{9.7}$$

α is a dimensionless parameter that depends on the bubble Reynolds number and has the value 1 when $Re_b = 0$ (Stokes regime) and asymptotes to the value 0 at high Re_b (potential flow). α also depends on the value of r. For values of r only slightly larger than R_b, α is close to unity.

α is related to the bubble Reynolds number by:

$$\alpha = \exp\left(-\frac{4Re_b^{0.72}}{45}\frac{r' - 1}{r'}\right) \tag{9.8}$$

where $r' = (r/R_b)$ and R_p is the radius of the particle.

The bubble Reynolds number $Re_b = (D_b U_b \rho_w / \mu_w)$ can be calculated from the terminal rise velocity given the bubble size and the load of solids that it carries (see Section 9.5).

9.2.2 A model for collision efficiencies

The collision efficiency is calculated as the fraction of the particles which are in the path of the bubble that actually collide with the bubble. In order to find which particles collide with the bubble, it is helpful to establish the streamlines for the particle motion since particles will deviate from the fluid streamlines that are shown in Figure 9.3. Particles deviate from the fluid streamlines because they fall relative to the fluid as a result of their greater density and because inertia prevents the particle from accelerating at the same rate as the fluid when the fluid deviates from a straight line path as it approaches a bubble.

The radial and azimuthal components of the fluid velocity are given by:

$$u_r = -\frac{1}{r^2 \sin \theta} \frac{\partial \psi}{\partial \theta} = -2 \frac{U_b R_b^2}{r^2} \cos \theta \mathscr{F}(r) \tag{9.9}$$

$$u_\theta = \frac{1}{r \sin \theta} \frac{\partial \psi}{\partial r} = \frac{U_b R_b^2}{r} \cos \theta \mathscr{F}'(r) \tag{9.10}$$

If the inertia of the particles is neglected, the particles move relative to the fluid at their terminal settling velocities, v_T. Terminal settling velocities can be calculated using the method described in Section 4.5. Then the components of the particle velocity vector are given by:

$$v_r = u_r - v_T \cos \theta \tag{9.11}$$

and

$$v_\theta = u_\theta + v_T \sin \theta \tag{9.12}$$

The stream function for the particle motion is therefore

$$\psi_p = U_b R_b^2 \sin^2 \theta \left(\mathscr{F}(r) + \frac{1}{2} \frac{v_T}{U_b} \left(\frac{r}{R_b} \right)^2 \right) \tag{9.13}$$

The streamlines for water calculated using Equation 9.6 and the trajectories of particles having terminal settling velocity equal to $0.1U_b$, calculated using Equation 9.13, are shown in Figure 9.4 for a bubble rising with a Reynolds number of 20.

The collision efficiency for particles of type ij is calculated as the flux of particles crossing an imaginary hemisphere of radius $R_b + R_p$ that shrouds the front hemisphere of the bubble divided by the flux of particles that cross a horizontal plane of area $\pi(R_b + R_p)^2$.

Therefore

$$E_{Cij} = \frac{-2\pi \int_0^{\frac{\pi}{2}} v_r|_{r=R_b+R_p} (R_b + R_p) \sin \theta (R_b + R_p) \, d\theta}{\pi(R_b + R_p)^2 (U_b + v_{Tij})}$$

$$= \frac{\dfrac{2U_b R_b^2 \mathscr{F}(R_b + R_p)}{(R_b + R_p)^2} + v_{Tij}}{U_b + v_{Tij}} \tag{9.14}$$

Collision efficiencies can be measured in the laboratory using single bubble flotation experiments with particles that are conditioned to be extremely hydrophobic. These particles have adhesion efficiencies of 1.0 and the measured collection efficiency is therefore equal to the collision efficiency. The collision

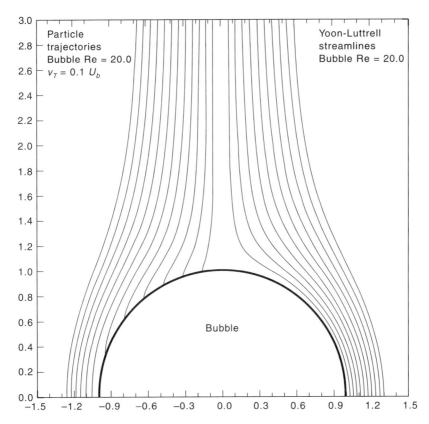

Figure 9.4 Streamlines for water (right-hand side) calculated using Equation 9.6 and streamlines for particles (left-hand side) having terminal settling velocity equal to $0.1U_b$ calculated using Equation 9.13

efficiencies predicted by Equation 9.14 are compared to experimental data in Figure 9.5. The data of Afruns and Kitchener (1976) were determined using quartz particles and that of Yoon and Luttrell (1989) using coal particles. The predicted efficiencies are close to the measured data confirming that the theory provides a reasonable model of the collision process although it over-estimates the collision efficiency by a factor of about 2 for bubbles approaching 1 mm in diameter.

Another important factor to note is that particles that do strike the bubble will do so at different latitudes on the bubble. This has important implications for the modeling of the attachment process, which is discussed in Section 9.2.3. The latitude angle at which the particle on streamline Ψ_0 strikes the bubble can be calculated using Equation 9.13 by setting $r = R_b + R_p$ and is given by:

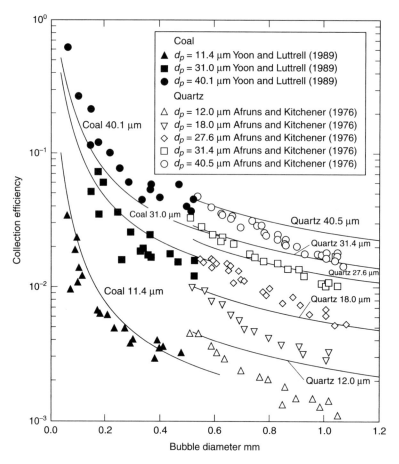

Figure 9.5 Comparison between measured and predicted collection efficiencies. The lines were calculated using Equation 9.14 for each of the two minerals

$$
\theta_c = \arcsin\left(\left(\frac{\Psi_0}{\mathscr{F}(R_b + R_p) + \frac{1}{2}\frac{v_T}{U_b}\left(\frac{R_b + R_p}{R_b}\right)^2}\right)^{\frac{1}{2}}\right) \tag{9.15}
$$

Particles of type *ij* can start their approach to a bubble at any distance R_o from the collision center line. Each particle of type *ij* will have a maximum distance R_{oij} from the center line that will result in a grazing trajectory. Providing that $R_o < R_{oij}$ a collision will occur. The collision angle will vary with R_o and since the location of each particle in the pulp is random, θ_c should be regarded as a random variable. Its associated probability distribution can be evaluated by noting that R_o is uniformly distributed in $[0, R_{oij}]$.

R_o can be related to ψ_{po} by noting that far out from the bubble

$$\sin \theta = \frac{R_o}{r} \tag{9.16}$$

so that Equation 9.13 gives

$$R_o = \left(\frac{2\psi_{po}}{U_b + v_{Tij}} \right)^{\frac{1}{2}} \tag{9.17}$$

and

$$R_{oij} = \left(\frac{2\psi_{poij}}{U_b + v_{Tij}} \right)^{\frac{1}{2}} \tag{9.18}$$

where ψ_{poij} is the value of the particle stream function for the streamline that represents the grazing trajectory for a particle of type ij at the equator of the bubble. This is evaluated using Equation 9.13 with $\theta = \pi/2$ and $r = R_b + R_{pij}$

$$\Psi_{oij} = \mathscr{F}(R_b + R_{pij}) + \frac{1}{2} \frac{v_{Tij}}{U_b} \left(\frac{R_b + R_{pij}}{R_b} \right)^2 \tag{9.19}$$

Substitution of Equations 9.17, 9.18 and 9.19 into Equation 9.15 gives a simple relationship between θ_c and R_o.

$$\sin \theta_c = \frac{R_o}{R_{oij}} \tag{9.20}$$

By noting the uniform probability distribution of the random variable R_o, the distribution density for the collision angle θ_c is:

$$f_{\theta_c}(\theta_c) = \cos \theta_c \tag{9.21}$$

The corresponding cumulative distribution function for θ_c is:

$$F_{\theta_c}(\theta_c) = \sin \theta_c \quad \text{for } 0 \le \theta_c \le \frac{\pi}{2} \tag{9.22}$$

9.2.3 Particle–bubble attachment

The attachment process requires significantly more complex modeling than the collision process, which, as shown in the previous section, is governed primarily by the fluid dynamics close to the bubble. The attachment process is governed by hydrodynamic and chemical factors which interact in complex ways that ultimately determine whether a particle will attach to the bubble or not.

When a particle collides with the bubble, the particle cannot immediately attach to the bubble because a thin film of liquid between the particle and the bubble must first drain. When the intervening film becomes sufficiently thin it can rupture allowing the particle to penetrate the skin of the bubble. The three-phase contact line that defines the penetration boundary of the bubble around the particle must then develop to a stable configuration that is governed primarily by the contact angle (both receding and advancing) of the solid and also by its shape. Although the actual rupture step is very rapid the film thinning and movement of the three-phase contact line are governed by kinetic processes and each requires a finite time. The time taken from the instant of collision to the establishment of a stable three-phase contact is called the induction time, which will be represented by t_{ind}. The induction time for a particle is determined primarily by its contact angle but the particle size and shape are also important. Other chemical factors such as the concentration of surfactants at the bubble surface and the interaction between collector adsorbed on the solid and frother on the bubble surface also play a role. Purely physical factors such as the precise orientation of the particle on first contact and the velocity profile close to the bubble surface and surrounding bubble all contribute to this enormously complex phenomenon. A detailed understanding of these effects has not yet been developed to the stage where induction times for irregular particles with heterogeneous surfaces can be confidently calculated. However, in general, particles with larger contact angles have shorter induction times than similar particles having smaller contact angles. This variation in induction time is the origin of the differential behavior of particles during flotation and consequently it must form the kernel of any quantitative model of flotation.

While the film thinning, film rupture and receding three-phase contact line are proceeding, another purely physical process is occurring. The particle is being carried downward over the surface of the bubble by the water as it moves past the bubble surface. Particles against or attached to the bubble are washed to the rear of the bubble. If a stable three-phase contact has been established before the fluid streamlines start to diverge from the bubble, successful attachment is achieved and the particle remains attached to the bubble and continues its journey over the bubble surface until it collides with other particles already attached to the bubble which covers its lower pole. The accumulating collection of particles on the lower surface gradually builds up until the whole of the lower hemisphere of the bubble is covered with adhering particles.

Particles that have not formed a stable three-phase contact by the time the streamlines start to diverge from the bubble surface at the equator are pulled away from the bubble surface and they do not attach. The time taken by a particle to slide over the bubble surface from its point of collision to the point of divergence is called the sliding time, t_s. The fundamental principle that governs the collision model of flotation is that a particle of size d_p and composition g will attach to a bubble of size R_b only if it experiences a successful collision and its induction time is less than the sliding time.

$$t_{ind} \leq t_s \qquad (9.23)$$

This should properly be referred to as the Sutherland principle in recognition of his pioneering attempt to put the understanding of flotation principles on a firm scientific basis. This model of the flotation process is illustrated in Figure 9.6.

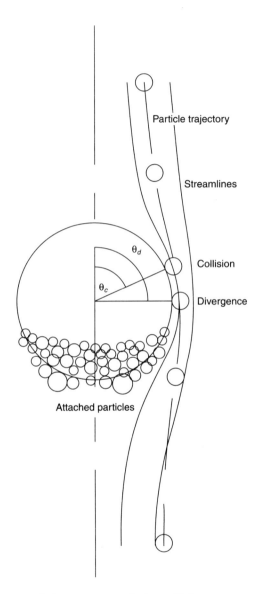

Figure 9.6 Collision and divergence angle for collisions

In order to use the principle embodied in Equation 9.23 to calculate the rate at which particles are captured by bubbles, it is necessary to calculate the sliding time and the induction time.

The sliding time for the particle can be calculated with reasonable accuracy from a knowledge of the hydrodynamics in the water close to the bubble surface. The tangential velocity in the water surrounding the bubble can be calculated using the stream function given in Equation 9.6:

$$u_\theta = \frac{1}{R_b \sin(\theta + \theta_p)} \frac{d\psi}{dr} = U_b \sin(\theta + \theta_p) \left(\frac{r}{R_b} - \frac{3}{4}\alpha + \frac{R_b^2}{2r^2} - \frac{3}{4}\frac{\alpha R_b^2}{r^2} \right) \quad (9.24)$$

where θ_p is the angle subtended at the bubble center by the radius of a particle that touches the bubble. The particle slides over the surface of the sphere approximately at the water velocity at one particle radius from the surface of the bubble, which for small particles is given by:

$$u_\theta = U_b \sin(\theta + \theta_p) \left[\frac{3}{2}(1 - \alpha + \alpha r_p') + \left(\frac{1}{2} - \frac{3}{4}\alpha \right) r_p'^2 \right]$$
$$= U_b^* \sin(\theta + \theta_p) \quad (9.25)$$

where

$$r_p' = \frac{R_p}{R_b} \quad (9.26)$$

$$U_b^* = U_b \left[\frac{3}{2}(1 - \alpha + \alpha r_p') + \left(\frac{1}{2} - \frac{3}{4}\alpha \right) r_p'^2 \right]$$

In addition the particle moves relative to the water at its terminal settling velocity so that the sliding time is:

$$t_s = \int_{\theta_c}^{\frac{\pi}{2}} \frac{R_b + R_b}{u_\theta + v_t \sin(\theta + \theta_p)} d\theta$$

$$= \frac{R_b + R_p}{U_b^* + v_t} \int_{\theta_c}^{\frac{\pi}{2}} \frac{d\theta}{\sin(\theta + \theta_p)} \quad (9.27)$$

$$= \frac{R_b + R_p}{2(U_b^* + v_t)} \ln \left[\frac{1 - \cos\left(\frac{\pi}{2} + \theta_p \right)}{1 + \cos\left(\frac{\pi}{2} + \theta_p \right)} \times \frac{1 + \cos(\theta_c + \theta_p)}{1 - \cos(\theta_c + \theta_p)} \right]$$

The maximum sliding time for a particle of size R_p is evaluated from Equation 9.27 at $\theta_c = 0$ and is given by:

$$t_{smax} = \frac{R_b + R_p}{2(U_b^* + v_t)} \ln \left[\frac{1 - \cos\left(\frac{\pi}{2} + \theta_p \right)}{1 + \cos\left(\frac{\pi}{2} + \theta_p \right)} \times \frac{1 + \cos\theta_p}{1 - \cos\theta_p} \right] \quad (9.28)$$

In industrial flotation machines it is not uncommon to find bubbles loaded so that a considerable fraction of their available surface area is covered by adhering particles. If this were not so, the recovery rate of floated material would be uneconomically low. The sweeping action on the adhering particles and the steady loading of the bubble surface from below is graphically illustrated in the sequence of photographs of a single bubble during its lifetime in a flotation pulp that is shown in Figure 9.2. When the layer of attached particles builds up past the equator of the bubble, the available sliding time is shortened as shown in Figure 9.7. When the bubble is more than 50% loaded, Equation 9.27 becomes

$$t_s = \frac{R_b + R_p}{2(U_b^* + v_t)} \ln\left[\frac{1 - \cos\theta_L}{1 + \cos\theta_L} \times \frac{1 + \cos(\theta_c + \theta_p)}{1 - \cos(\theta_c + \theta_p)}\right] \qquad (9.29)$$

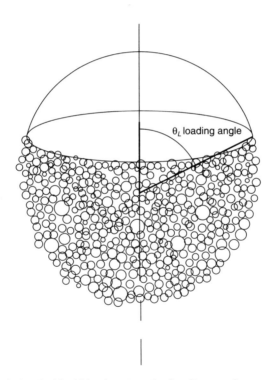

θ_L loading angle

Figure 9.7 Heavily loaded bubble showing the loading angle

L is the fractional loading on the bubble.

$$L = \frac{\text{Fraction of bubble surface covered by adhering particles}}{\text{Total bubble surface area}}$$

$$\qquad (9.30)$$

$$L = \frac{1}{2}(1 + \cos\theta_L)$$

θ_L is called the loading angle.

The corresponding value for t_{smax} is:

$$t_{smax} = \frac{R_b + R_p}{2(U_b^* + v_T)} \ln\left[\frac{1-L}{L} \times \frac{1 + \cos(\theta_L + \theta_p)}{1 - \cos(\theta_L + \theta_p)}\right] \qquad (9.31)$$

While it is possible to calculate the sliding time at least approximately for particles of arbitrary size and density when the bubble size is known, the same cannot be said for the induction time. The sliding time is governed primarily by physical factors and the hydrodynamics in the water phase close to the bubble. The induction time, on the other hand, is influenced strongly by the chemical conditioning of particle surface. In particular the contact angles, advancing and receding, play an important role. In addition the precise shape of the particle as well as the roughness of its surface and the mineralogical composition and texture of its surface all contribute strongly to the time that it takes for stable attachment to occur.

Three stages in the attachment process have been identified: thinning of the water film between the particle and the bubble, actual rupture of the bubble skin when the thinning film reaches an unstable condition and finally the retraction of the water from the particle surface to form a stable three-phase contact perimeter. Of these subprocesses only the first and third contribute to the induction time. The rupture of the film is much quicker and the rupture time is usually neglected by comparison to the time taken for film drainage and the formation of the stable three-phase contact perimeter.

In spite of considerable research effort, there are no quantitative models available that allow the calculation of the induction time for a specific particle. However, it can be assumed that the induction time is dominated by the particle characteristics such as advancing and receding contact angles, mineralogical texture of the surface, particle shape and surface roughness. The induction time is not significantly influenced by the hydrodynamic conditions of the collision nor by the bubble size. This is not completely true because the exact orientation of the particle relative to the bubble surface on first contact is known to influence the film drainage process. In spite of this it is postulated that the induction time is entirely a particle-specific constant. Each particle in the flotation cell has its own unique value of induction time, which remains the same for every collision that the particle makes with a bubble during its lifetime in the flotation environment. At least any variation in the induction time that a particular particle experiences from collision to collision will reflect only small statistical fluctuations around an average value.

This postulate of particle specificity for the induction time leads to an important conclusion relating to the kinetics of the overall flotation process. If any particle has an induction time larger than the sliding time associated with a direct center line collision with the largest available bubble, it will never float no matter how long it remains in the flotation cell. This gives rise to the concept of a non-floatable component for every type of particle. This behavior is commonly observed during kinetic studies on flotation systems and has been routinely applied to the analysis of flotation kinetics since at least the 1950s and probably much earlier than that.

Non-floatable particles will be found in virtually any flotation environment from precisely controlled Hallimond tubes through to large industrial flotation cells. The only exception will be in situations where the particles are conditioned to have very hydrophobic surfaces so that they all have induction times short by comparison with available sliding times. Under these conditions particle capture is dominated by the collision process and this is often exploited to measure the collision efficiency experimentally.

The collision and sliding phenomena that are described above are illustrated graphically in the sequence of photographs in Figure 9.2, which show a single bubble during its sojourn in a fairly typical flotation environment. These photographs show clearly that the particles slide over the front surface of the bubble and accumulate in the lower hemisphere. The close packing of the particles reflects their ability to form stable particle–particle agglomerates because of their hydrophobic character. The layer of packed particles gradually builds up on the bubble and eventually encroaches on the upper hemisphere where it reduces the sliding time as the load on the bubble increases. Once the particle layer crosses the equator of the bubble, the bubble loading starts to impact the rate of transfer of particles to the bubble phase and this must be taken into account when developing models based on the flotation kinetics.

9.3 Distributed rate constant kinetic model for flotation

The kinetic model for flotation is based on Equation 9.3. This equation shows clearly that the rate of transfer of particles of type ij from the pulp phase to the bubble phase is first order with respect to the concentration of these particles in the pulp phase. This observation leads to a simple formulation of the kinetic law as

$$\text{Rate of transfer of particles of type } ij = K_{ij}C_{ij} \tag{9.32}$$

where

$$K_{ij} = \frac{3}{2}(D_p + d_{pi})^2 (U_b + v_{Tij}) \times \frac{G_v \tau_b}{D_b^3} \times E_{Cij} E_{Aij} (1 - E_{Dij})$$

$$= \frac{1}{4}\left(1 + \frac{d_{pi}}{D_b}\right)^2 (U_b + v_{Tij}) \times S_v \times E_{Cij} E_{Aij} (1 - E_{Dij}) \tag{9.33}$$

$$= \kappa_{ij} S_v$$

and

$$S_v = \frac{6G_v \tau_b}{D_b} \quad \text{m}^2 \text{ bubble surface/m}^3 \text{ cell} \tag{9.34}$$

is the total bubble surface area in contact with the pulp phase per unit volume of the cell. It is easy to see from Equation 9.33 that K_{ij} has units of s^{-1}. K_{ij} is a complex function of particle size and particles density because of the complex dependence of E_{Cij} and E_{Aij} on the size of the particles. κ_{ij} is a mass transfer coefficient for transfer of particles from the pulp to the surface of the bubbles and it can be scaled independently of the bubble surface area. κ_{ij} has units of m/s.

Equation 9.33 appears to be a perfectly conventional linear kinetic law but the Sutherland principle imparts a special character to K_{ij}. A definite fraction of the particles of type ij (i.e. particles characterized by size d_{pi} and mineralogical grade g_j) have $K_{ij} = 0$ because $t_{ind} > t_{smax}$ for these particles and therefore $E_{Aij} = 0$. This is the non-floatable fraction for this particle type.

The size of the non-floatable fraction depends on the distribution of induction times across the population of particles that are characterized by size d_{pi} and mineralogical composition g. There is apparently no experimental data from which these distributions can be estimated although experimental techniques have been developed for the measurement of induction times. Since the induction time cannot be negative, it is reasonable to postulate that its value has a gamma distribution defined by the probability density function:

$$f_{ij}(t_{ind}) = \frac{t_{ind}^{\beta_{ij}-1} \exp\left(-\dfrac{t_{ind}}{\tau_{ij}}\right)}{\tau_{ij}^{\beta_{ij}} \Gamma(\alpha_{ij})} \tag{9.35}$$

which has two parameters β_{ij} and τ_{ij} that vary with particle size and particle composition.

The fraction of the particles in class ij that are non-floatable is usually specified in terms of the ultimate recovery, $\Re_{ij}$, of particles of this type

$$\Re_{ij} = \text{Prob}\{t_{ind} < t_{s\,max}\} \tag{9.36}$$

where $t_{s\,max}$ is the longest sliding time available and is given by Equation 9.28.

Using Equations 9.35 and 9.36, the ultimate recovery of particles of type ij is given by

$$\Re_{ij} = \frac{\Gamma_z(\beta_{ij})}{\Gamma(\beta_{ij})} \tag{9.37}$$

Where Γ_z is the incomplete gamma function and

$$z = \frac{t_{smax}}{\tau_{ij}} \tag{9.38}$$

The parameters α_{ij} and τ_{ij} in the distribution of induction times are related uniquely to the mean and variance of the distribution.

$$\tau_{ij} = \frac{\text{Variance}}{\text{Mean}} = \frac{\sigma_{ind}^2}{\bar{t}_{ind}} \tag{9.39}$$

$$\beta_{ij} = \frac{\text{Mean}}{\tau_{ij}} = \frac{\bar{t}_{\text{ind}}^2}{\sigma^2} \qquad (9.40)$$

$\bar{t}_{\text{ind}}$ is the average of the induction time distribution and σ_{ind}^2 its variance.

The specific flotation constant for the floatable component of particles of type ij in given by Equation 9.33 with the appropriate value of the attachment efficiency E_{Aij}.

The attachment efficiency for the floatable fraction of particles of type ij is given by

$$E_{Aij} = \text{Prob}\{t_s - t_{\text{ind}} > 0\} \qquad (9.41)$$

Equation 9.41 contains two random variables t_{ind} and t_s and the indicated probability can be evaluated from their respective density functions.

The sliding time t_s inherits its randomness from the collision angle θ_c which in turn inherits its randomness from the position of the particles relative to the axis of the bubble motion. θ_c has probability distribution function given by Equation 9.22 and distribution density function:

$$f_{\theta_c}(\theta_c) = \cos\theta_c \qquad (9.42)$$

t_s is related to θ_c by Equation 9.27 and its distribution density function is:

$$f_{t_s} = \frac{f_{\theta_c}(\theta_c)}{\left| \dfrac{dt_s}{d\theta_c} \right|} \qquad (9.43)$$

Thus

$$E_{Aij} = \int_0^{t_{s\max}} \int_0^{t_s} f_{t_s}(t_s) \times f_{ij}(t_{\text{ind}}) \, dt_{\text{ind}} \, dt_s$$

$$= \int_0^{\frac{\pi}{2}} f_{\theta_c}(\theta_c) \times F_{ij}(t_s) \, d\theta_c \qquad (9.44)$$

$$= \int_0^{\frac{\pi}{2}} \cos\theta_c \times F_{ij}(t_s) \, d\theta_c$$

where f_{ij} is the distribution density function for the induction time given by Equation 9.35 and F_{ij} is the corresponding distribution function. In practical applications the integral in Equation 9.44 must be evaluated numerically but this is comparatively easy to do.

9.4 Bubble loading during flotation

Up to now the discussion of the kinetic model for flotation has been presented from the point of view of the particles. Some additional considerations emerge when the process is viewed from the point of view of the bubble. As the bubble rises through the pulp phase, it adds to its load of particles and the lower hemisphere of the bubble becomes increasingly covered with accumulated attached particles as a result of successful collisions. The sequence of photographs in Figure 9.2 shows this accumulation process in graphic detail.

This loading process can be quantitatively modeled using the kinetic model of particle capture that is described in Section 9.3. As a bubble rises through the well-mixed pulp it collides with all types of particles, some of which are captured according to the model that has been developed. The different types of particle will be captured at different rates depending on their size and hydrophobicity which determine their specific flotation rate constants. The bubble accordingly accumulates a load that has a particle composition and size distribution that is distinctly different to the average particle composition of the pulp. However, during the lifetime of a particular bubble the pulp environment does not change so that the bubble becomes increasingly covered with a load of particles of size and composition distribution that does not change with bubble residence time.

The surface area of the bubble that is occupied by an adhering particle depends primarily on the shape of the particle but also depends on the packing density of the particles on the curved bubble surface. Hydrophobic particles form agglomerates in suspension even when no bubble surface is present and this phenomenon can increase the bubble surface loading. Evidence of these agglomerates can be seen in the photograph of a heavily loaded bubble in Figure 9.8. The lower hemisphere of the bubble is covered by several

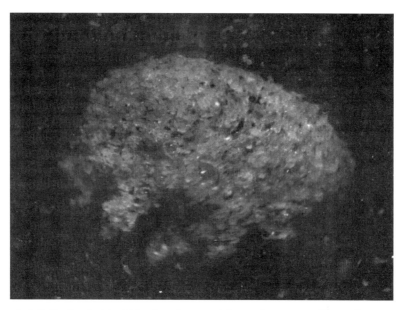

Figure 9.8 Fully loaded bubble showing particle agglomerates. Photo from King *et al.* (1974)

layers of particles where evidently not every particle is in contact with the bubble surface.

A model for bubble loading by particles on its surface can be based on the specific coverage achieved by each type of particle.

Let η_{ij} be the load per unit volume of bubble on a bubble fully loaded with particles of type ij

$$\eta_{ij} = \frac{(1 - \varepsilon_{ij})3(R_b + R_{pi})^2 R_{pi}\rho_{ij} + \eta_0}{R_b^3} \quad \text{kg/m}^3 \text{ bubble volume} \quad (9.45)$$

ε_{ij} is the packing voidage for particles of type ij, ρ_{ij} is the density of particles of type ij and η_0 is the load on a very small bubble that is composed largely of agglomerates of hydrophobic particles.

A fully loaded bubble in a flotation pulp carries a load given by

$$L_{\max} = V_b \sum_{ij} m_{ij}\eta_{ij} \quad (9.46)$$

where V_b is the volume of the bubble and m_{ij} is the mass fraction of particles of type ij in the bubble load.

$$L_{\max} = 4\pi \sum_{ij} (1 - \varepsilon_{ij})m_{ij}R_{pi}\rho_{ij}(R_b + R_{pi})^2 + \eta_0 \quad (9.47)$$

$L_{\max}$ can be measured by capturing single completely loaded bubbles and weighing the bubble load. For very small bubbles the maximum load is approximately independent of the bubble volume and for bubbles that are much larger than the particle size, $L_{\max}$ varies approximately as $V_b^{2/3}$. If the particles are all of the same size and density

$$L_{\max} = a(R_b + R_{pi})^2 + \eta_0 \quad (9.48)$$

Some data from captured fully loaded bubbles are given in Figure 9.9.

The development of the bubble load L is described by the following differential equation as the bubble moves through the pulp phase and its residence time increases:

$$\frac{6G_v\tau_b L_{\max}}{D_b} \frac{dL}{d\tau} = \sum_{ij} K_{ij}C_{ij} \quad (9.49)$$

with K_{ij} given by Equation 9.33 and τ representing the bubble residence time.

During the first few seconds of the loading process $L < 0.5$ and the right hand side of Equation 9.49 is constant and the bubble load increases linearly with bubble residence time. As the load on the bubble increases, the rate of loading decreases as a result of two effects. As the solid load carried by the bubble increases, its rise velocity decreases because of its increased mass. Since K_{ij} is proportional to U_b according to Equation 9.33, it decreases as the rise velocity decreases. When the load on the bubble exceeds 50%, K_{ij} decreases with increase in bubble load because the sliding time, and therefore E_{Aij},

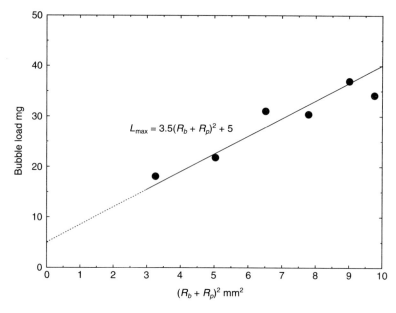

Figure 9.9 Measured load on fully loaded single bubbles. Data from King *et al.* (1974)

decreases according to Equation 9.27. The sliding time eventually becomes zero as the bubble approaches complete loading.

Under these conditions of inhibited flotation the bubble transports a load of particles to the froth phase that is limited either by the surface capacity of the bubbles or by the lifting power of the bubble. It should however always be considered in industrial flotation applications because economic considerations usually dictate that industrial flotation cells be operated at significant bubble loadings.

9.5 Rise times of loaded bubbles

In order to quantify the effect of bubble loading on the specific rate of flotation it is necessary to calculate the bubble rise velocity as a function of the load that is carried by the bubble. Light particles rise in water at terminal velocities that are lower than those calculated from the drag coefficient for heavy particles that sink. Deformable bubbles have terminal rise velocities that are still lower. Very small bubbles rise in water as though they behaved as rigid spheres having the density of air. Larger bubbles deform while they rise and become ellipsoidal in shape. This is evident in the photographs in Figure 9.2. Still larger bubbles have a spherical cap shape. The drag coefficient varies accordingly.

In spite of these anomalies, the terminal rise velocities of bubbles can be accurately calculated using the drag coefficient for lighter-than-water solid spheres provided that the bubble Reynolds number is calculated using the horizontal projection of the bubble diameter. The diameter of the horizontal projection of the bubble can be calculated from the effective volume diameter using the empirical relationship:

$$\frac{D_{be}}{D_{bh}} = \frac{1 + 0.60Ta^3}{1 + Ta^3} \tag{9.50}$$

where

$$Ta = Re_{be}Mo^{0.23} \tag{9.51}$$

$$Re_{be} = \frac{D_{be}U_b\rho_w}{\mu} \tag{9.52}$$

$$D_{be} = \left(\frac{6V_b}{\pi}\right)^{\frac{1}{3}} \tag{9.53}$$

and V_b is the bubble volume. *Mo* is the Morton number and is defined as

$$Mo = \frac{g\mu^4}{\rho\sigma^3} \tag{9.54}$$

where σ is the surface tension of the solution at the bubble surface.

The terminal rise velocity of a partly loaded bubble can be calculated from

$$
\begin{aligned}
U_b &= \left(\frac{8(V_b\rho_w - m_b)g}{C_D\pi D_{bh}^2\rho_w}\right)^{\frac{1}{2}} \\
&= \frac{D_{be}}{D_{bh}}\left(\frac{4(\rho_w - LL_{max})gD_{be}}{3C_D\rho_w}\right)^{\frac{1}{2}}
\end{aligned}
\tag{9.55}
$$

where C_D is the drag coefficient for a rising bubble.

Experimental data measured on rising bubbles in water containing surfactant reveals that C_d shows the same variation with bubble Reynolds number as solid lighter-than-water spheres provided that due allowance is made for the deformation of the bubble. The data is shown in Figures 9.10 and 9.11 where the measured drag coefficient is plotted against the bubble Reynolds number calculated using the horizontally projected bubble velocity

$$Re_{bh} = \frac{D_{bh}U_b\rho_w}{\mu_w} \tag{9.56}$$

The line in Figures 9.10 and 9.11 represents the drag coefficient for rising solid spheres and is given by:

$$
\begin{aligned}
C_D &= 0.28\left(1 + \frac{9.06}{Re_{bh}^{1/2}}\right)^2 \qquad &\text{for } Re_{bh} \leq 108 \\
&= 0.98 &\text{for } Re_{bh} > 108
\end{aligned}
\tag{9.57}
$$

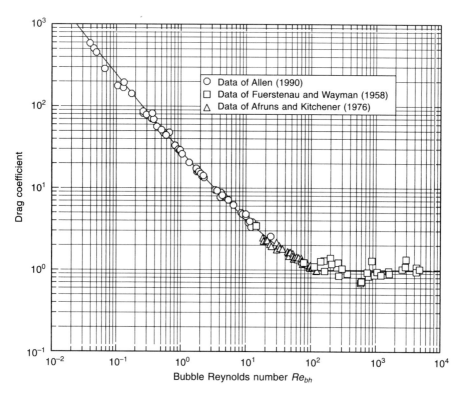

Figure 9.10 Drag coefficient for rising bubbles in viscous fluids. Experimental data from Allen (1900), Fuerstenau and Wayman (1958) and Afruns and Kitchener (1976). Solid line represents data for rising solid spheres

The corresponding equation for the line in Figure 9.11 is:

$$C_D = 0.28\left(\frac{(1 + 0.0921\Phi^{1/2})^{1/2} + 1}{(1 + 0.0921\Phi^{1/2})^{1/2} - 1}\right)^2 \quad \text{for} \quad \Phi < 1.14 \times 10^4 \qquad (9.58)$$

Figure 9.11 is useful because it can be used to calculate the terminal rise velocity of the bubble without trial and error methods because at terminal settling velocity.

$$C_D Re_{bh}^2 = \frac{4}{3}\frac{(\rho_w - LL_{max})\rho_w g D_{be}^3}{\mu^2} \qquad (9.59)$$

This can be calculated without knowing the rise velocity and the value of C_D can be obtained direct from Figure 9.11. This provides a method to calculate the terminal rise velocity of the bubble as its load of solid increases.

The application of these methods is described using the following illustrative examples.

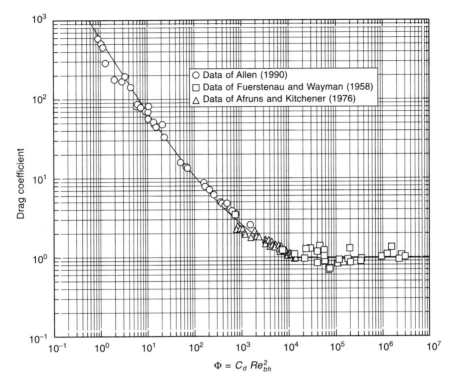

Figure 9.11 Drag coefficient for rising bubbles in viscous fluids. Experimental data from Allen (1900), Fuerstenau and Wayman (1958) and Afruns and Kitchener (1976). Solid line represents data for rising solid spheres. This coordinate system is most useful for the calculation of the terminal rise velocity of bubbles

---

Illustrative example 9.1

Calculate the terminal rise velocity of the following bubbles in a dilute aqueous solution of butanol having surface tension 0.052 N/m. The density of the water is 999.1 kg/m^3, and its viscosity 0.001 kg/ms.

(a) Small bubble of volume effective diameter = 0.5 mm.

$$\Phi = C_D Re_{bh}^2 = \frac{4}{3} \frac{\rho_w^2 g D_{be}^3}{\mu^2}$$

$$= \frac{4}{3} \frac{999.1^2 \times 9.81 \times (0.5 \times 10^{-3})^3}{(0.001)^2}$$

$$= 1.63 \times 10^3$$

The value of C_D can now be evaluated from Figure 9.11 or Equation 9.58

$$C_D = 0.28 \left(\frac{1 + 0.0921 \times (1.63 \times 10^3)^{1/2})^{1/2} + 1}{(1 + 0.0921 \times (1.63 \times 10^3)^{1/2})^{1/2} - 1} \right)^2$$

$$= 2.05$$

Bubbles of this small size will maintain a spherical shape while rising in water, thus try $D_{be}/D_{bh} = 1$ as a first estimate.

From Equation 9.55

$$U_b = \left(\frac{4 \times 9.81 \times 0.5 \times 10^{-3}}{3 \times 2.05} \right)^{1/2} = 0.056$$

Check the value of D_{be}/D_{bh}:

$$Mo = \frac{9.81 \times (0.001)^4}{999.1 \times (0.052)^3} = 6.98 \times 10^{-11}$$

$$Re_{be} = \frac{0.5 \times 10^{-3} \times 0.056 \times 999.1}{0.001}$$

$$= 28.0$$

$$Ta = 28.0 \times (6.98 \times 10^{-11})^{0.23}$$

$$= 0.129$$

From Equation 9.50

$$\frac{D_{be}}{D_{bh}} = \frac{1 + 0.6 \times 0.129^3}{1 + 0.129^3} = 0.999$$

which confirms the estimate.

The calculated value can be compared to the value of 0.042 m/s measured by Allen for a 0.48 mm diameter bubble in water.

(b) Intermediate bubble of effective volume diameter = 2 mm.

$$\Phi = C_D Re_{bh}^2 = \frac{4}{3} \frac{999.1^2 \times 9.81 \times (2 \times 10^{-3})^3}{(0.001)^2}$$

$$= 1.044 \times 10^5$$

Since $\Phi > 1.14 \times 10^4$, $C_D = 0.98$

A bubble of 2mm diameter in water is elliptical in shape, and the value of the ratio D_{be}/D_{bh} must be calculated. This requires some iterative calculation. Start the calculation at an intermediate value of $D_{be}/D_{bh} = 0.8$.

$$U_b = \frac{D_{be}}{D_{bh}} \left(\frac{4 \times 9.81 \times 2 \times 10^{-3}}{3 \times 0.98} \right)^{1/2}$$

$$= 0.8 \times 0.163 \text{ m/s}$$

The diameter ratio must now be refined.

$$Re_{be} = \frac{2 \times 10^{-3} \times 0.8 \times 0.163 \times 999.1}{0.001}$$

$$= 261$$

$$Ta = 261 \times (6.98 \times 10^{-11})^{0.23}$$

$$= 1.204$$

$$\frac{D_{be}}{D_{bh}} = \frac{1 + 0.6 \times 1.204^3}{1 + 1.204^3} = 0.746$$

Refining the value of U_b using this value for the diameter ratio gives

$$\frac{D_{be}}{D_{bh}} = 0.76$$

and

$$U_b = 0.76 \times 0.163 = 0.124 \text{ m/s}$$

This answer can be compared to the value of 0.136 m/s measured by Fuerstenau and Wayman (1958) for a 1.97 mm bubble in water.

(c) Large bubble having volume effective diameter = 12 mm.
Bubbles as large as this will have a spherical cap shape, and therefore C_D = 0.98 and the diameter ratio has the limiting value 0.6

$$U_b = 0.6\left(\frac{4 \times 9.81 \times 12 \times 10^{-3}}{3 \times 0.98}\right)^{1/2}$$

$$= 0.24 \text{ m/s}$$

This can be compared with the value 0.236 m/s measured by Fuerstenau and Wayman for a bubble having a volume effective diameter of 12.2 mm.

Illustrative example 9.2
Calculate the terminal rise velocity of a 0.5 mm diameter air bubble that is carrying 5 galena particles that may be taken as cubes of side 100 μm. The density of galena is 7500 kg/m^3.
The specific load carried by the bubble is

$$\frac{5 \times (100 \times 10^{-6})^3 \times 7500}{\frac{\pi}{6}(0.5 \times 10^{-3})^3} = 573.0 \text{ kg/m}^3$$

$$\Phi = C_D Re_{bh}^2 = \frac{4}{3} \frac{(999.1 - 573.0)999.1 \times 9.81 \times (0.5 \times 10^{-3})^3}{(0.001)^2}$$

$$= 6.96$$

$$C_D = 0.28 \left(\frac{1 + 0.0921 \times 696^{1/2})^{1/2} + 1}{1 + 0.0921 \times 696^{1/2})^{1/2} - 1} \right)^2$$

$$= 11.21$$

$$U_b = \left(\frac{4(999.1 - 573.0)9.81 \times 0.5 \times 10^{-3}}{3 \times 11.21 \times 999.1} \right)^{1/2}$$

$$= 0.016 \text{ m/s}$$

which is about $\frac{1}{3}$ of the rise velocity of an unloaded bubble.

Illustrative example 9.3
Calculate the specific flotation rate constant for 100 μm galena particles on 0.5 mm diameter bubbles in a continuous flotation cell that is aerated at 0.013 m^3 air/m^3 pulp. The bubble residence time is 5 s. Consider the case where the average induction time for the particles is 100 ms. Calculate the fraction of the non-floatable component for these particles. Consider the case when the bubble is carrying 5 galena particles $\sigma^2_{ind} = 90$.

Solution
Calculate the terminal settling velocity of the galena particles, which may be assumed to be cubic in shape. See illustrative example 4.4.

$$d_e = \left(\frac{6 v_p}{\pi} \right)^{1/3} = \left(\frac{6 \times 10^{-12}}{\pi} \right)^{1/3} = 1.241 \times 10^{-4} \text{ m}$$

$$\psi = \frac{\pi d_e^2}{a_p} = \frac{\pi (1.241 \times 10^{-4})^2}{6 \times 10^{-8}} = 0.806$$

$$v_T = 0.022 \text{ m/s}$$

From illustrative example 9.2, the bubble rise velocity = 0.030 m/s. Calculate E_c:

$$r'_p = \frac{R_p}{R_b} = \frac{0.05}{0.25} = 0.2$$

$$Re_b = \frac{D_b U_b \rho_w}{\mu_w} = \frac{0.5 \times 10^{-3} \times 0.030 \times 999.1}{0.001} = 14.9$$

From Equation 9.8

$$\alpha = \exp \left(-\frac{4 \times 14.9^{0.72}}{45} \cdot \frac{0.2}{1.2} \right)$$

$$= 0.901$$

$$U_b^* = 0.030\left[\frac{3}{2}(1 - 0.901 + 0.901 \times 0.2) + \left(\frac{1}{2} - \frac{3}{4} \times 0.901\right)0.2^2\right]$$

$$= 0.0124$$

$$E_c = 3(1 - \alpha)r_p' + \frac{3}{2}\alpha r_p'^2$$

$$= 3(1 - 0.901)0.2 + \frac{3}{2}0.901 \times 0.2^2$$

$$= 0.114$$

$$m_p = 5 \times (100 \times 10^{-6})^3 \times 7500$$

$$= 3.75 \times 10^{-8}$$

$$= 0.0375 \times 10^{-6}$$

$$V_b = \frac{\pi}{6}(0.5 \times 10^{-3})^3$$

$$= 6.545 \times 10^{-11}$$

Assuming the cubic particles pack closely together on the bubble, they occupy a total area = 5×10^{-8} m². The total surface area of the bubble is $\pi(0.5 \times 10^{-3})^2$ = 7.85×10^{-7} m².

$$\text{Fractional load on bubble} = L = \frac{5 \times 10^{-8}}{7.85 \times 10^{-7}} = 0.064$$

Since this is less than half load, t_{smax} is calculated from Equation 9.28.

$$\theta_p = \frac{R_p}{R_b + R_p}$$

$$= \frac{5.0 \times 10^{-6}}{0.25 \times 10^3 + 5.0 \times 10^{-6}}$$

$$= 0.167 \text{ radians}$$

$$t_{s\,max} = \frac{0.25 \times 10^{-3} + 50 \times 10^{-6}}{2(0.0124 + 0.022)}\ln\left[\frac{1 - \cos\left(\dfrac{\pi}{2} + 0.167\right)}{1 + \cos\left(\dfrac{\pi}{2} + 0.167\right)} \times \frac{1 + \cos(0.167}{1 - \cos(0.167)}\right]$$

$$= 4.35 \times 10^{-3} \times \ln[1.40 \times 142.8]$$

$$= 0.023 \text{ s}$$

E_A is evaluated by numerical integration of Equation 9.44.

$$E_A = \int_0^{\frac{\pi}{2}} \cos\theta_c \times F_{ind}(t_s)\, d\theta_c = 0.636$$

The specific rate constant for transfer from pulp to bubble phase can be calculated from Equation 9.33.

$$\kappa = \frac{3}{2}(1 + r')^2 (U_b + v_T) \times E_C \times E_A$$

$$= \frac{3}{2}(1 + 0.2)^2 (0.030 + 0.022) \times 0.114 \times 0.636$$

$$= 8.14 \times 10^{-3} \text{ m/s}$$

$$K = \frac{G\tau_b}{D_b} \kappa$$

$$= \frac{0.013 \times 5}{0.5 \times 10^{-3}} \times 8.14 \times 10^{-3} = 1.06 \text{ s}^{-1}$$

These calculated rate constants apply only to those particles in the pulp that have an induction time smaller than t_{smax}. The remainder make up the non-floatable component. The fraction of pyrite particles that are non-floatable is calculated from the Equation 9.37.

$$\beta = \frac{\text{Average induction time}^2}{\text{Variance of induction time}}$$

$$= \frac{(100 \times 10^{-3})^2}{90 \times 10^{-3}} = 0.111$$

$$\tau = \frac{90 \times 10^{-3}}{100 \times 10^{-3}} = 0.9 \text{ s}$$

$$z = \frac{t_{smax}}{\tau} = \frac{0.023}{0.9} = 0.0255$$

$$\text{Non-floatable fraction} = 1 - \frac{\Gamma_z(0.111)}{\Gamma(0.111)} = 0.299$$

These calculations can be repeated throughout the range of possible values of the parameter $\bar{t}_{ind}$ and for different particle loads on the bubble. The results are shown in Figure 9.12. The decrease in flotation rate constant with increase in the particle load is clearly evident in Figure 9.12. While the load is less than 50%, both the collision and attachment efficiencies remain constant because the colliding particle does not receive any interference from the adhering particles which are all on the lower hemisphere of the bubble. The decrease in the rate constant is due entirely to the reduction in the bubble rise velocity as the load of particles increases. Once the load exceeds 50%, the attachment efficiency decreases as the available free bubble surface on the upper hemisphere reduces and the rate constant decreases more rapidly. The rate constant reduces to close to zero when the load reaches a mass that the bubble is unable to lift and the number of collisions is reduced to a small number that is governed by the settling velocity of the solid particles in the pulp.

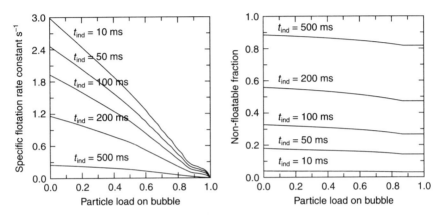

Figure 9.12 Calculated values of the specific flotation rate constant and the corresponding non-floatable fraction for 100 μm galena particles on 0.5 mm diameter bubbles. The parameter is the average of the induction time distribution for the particles

Since there are bubbles in all states of loading in the flotation cell, an average bubble loading must be calculated to fix the value of K_{ij} for the pulp volume as a whole. A method for doing this is described in connection with the simplified model in Section 9.8.

9.6 Particle detachment

In the previous sections it has been assumed that if a particle collides with and adheres to a bubble it will enter the froth phase when the carrying bubble breaks through the pulp–froth interface. This is not necessarily so because a particle can become detached from a bubble due to the effects of stresses that are induced by the turbulence in the flotation cell. The detachment process is complex and no comprehensive analysis is available. Two forces are dominant in causing a particle to become detached from a bubble: the weight of the particle and the inertia of the particle during the acceleration of the bubble that is induced by turbulent eddies in the fluid. The distortion of the bubble due to the weight of the particle is illustrated in Figure 9.13. The bubble is distorted and the three-phase contact line is pulled back over the particle until a position of equilibrium is established or until the bond between particle and bubble is broken. In the ideal situation that is illustrated in

Figure 9.13, the separating force F_g is balanced by the vertical component F_c of the surface tension force F_σ. If the separating force exceeds the maximum value that F_c can achieve, the particle will be detached. The maximum value of F_c is obtained as follows:

$$F_c = \sigma 2\pi R_p \sin\beta \, \sin\varphi \qquad (9.60)$$

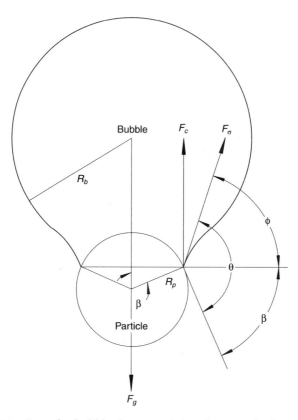

Figure 9.13 Distortion of a bubble due to weight of an attached particle. Adapted from Drzymala (1994)

This must be maximized subject to the condition that $\beta + \varphi = \theta$ where θ is the contact angle. Setting the derivative of F_c equal to zero shows that F_c is maximized when

$$\varphi = \beta = \frac{\theta}{2} \qquad (9.61)$$

$$F_{c\max} = 2\pi R_p \, \sigma \sin^2\left(\frac{\theta}{2}\right) \qquad (9.62)$$

The largest particle that can attach to the bubble in a quiescent pulp is given by

$$\frac{\pi}{6}d_{p\max}^3 g(\rho_s - \rho_w) = \pi d_{p\max}\sigma\sin^2\left(\frac{\theta}{2}\right)$$ (9.63)

$$d_{p\max} = \left(\frac{6\sigma}{g(\rho_s - \rho_w)}\right)^{\frac{1}{2}}\sin\left(\frac{\theta}{2}\right)$$ (9.64)

The additional forces caused by turbulence can detach a smaller particle and an approximate analysis by Woodburn *et al.* (1971) leads to the following model for the detachment efficiency.

$$E_D = \left(\frac{d_p}{d_{p\max}}\right)^{1.5}$$ (9.65)

9.7 The froth phase

9.7.1 Motion of the froth

The kinetic models for the collection of solid particles in the pulp are not adequate to describe the overall behavior of the flotation cell because the froth on the top of the cell exerts a strong influence on the kinetic behavior of the flotation system. The processes that take place in the froth phase are complex and they ultimately determine whether a particle that is captured by a bubble in the pulp phase will be recovered in the concentrate stream from the flotation cell. Two aspects of froth behavior must be considered: the motion of the froth from the surface of the pulp towards the upper surface of the froth and the concentrate overflow weir and the processes that occur within the froth while it is flowing. These processes include bubble bursting, particle detachment and drainage of water and solids from the froth. It is useful to consider the motion of the froth first and then superimpose the internal subprocesses on the basic flow pattern.

Bubbles, carrying their loads of attached particles, leave the pulp phase and pass into the froth phase at the pulp–froth boundary at the top of the pulp. The bubbles do not burst immediately and they congregate together and float on the surface of the pulp. The bubbles are pushed upward steadily by the arrival of new bubbles from the pulp phase. Provided that the bubbles are not too large, they retain their spherical shape as they pass from the pulp into the lower layer of the froth phase. They tend to arrange themselves in a hexagonal close-packed arrangement with the intervening spaces between the bubbles filled with liquid that is carried upward with the bubble from the pulp phase. This interstitial liquid entrains significant quantities of particulate material from the pulp and the entrained material can include significant quantities of hydrophilic non-floating particles. The composition of the entrained solids is determined primarily by the composition of the solid that is suspended in topmost regions of the pulp phase.

The interstitial water drains quite rapidly from the froth and the bubbles move closer together and deform steadily to take up polyhedral shapes as the films that separate any two bubbles become thinner and more or less flat. The few layers of spherical bubbles immediately above the pulp surface can usually be neglected when considering the froth phase as a whole unless the froth layer is very thin. As the interstitial water drains the structure becomes a true froth with a characteristic geometrical structure that is illustrated in two dimensions in Figure 9.14. Three lamellar films that separate the bubbles meet to form edges called Plateau borders that form a network of channels through which water can drain from the froth under gravity. The radius of curvature of the lamellar films is considerably larger than the radii of the walls of the Plateau borders, as illustrated in Figure 9.14. This means that the pressure inside a Plateau border is less than the pressure inside the laminar films to which it is attached. Consequently water drains from the lamellar films into Plateau borders and the films become steadily thinner. The plateau borders vary in length, cross-sectional area and orientation throughout the froth. The films that separate the bubbles contain all particles that were attached to a bubble when the bubble passed from the pulp phase into the froth.

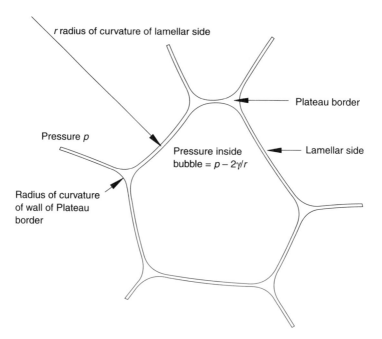

Figure 9.14 Geometrical structure of a single bubble in a froth shown here in two dimensions

The bubbles continue to move upward toward the top surface of the froth and laterally toward the concentrate overflow weir from which the particle-laden froth is recovered into the concentrate. Some of the bubbles reach the top surface of the froth where they burst and so never reach the concentrate overflow. The motion of the froth phase can be modeled as a two-dimensional potential flow problem. This is illustrated in Figures 9.15 and 9.16, which show a vertical cross-section through the froth phase on top of a typical flotation cell. The froth is assumed to be incompressible and therefore streamlines for the flow pattern can be calculated as constant values of the stream function ψ which satisfies the Laplace equation

$$\frac{\partial^2 \psi}{\partial x^2} + \frac{\partial^2 \psi}{\partial y^2} = 0 \tag{9.66}$$

subject to suitable boundary conditions. Because there is almost no macroscale turbulence in the froth, the streamlines represent the path taken by a bubble through the froth. Streamlines that terminate on the top surface of the froth represent bubbles that burst on the surface while streamlines that terminate on the vertical plane above the overflow weir represent bubbles that pass into the concentrate stream. It is not difficult to calculate the streamline trajectories for simple froth geometries such as those shown in Figures 9.15 and 9.16 because solutions to the Laplace equation for these geometries are readily available. Carslaw and Jaeger (1959) is a good source for these solutions. It is necessary only to specify the boundary conditions for the particular cell geometry that is to be studied.

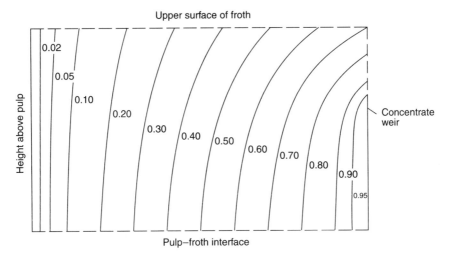

Figure 9.15 Streamlines in flotation froth calculated from Equation 9.72. The froth is relatively unstable with stability coefficient 1 - α = 0.7. Solid boundary lines are impermeable to air which can flow only through boundaries represented by broken lines

The boundary conditions are formulated in terms of the gas flows across the various boundaries of the froth. The gas flow across the pulp–froth interface

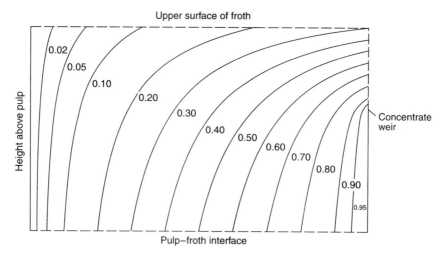

Figure 9.16 Streamlines in flotation froth calculated from Equation 9.72. The froth is relatively stable with stability coefficient $1 - \alpha = 0.3$. Solid boundary lines are impermeable to air which can flow only through boundaries represented by broken lines

is related to the aeration rate in the flotation cell G m^3/s. Consider a flotation cell with a rectangular cross-section and a side-to-side dimension w meters and a front-to-back dimension b meters. The superficial gas velocity through the pulp–froth interface is G/wb and the linear aeration rate is G/w per unit length of interface. The total gas flow through the interface that is bounded by the back wall of the cell and an imaginary line a distance x from the back wall is Gx/b. Thus, if the value of the stream function against the back wall is given the reference value 0, the value of the stream function at a distance x from the back wall is $\psi = (Gx/b)$. The stream function can be normalized as:

$$\Psi = \frac{\psi}{G} \tag{9.67}$$

and Ψ also obviously satisfies Equation 9.66. The dimensions of the froth are normalized with respect to the cell dimension b.

$$x' = \frac{x}{b}$$
$$y' = \frac{y}{b} \tag{9.68}$$

On the pulp–froth interface the boundary value for the stream function Ψ is given by $\Psi = x'$.

   Similarly the stream function is zero along the entire back wall because no air passes through that plane. No gas passes through the front wall beneath the overflow weir. An important feature of the pattern of streamlines in the froth is the fractional split of the gas flow between that contained in the concentrate froth and that which leaves the froth through the top surface where bubbles burst and release gas. If a fraction $1 - \alpha$ of the froth passes over the weir, the streamline that passes through the top right hand corner of the froth is associated with the stream function with normalized value α because this streamline separates the recovered froth from the froth that terminates its life by bursting on the top surface. Thus the boundary condition along the top surface of the froth is

$$\Psi = \alpha x' \tag{9.69}$$

and along the front wall

$$\Psi = 1 \quad \text{for } 0 < y \le Wh \tag{9.70}$$

and

$$\Psi = 1 - (1 - \alpha)\frac{y - Wh}{H - Wh} \quad \text{for } Wh < y < H \tag{9.71}$$

where H represents the total height of the froth layer and Wh represents the height of the weir above the pulp–froth interface.

The parameter α is determined largely by the stability of the froth. Stable froths tend to rise higher above the pulp thus forcing more froth over the weir into the concentrate. Unstable froth tends to break more easily and consequently more gas leaves the froth through the top surface. Lower values of α represent more stable froths. Obviously the mobility of the froth also affects the value of parameter α since greater mobility of the froth phase will allow more to be recovered over the weir.

Carslaw and Jaeger provide a convenient steady state solution to Equation 9.66 in a rectangular region with three sides having $\Psi = 0$ and an arbitrary function along the fourth side. The solution to the present problem can be obtained by superimposing three separate solutions each with the appropriate boundary function specified for the pulp–froth interface, the top surface of the froth or the front wall and concentrate overflow.

The general solution has the form

$$\Psi(x', y') = \Psi_1(x', y') + \Psi_2(1 - x', H' - y') + \Psi_3(y', 1 - x') \tag{9.72}$$

where

$$\Psi_i(x', y') = \sum_{n=1}^{\infty} a_n^{(i)} \sin n\pi x' \frac{1 - \exp(-2n\pi(H' - y'))}{1 - \exp(-2n\pi H')} \exp(-n\pi y') \tag{9.73}$$

represents the solution for boundary i and the coefficients $a_n^{(i)}$ for each solution are given by

$$a_n^{(1)} = -\frac{2}{n\pi}(-1)^n \qquad \text{(pulp–froth interface)}$$

$$a_n^{(2)} = \frac{2\alpha}{n\pi} \qquad \text{(upper surface of froth)} \tag{9.74}$$

$$a_n^{(3)} = \frac{2}{n\pi}\left(1 - \alpha(-1)^n + \frac{1-\alpha}{n\pi(1-Wh')}\sin(n\pi Wh')\right) \text{ (front edge)}$$

The contours for constant values of Ψ (the streamlines) calculated from Equation 9.72 are illustrated in Figures 9.15 and 9.16. These figures give a visual picture of the flow of the froth.

9.7.2 Dynamic processes in the froth phase

As individual bubbles in the froth move upward towards the free surface and laterally toward the overflow weir, a number of dynamic processes occur. The most obvious of these is the coarsening of the froth structure due to rupture of the lamellar films between the bubbles. A rupture of this kind gives rise to a sudden increase in the bubble volume and the froth structure becomes increasingly coarse. This phenomenon is easily observable on any flotation cell and there are a number of studies underway aimed at diagnosing the behavior of the froth from the size of bubbles in the top layer which can be measured fairly easily by image analysis techniques.

The rupturing of a film in the froth leads to the partial detachment of the hydrophobic particles that were present in the film before rupture. The water and particles contained in the ruptured lamellar film are transferred to the plateau borders that are connected to the film. It is reasonable to suppose that some of the particles will, however, remain attached at the air–water interface and will simply move to a neighboring lamellar film. Particles that have a high hydrophobicity (high contact angle) will have a greater tendency to remain attached while particles with lower hydrophobicity will tend to transfer to the Plateau borders. This is the origin of the upgrading that can occur in the froth, a phenomenon that has been observed in experiments.

The water in the Plateau borders together with its load of suspended solids drains under gravity through the network of Plateau borders so the froth becomes steadily denuded of both water and particles as it moves toward the overflow weir.

Hydrophilic particles that entered the froth by entrainment are unlikely to become attached to bubbles in the froth and they are free to drain from the froth with the water. Particles also settle relative to the water because of their greater density and this adds to the downward flux of particles in the Plateau borders.

Particles that are entrained in the water in the lamellar films that separate the bubbles do not drain from the lamellar films to the Plateau borders because they are hindered by other particles that are attached to the bubble surfaces. This restriction was first suggested by Gaudin who provided a sketch to illustrate the idea.

The rate of breakage of lamellar films is governed in a complex fashion by the surface tension of the fluid, the concentration and type of surfactants, the thickness of the film, the presence of particles, particularly those having high contact angles, the differential pressure across the film and the size of the bubble, and therefore the area of the film. Dippenaar (1982) has demonstrated that hydrophobic particles with high contact angles lead to fairly rapid film rupture but there is also evidence that hydrophobic particles with lower contact angles have a stabilizing effect on the froth. This can sometimes be observed in flotation plants where the froths produced in cleaner cells are often more stable than those produced in rougher or scavenger cells.

The dynamic behavior of the froth and its evolution with time is governed by three rate processes: the rate of drainage of water from the lamellar film into the Plateau borders, the rate of breakage of the lamellar films, and the rate of drainage of water and suspended solids from the Plateau borders. In addition, the compositions of the particle populations in the lamellar films and the Plateau borders at any point in the froth phase are governed by the differential rate of detachment of the different types of particle from the lamellar films into the Plateau borders.

These individual rate processes are difficult to observe and study in the laboratory so that almost no reliable information is currently available on appropriate model structures that can be used for their description in quantitative terms. It is however useful to formulate a simple model for each process and to superimpose these on the model of the froth motion that is described in Section 9.7.1. This leads to an overall kinetic model for the flotation process that captures the elements of the all important subprocesses that occur in the froth and which is capable of calibration against operating data. This model for the froth phase can be combined with the kinetic models for the pulp phase to generate a viable model structure that can be used for simulation of operating flotation cells and flotation plants.

9.7.3 Rate of breakage of lamellar films

The rate of breakage of the lamellar films that separate bubbles in the froth is a complex process that depends on many factors. The two internal properties of the froth that are most directly affected by the rupturing of the lamellar films are the surface area per unit volume of froth and the amount and composition of the solids that remain attached to the bubble surfaces. The amount of water and solid and the composition of the solid in the Plateau borders is also determined by the film breakage process because each rupture event adds both water and, selectively, solid particles to the Plateau borders.

The rate at which lamellar films break varies throughout the froth layer but the factors that determine the rate of breakage are not well enough understood at the present time to attempt anything but the simplest simulation of this subprocess and its effect on the performance of the flotation cell. Accordingly a simple model for the breakage of lamellar films is used here. A method is now developed to calculate amount and composition of the solid concentrate and the amount of water that is recovered over the weir.

The method of solution is based on the observation that individual bubbles pass through the froth entirely on a single streamline. Thus the dynamic changes that occur during the time that a bubble is in the froth can be evaluated by integration along a streamline. This can be illustrated by calculating bubble surface area per unit volume of froth from the time that a bubble enters the froth until it leaves in the concentrate or bursts on the surface of the froth. Let S_F represent bubble surface area per unit volume of froth. A differential balance on the surface area in the froth gives

$$u_x \frac{\partial S_F}{\partial x} + u_y \frac{\partial S_F}{\partial y} = -R_B \tag{9.75}$$

where u_x and u_y are the local components of the froth motion and R_B is the rate of breakage of lamellar films per unit volume of froth. The incompressibility of the froth requires

$$\frac{\partial u_x}{\partial x} + \frac{\partial u_y}{\partial y} = 0 \tag{9.76}$$

The solution to Equation 9.75 can be generated by integrating along a streamline since all streamlines are characteristic lines for this equation. The local components of the velocity vector can be obtained by differentiation of the stream function.

$$u_x = \frac{1}{w} \frac{\partial \psi}{\partial y} = \frac{G}{wb} \frac{\partial \Psi}{\partial y'} = \frac{G}{wb} u_x' = U_G u_x' \tag{9.77}$$

and
$$u_y = -\frac{1}{w} \frac{\partial \psi}{\partial x} = -\frac{G}{wb} \frac{\partial \Psi}{\partial x'} = \frac{G}{wb} u_y' = U_G u_y' \tag{9.78}$$

where $U_G = (G/wb)$ is the superficial gas velocity across the pulp–froth interface. Along a streamline

$$\frac{dS_F}{dy} = -\frac{R_B}{u_y} \tag{9.79}$$

which is easy to integrate numerically starting at $S_F = (6/D_b)$ at $y = 0$. D_b is the diameter of the bubble as it leaves the pulp phase. It is convenient to normalize S_F with respect to the surface area per unit volume of a single spherical bubble as it emerges from the pulp phase and R_B with respect to the surface area production rate in the flotation cell $6G/D_b$

$$S_F' = S_F \frac{D_b}{6} \tag{9.80}$$

$$R_B' = R_B \frac{D_b wb^2}{6G} \tag{9.81}$$

Then Equation 9.79 becomes

$$\frac{dS_F'}{dy'} = -\frac{R_B'}{u_y'} \tag{9.82}$$

with $S_F' = 1$ at $y' = 0$.

Each bubble burst alters the amount of solid that is attached to the bubble surfaces that bound the lamellar films because there is a strong tendency for the attached solid to become detached and to enter the water in the Plateau borders as suspended solid. Not all of the adhering solid will become detached and some will remain attached at the air–water interface and will slide along the surface and join the attached load on a neighboring lamellar film. This phenomenon is described by a solid attachment coefficient σ_{ij} which is the fraction of particles of type ij that remain attached during a bursting event. σ_{ij} depends strongly on the size and hydrophobicity of the particle. Highly hydrophobic particles have a greater tendency to remain attached and this is the primary mechanism by which the grade of mineral increases in the froth as the froth moves up from the pulp surface. This phenomenon has been observed in practice and is sometimes exploited to improve the grade of the concentrate in industrial flotation cells. The composition of the solid attached to bubble surfaces is governed by the equation

$$u_x \frac{\partial S_F \Gamma_{ij}}{\partial x} + u_y \frac{\partial S_F \Gamma_{ij}}{\partial y} = -(1 - \sigma_{ij})\Gamma_{ij}R_B \tag{9.83}$$

where Γ_{ij} is the amount of solid of type ij attached per unit area of bubble surface. σ_{ij} is the fraction of type ij that remains attached during the rupture of a single lamellar film. The streamlines are characteristics for Equation 9.83 and along a streamline

$$\frac{dS_F \Gamma_{ij}}{dy} = -\frac{(1 - \sigma_{ij})\Gamma_{ij}R_B}{u_y} \tag{9.84}$$

which must be integrated from the initial condition $\Gamma_{ij} = m_{ij}LL_{max}$ at $y = 0$. LL_{max} is the load carried by the bubbles when they leave the pulp phase and m_{ij} is the mass fraction of particles of type ij in the bubble load. Γ_{ij} is normalized by

$$\Gamma'_{ij} = \frac{\Gamma_{ij}}{m_{ij}LL_{max}} \tag{9.85}$$

and Equation 9.84 is normalized

$$\frac{d\Gamma'_{ij}}{dy'} = \sigma_{ij} \frac{\Gamma'_{ij}R'_B}{S'_F u'_y} \tag{9.86}$$

S'_F and Γ'_{ij} can be calculated along a characteristic streamline using Equations 9.82 and 9.86 provided that a model for bubble bursting rate can be formulated. The rate at which laminar films rupture is assumed here to be constant.

The residence time of any bubble in the froth is obtained by integrating along a streamline

$$\frac{d\tau'}{dy'} = \frac{1}{u'_y} \tag{9.87}$$

where τ' is the bubble residence time normalized with respect to the superficial velocity of the gas.

$$\tau' = \frac{\tau U_G}{b} \tag{9.88}$$

Equations 9.82 and 9.86 can be integrated simultaneously to give explicit expressions for S_F' and Γ_{ij}' throughout the froth in terms of the residence time as a parameter.

$$S_F' = 1 - R_B' \tau' \tag{9.89}$$

$$\Gamma_{ij}' = (S_F')^{-\sigma_{ij}} = (1 - R_B' \tau')^{-\sigma_{ij}} \tag{9.90}$$

These expressions are necessary for calculating the conditions in the network of Plateau borders throughout the froth as descibed in Section 9.7.4.

9.7.4 Froth drainage through plateau borders

Flow of slurry in the network of Plateau borders varies throughout the froth. The rate at which water and solids drain through the network is larger at the bottom of the froth because of the accumulated water and solids due to breakage of lamellar films as the froth flows toward the surface or the concentrate weir. The water and solids that drain from the lamellar films and which leaves the films because of the breakage mechanism that is described in Section 9.7.3 causes a steady increase in the flow through the Plateau borders which increase in cross-sectional area to accommodate the additional flow. The slurry that drains through the Plateau borders is driven by gravity and consequently moves relative to the moving froth.

The path that any element of draining slurry takes through the froth can be calculated from the streamlines established for the froth and therefore for the network of Plateau borders which moves integrally with the froth. The velocity at which the slurry drains relative to the network can be approximated by assuming laminar flow under gravity in the Plateau borders. A short segment of a channel within the network of Plateau borders might appear as shown in Figure 9.17. In a channel inclined at angle θ to the vertical the average velocity relative to the channel wall is:

$$V(\theta) = \frac{g\rho\cos\theta D_H^2}{32\mu} \tag{9.91}$$

where D_H is the hydraulic diameter of the channel. The Plateau borders are assumed to have no preferred orientation so that θ is uniformly distributed in the interval $[0, \pi/2]$. The average draining velocity is

$$\overline{V} = \frac{2}{\pi} \int_0^{\pi/2} \frac{g\rho D_H^2}{32\mu} \cos\theta \, d\theta$$

$$= \frac{g\rho D_H^2}{16\pi\mu} \tag{9.92}$$

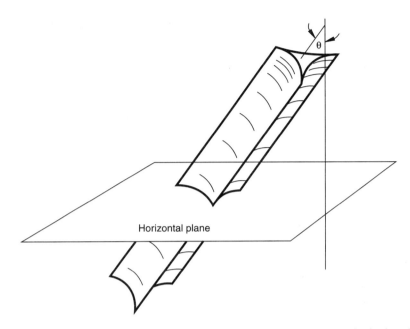

Figure 9.17 Segment in the network of Plateau borders through which the slurry drains under gravity

If D_H remains constant throughout the froth, $\overline{V}$ may be taken as constant and the components of the vector of draining water velocities in the froth is:

$$v_x = u_x$$

$$v_y = u_y - \overline{V}$$
(9.93)

The characteristic trajectories for draining slurry are defined by:

$$\frac{dx}{dy} = \frac{v_x}{v_y} = \frac{u_x}{u_y - \overline{V}}$$
(9.94)

which are integrated starting at $y = H$.

For convenience, Equation 9.94 can be written in terms of normalized variables:

$$\frac{dx'}{dy'} = \frac{u'_x}{u'_y - \overline{V}'}$$
(9.95)

where

$$\overline{V}' = \frac{\overline{V}}{\overline{u}_G}$$
(9.96)

A typical set of trajectories calculated from Equation 9.95 are shown in Figure

9.16. Trajectories that pass over the concentrate weir represent water and solids that are recovered by entrainment. Those trajectories that end on the pulp–froth interface represent slurry that is returned from the froth to the pulp phase.

The quantity of slurry that drains through the network of Plateau borders increases steadily from the top of the froth because the bursting bubbles continually add water to the drainage channels, which become increasingly larger in cross-section to accommodate the increase in the rate of drainage. If ε represents the volume of the Plateau borders per unit volume of froth, the differential volume balance for the draining slurry is

$$u_x \frac{\partial \varepsilon}{\partial x} + (u_y - \overline{V}) \frac{\partial \varepsilon}{\partial y} = \frac{1}{2} R_B \delta \tag{9.97}$$

where δ is the average thickness of the lamellar films in the froth. Equation 9.97 must be solved with boundary conditions specified along the top surface of the froth, which reflect the release of slurry due to the bursting of the bubbles on that surface. All lamellar films break that are associated with the gas that escapes through the top surface of the froth and the slurry produced enters the Plateau borders and starts to drain down the network of channels. Any wash water that is added to the top surface of the froth also adds to the flow at the upper surface.

$$\overline{V}\varepsilon = \frac{1}{2} S_F \delta \alpha U_G + \text{wash water rate} \quad \text{at } y = H$$

$$\varepsilon = 3 \frac{S_F' \delta' \alpha}{\overline{V}'} + \frac{\text{wash water rate}}{\overline{V}} \tag{9.98}$$

The drainage trajectories shown in Figure 9.18 are characteristic lines for the solution of Equation 9.97. Along any trajectory

$$\frac{d\varepsilon}{dy} = \frac{1}{2} \frac{R_B \delta}{u_y - \overline{V}} \tag{9.99}$$

In terms of normalized variables

$$\frac{d\varepsilon}{dy'} = \frac{3 R_B' \delta'}{u_y' - \overline{V}'} \tag{9.100}$$

where $\delta' = (\delta / D_b)$.

A differential balance for the concentration C_{ij} of solids of type ij in the slurry in the Plateau borders is

$$u_x \frac{\partial \varepsilon C_{ij}}{\partial x} + (u_y - \overline{V} - v_{Tij}) \frac{\partial \varepsilon C_{ij}}{\partial y} = (1 - \sigma_{ij}) R_B \Gamma_{ij} \tag{9.101}$$

where v_{Tij} is the terminal settling velocity in the narrow channels of particles of type ij. Boundary conditions are:

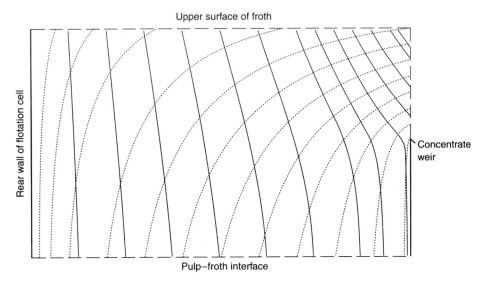

Figure 9.18 Drainage trajectories for slurry in the network of Plateau borders

$$C_{ij} = \frac{2\Gamma_{ij}}{\delta} \quad \text{at } y = H \tag{9.102}$$

Characteristic lines for this equation can be calculated by integrating:

$$\frac{dx'}{dy'} = \frac{u'_x}{u'_y - \overline{V}' - \dfrac{v_{Tij}}{U_G}} \tag{9.103}$$

These lines have the same general shape as the drainage characteristics but are somewhat steeper, becoming almost vertical throughout most of the froth for particles that settle rapidly.
    Along these characteristic lines

$$\frac{d\varepsilon C_{ij}}{dy} = \frac{(1 - \sigma_{ij})R_B\Gamma_{ij}}{u_y - \overline{V} - v_{Tij}} \tag{9.104}$$

In terms of normalized variables:

$$\frac{d\varepsilon C'_{ij}}{dy'} = \frac{(1 - \sigma_{ij})R'_B\Gamma'_{ij}}{u'_y - \overline{V}' - \dfrac{v_{Tij}}{U_G}} \tag{9.105}$$

with

$$C'_{ij} = \frac{C_{ij}D_b}{6m_{ij}LL_{max}} \tag{9.106}$$

Equation 9.105 must be solved after the surface concentration Γ_{ij} has been evaluated at every point in the froth using Equation 9.90.

The total concentration of particles of type ij in a sample drawn from any point in the froth is given by

$$\text{Froth sample concentration} = \varepsilon\, C_{ij} + S_F \Gamma_{ij} \quad \text{kg/m}^3 \qquad (9.107)$$

Samples can be drawn from different points in the froth on operating cells and compared with the model predictions using this equation.

It is now possible to evaluate the fractional recovery of particles of type ij particles by the froth. This quantity is called the froth transmission coefficient γ_{ij} and it is an important link between the pulp phase kinetics and the recovery of solids. Its use in a simulation model is discussed in Section 9.8.1. The froth transmission coefficient can be evaluated by integrating the flow of particles, both those attached to lamellar films and those entrained in the Plateau borders, in the streams that pass across the concentrate weir.

$$\gamma_{ij} = \frac{w \displaystyle\int_{Wh}^{H} (u_x S_F \Gamma_{ij})_{x=b}\, dy + w \displaystyle\int_{Wh}^{H} (u_x \varepsilon C_{ij})_{x=b}\, dy}{\dfrac{6}{D_b} g m_{ij} L L_{\max}} \qquad (9.108)$$

$$= \int_{Wh'}^{H'} u'_x (S'_F \Gamma'_{ij} + \varepsilon C'_{ij})_{x'=1}\, dy'$$

Likewise, the recovery of water to the concentrate stream can be calculated from:

$$\frac{Q_w^C}{G} = \int_{Wh'}^{H'} u'_x (3 S'_F \delta' + \varepsilon)\, dy' \qquad (9.109)$$

The calculated values of the froth transmission coefficient using Equation 9.108 are shown as a function of the adhesion coefficient σ_{ij} and the stability factor for the froth in Figure 9.19. The flowrate of water over the concentrate weir calculated from Equation 9.109 is given in Table 9.1.

Table 9.1 Calculated values of the recovery of water through the froth phase

Stability coefficient $1 - \alpha = 0.7$		Stability coefficient $1 - \alpha = 0.2$	
Breakage rate R_B $(\text{m}^2/\text{m}^3\ \text{s})$	Concentrate water flow $(\text{m}^3\ water/\text{m}^3\ air)$	Breakage rate R_B $(\text{m}^2/\text{m}^3\ \text{s})$	Concentrate water flow $(\text{m}^3\ water/\text{m}^3\ air)$
0.0	0.0326	0.0	0.0279
0.2	0.0292	0.5	0.0200
0.4	0.0259	0.7	0.0168
0.6	0.0225	0.9	0.0137

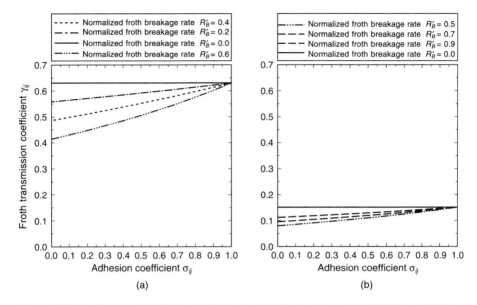

Figure 9.19 Froth transmission coefficient calculated using Equation 9.108 for a flatation froth having a weir height equal to 51.4% of the total froth height. (a) Relatively stable froth with $\alpha = 0.3$ and froth stability coefficient = 0.7 and (b) relatively unstable froth with $\alpha = 0.8$ and froth stability coefficient = 0.2

The data in Figure 9.19(b) shows that the value of the froth transmission coefficient can be quite small if the froth is not stable. These low values of γ_{ij} have often been observed in industrial flotation cells. The application of the froth transmission coefficient is described in Section 9.8.1. The froth transmission coefficient has an upper limit that is determined by the froth stability and the relative height of the weir. In practice the froth stability can be controlled by the nature and concentration of frother and the weir height can be varied by control of the pulp level in the flotation cell. These are control actions that are often exploited in industrial practice.

9.8 Simplified kinetic models for flotation

Many batch flotation tests that have been conducted in the laboratory have indicated that the kinetic model for flotation does describe the essential nature of the flotation process at least in a well stirred flotation environment where the solid particles are kept in suspension and are available for capture by the bubbles. The analysis presented in Section 9.2 indicates that the rate at which particles are captured on to the bubble surfaces is proportional to the concentration of the particles in the pulp phase. Particles of different type will have different specific capture rates primarily because of the variation of floatability due to variation in the contact angle exhibited by the particle surfaces. There is abundant evidence in the literature that the particle size also influences the rate of capture of particles. This is reflected in the model for the specific flotation constant K_{ij} that is described in Section 9.3 which leads to Equations 9.32 and 9.33 for the rate of transfer of particles from the pulp phase to the bubble phase.

$$\text{Rate of transfer of particles of type } ij = K_{ij}C_{ij} = \kappa_{ij}SC_{ij} \qquad (9.110)$$

In this equation the indices refer to the particle size and particle composition type respectively. K_{ij} and κ_{ij} are complex functions of the particle size, bubble size, bubble loading and of the attachment induction time which is apparently dominated by the contact angle that is exhibited by a particle in the environment of the flotation cell.

Many experimental studies, particularly in laboratory batch flotation cells, have shown that it is unusual to find K_{ij} or κ_{ij} to be constant in a single test. This is illustrated in Figure 9.20 where the fraction of mineral remaining as a function of time in a batch flotation test is shown. The mineral apatite was essentially completely liberated from the gangue minerals in this experiment so particles of mixed composition are not a significant factor in this case.

The batch test is described by:

$$\frac{dC_{ij}}{dt} = -K_{ij}C_{ij} \qquad (9.111)$$

which integrates to:

$$\frac{C_{ij}(t)}{C_{ij}(0)} = \exp(-K_{ij}t) \qquad (9.112)$$

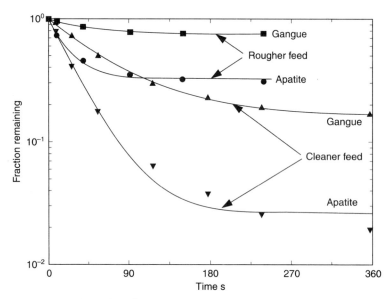

Figure 9.20 Typical results of batch flotation tests. Data from King (1978)

If the process were governed entirely by Equation 9.111 the data in Figure 9.20 would plot as a straight line according to Equation 9.112. It clearly does not and the specific rate of flotation of both minerals has decreased during the course of the test and both eventually stop floating. This kind of behavior can be described by assuming that a fraction of each mineral is non-floatable and has $K_{ij} = 0$. This fraction is related to the ultimate recovery by $\Re_{ij}$ and Equation 9.112 becomes:

$$\frac{C_{ij}(t)}{C_{ij}(0)} = \Re_{ij} \exp(-K_{ij}t) + (1 - \Re_{ij})$$

$$= 1 - \Re_{ij}(1 - \exp(-K_{ij}t)) \tag{9.113}$$

This is often written in terms of the recovery, R_{ij}, of the mineral species

$$R_{ij} = \frac{C_{ij}(0) - C_{ij}(t)}{C_{ij}(0)} = \Re_{ij}(1 - \exp(-K_{ij}t)) \tag{9.114}$$

If both K_{ij} and $\Re_{ij}$ in Equation 9.114 are regarded as adjustable parameters, this simple model can be made to fit the data well as can be seen for the data collected from the rougher feed in Figure 9.20. The lines on the graph show the best fit of Equation 9.114 to the data. Many such sets of data collected from a wide range of batch flotation cells can be found in the literature.

There are a number of issues raised by data such as that shown in Figure 9.20. Firstly the feed material had quite a broad size distribution as shown in Figure 9.21. The analysis presented in Section 9.2 shows that the collision and attachment probabilities, and therefore the specific flotation rate constant, are functions of the particle size. The curvature in Figure 9.20 could be due to the effect of the particle size distribution. This could occur because the particles having sizes favorable for flotation will float relatively rapidly leaving the slower floating sizes in the pulp with the consequent steady decline in the average specific rate of flotation. In fact this is the cause of the apparent lack of fit to the simple model shown for the cleaner feed. Another possible explanation for the steady decline in the specific rate of flotation is the possibility that the froth phase progressively loses its ability to transmit to the overflow lip the particles that are attached to bubbles. Factors such as a steady decline in frother concentration during the test contribute to this phenomenon, which is often observed as a steady decline in the froth collection rate during the course of a batch flotation test.

Some authors, notably Imaizumi and Inoue (1965), have postulated that the rate constant is distributed over a broad continuum of values. In practice this has rarely been found to be an effective model for describing actual flotation data mostly because the continuous distribution would have to be bimodal with a definite concentration around the value 0 representing the non-floatable components.

The differences exhibited by the two experiments shown in Figure 9.20 is instructive. The data labeled rougher feed represents the raw flotation feed

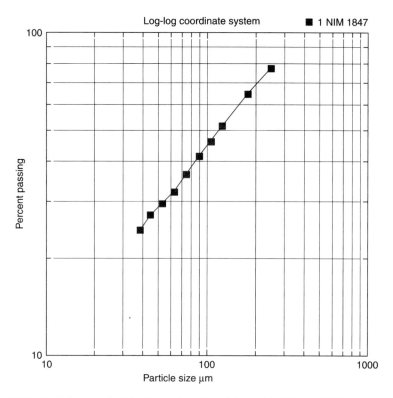

Figure 9.21 Particle size distribution of material tested in Figure 9.20

produced by milling. The data labeled cleaner feed shows the results of floating the accumulated concentrate from the rougher float. It is immediately obvious that the non-floatable component for both the mineral and the gangue is reduced considerably in the cleaner feed. In the case of apatite it is reduced to about 2%. This is consistent with the postulate that non-floatable material should never get into the concentrate. Any small amount in the rougher concentrate can be attributed to entrainment.

If this simple kinetic model is to be useful for practical application to real flotation systems, the effect of varying particle size must be included. A simple approach that has proved to be effective in allowing for the effect of particle size is to split the specific flotation rate constant K_{ij} into three factors using Equation 9.33 as a guide:

$$K_{ij} = k_j \Phi_j(d_{pi}) S_{av} \tag{9.115}$$

$\Phi_j(d_{pi})$ includes all the effects of particle size but it can also depend on the particle type. S_{av} is the available bubble surface area averaged over the entire bubble population in the flotation cell. k_j is a residual constant that is specific to the particle type and is independent of particle size.

The average available surface area S_{av} can be modeled using a method proposed by Pogorely (1962) who noted that the rate at which the available surface area on a bubble decreases can be modeled by Equation 9.49:

$$V_c A \frac{dS}{d\tau} = MAS \sum_j k_j \Phi_j (d_{pi}) p_{ij} = MAS\bar{k} \qquad (9.116)$$

where M is the mass of solids in the cell and p_{ij} is the fraction of the solids in the cell that is in composition class j and size class i. V_c is the volume of the flotation cell and A is the bubble surface area per unit volume.

The available surface area after the bubble has been in the cell for a time τ is obtained by integration of Equation 9.116

$$
\begin{aligned}
S &= \exp\left(-\frac{M\tau}{V_c} \sum_j k_j \Phi_j (d_{pi}) p_{ij}\right) \\
&= \exp\left(-\frac{M\tau\bar{k}}{V_c}\right)
\end{aligned}
\qquad (9.117)
$$

The average available area in the cell is obtained by averaging S over the interval from 0 to τ_b where τ_b is the time that a bubble spends in the cell.

$$S_{av} = \int_0^{\tau_b} S_v S d\tau = \frac{S_v V_c}{M\tau_b \bar{k}}\left(1 - \exp\left(-\frac{M\tau_b \bar{k}}{V_c}\right)\right) \qquad (9.118)$$

9.8.1 Application to flotation cells in complex flowsheets

The rate of transfer of particles of type ij to the froth phase in a perfectly mixed flotation cell is

$$\text{Rate of transfer} = k_j \Phi_j (d_{pi}) S_{av} M p_{ij} \quad \text{kg/s} \qquad (9.119)$$

A material balance for particles of type ij is:

$$
\begin{aligned}
W^F p_{ij}^F &= W^C p_{ij}^C + W^T p_{ij}^T \\
&= \gamma_{ij} k_j \Phi_j (d_{pi}) S_{av} M p_{ij}^T + W^T p_{ij}^T
\end{aligned}
\qquad (9.120)
$$

where γ_{ij} is the froth transmission coefficient for particles of type ij.

$$W^T p_{ij}^T = \frac{W^F p_{ij}^F}{1 + \gamma_{ij} k_j \Phi_j (d_{pi}) S_{av} \dfrac{M}{W^T}} \qquad (9.121)$$

The residence time of the solids in the cell is based on the flowrate of tailings

$$\theta_T = \frac{M}{W^T} \qquad (9.122)$$

which is usually approximated as

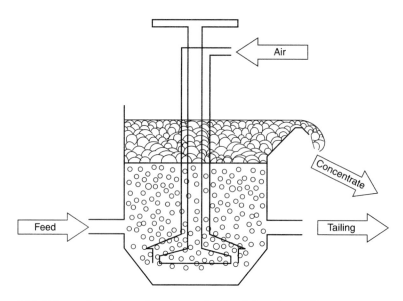

Figure 9.22 Schematic of a continuous well-mixed flotation cell

$$\theta_T = \frac{V_p}{Q^T} \tag{9.123}$$

where V_p is the volume of pulp in the cell and Q_T is the volumetric flowrate of the tailing stream equal to the sum of the water and solids volumetric flowrates.

$$Q^T = Q_s^T + Q_w^T$$

$$= \sum_{ij} \frac{Wp_{ij}^T}{\rho_{sj}} + Q_w^T \tag{9.124}$$

where ρ_{sj} is the density of the solid in composition class j.

Equation 9.121 becomes:

$$W^T p_{ij}^T = \frac{W^F p_{ij}^F}{1 + \gamma_{ij} k_j \Phi(d_{pi})S_{av}\theta_T} \tag{9.125}$$

which is the working equation used to simulate the operation of a single continuous flotation cell. This equation requires iterative solution because both S_{av} and θ_T depend on the solution p_{ij}^T through Equations 9.118, 9.123, 9.124 and 9.125.

In practice each particle type has more than one associated specific rate constant. Usually there are two – one for the floatable fraction and one, equal to zero, for the non-floatable fraction. Occasionally, more than two rate constants are associated with a particular type although this is unusual in practice. The discrete distribution of particle types is accordingly extended into a third dimension and we write p_{ijk} to represent the fraction of the particle population in size class i, composition class j, and flotation rate class k. There is a unique value of the specific rate constant k associated with each k class. When applying the model to a complete flotation circuit, it is postulated that any particles that enter the plant in flotation rate class k will remain in that class throughout their sojourn in the plant. It is possible that the value of the flotation rate constant that is associated with a specific k class will change its value at a specific point in the plant. For example the addition of collector at the head of a scavenger or cleaner bank will usually lead to an increase in the value of the rate constant for that bank of cells.

9.8.2 The effect of particle size on the rate of flotation

The function $\Phi_j(d_{pi})$ that is introduced into Equation 9.115 must account for the effects of particle size on the collision, attachment and detachment probabilities. A number of empirical expressions appear in the literature for the separate probabilities and a composite function that attempts to account for all three simultaneously is given by

$$\Phi_j(d_{pi}) = 2.33 \left(\frac{\varepsilon}{d_{pi}^2} \right)^{1/2} \exp\left(-\frac{\varepsilon}{d_{pi}^2} \right) \left(1 - \left(\frac{d_{pi}}{d_{pmax}} \right)^{1.5} \right) \tag{9.126}$$

This equation has two constants ε and $d_{p\,max}$. ε is related to the level of turbulence in the flotation pulp and $d_{p\,max}$ is the size of the largest particle that can be floated without detachment from the bubble. Although the theoretical foundation for this equation is only approximate, it has been found to be reasonably successful in describing the measured dependence of the flotation rate constant on particle size. The parameters in Equation 9.126 can be estimated from recovery-by-size data collect from a simple batch flotation test. Substituting K_{ij} from Equation 9.115 into Equation 9.112 gives:

$$R_{ij} = 1 - \exp(-k_j S_{av} \Phi_j(d_{pi}) t) \tag{9.127}$$

A typical set of data for galena is shown in Figure 9.23. The lines on the graph were calculated using Equation 9.127 with $\varepsilon = 112\ \mu m^2$ and $d_{p\,max} = 92\ \mu m$. In spite of the scatter in the measured data, the variation of recovery with particle size is reasonably well represented by Equation 9.127.

The measured recovery at very fine sizes should be corrected for the effects of entrainment which can account for a substantial fraction of the recovered particles at these sizes. This correction can be easily made for minerals that have no natural floatability. A blank test is run without collector so that all the recovery is due to entrainment. An example is shown in Figure 9.24. The measured entrained recovery was subtracted from the total recovery to generate

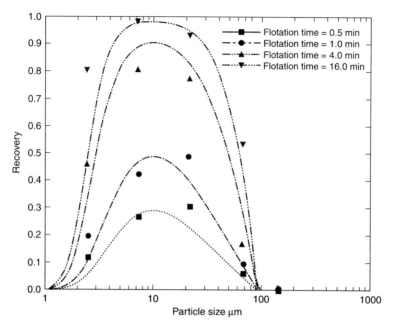

Figure 9.23 Recovery of galena in a batch flotation cell as a function of particle size. Data from Trahar (1976). Lines are calculated using Equation 9.127 with $\varepsilon = 112.5$ and $d_{p\,max} = 91.9$

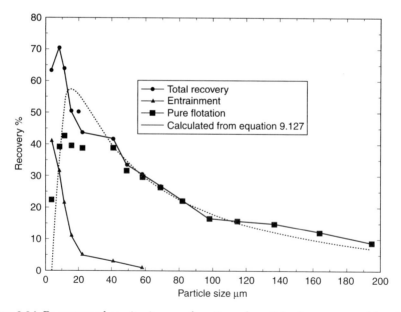

Figure 9.24 Recovery of cassiterite as a function of particle size measured in a batch flotation cell. Data from de Ruiter (1979). The broken line was calculated using Equation 9.127

the true flotation recovery, which is represented by square symbols in the figure. The true flotation recovery is compared to that calculated from Equation 9.127 in the figure. The function $\Phi_j(d_{pi})$ given by Equation 9.126 with $\varepsilon = 130$ μm^2 and $d_{p\,max} = 400$ μm fits the data well.

The parameter ε in Equation 9.126 is comparatively easy to estimate because it is related directly to the particle size, d_{pm} at which the recovery is a maximum through the equation

$$\varepsilon = 0.5 d_{pm}^2 \tag{9.128}$$

The constant 2.33 in Equation 9.126 normalizes the function so that $\Phi_j(d_{pm}) = 1.0$.

9.8.3 The water balance in an operating flotation cell

The water balance across the flotation cell can be established in a number of ways. The rate of water transfer by bubbles into the froth phase together with a model for froth breakage and drainage is the most reliable model for this purpose but the model described in Section 9.7 has not yet been developed to the stage where the parameters can be related for prediction purposes to the thermodynamic properties of the flotation environment. However the stability coefficient $1 - \alpha$, the froth breakage rate R_B and the lamellar film thickness δ can be estimated from measurements of the water recovery rate at different froth heights. Two alternative methods have been found to be useful in practice: the total solids content of either the concentrate or tailings stream can be specified.

If the solid content of the concentrate, s^C, is specified the calculation of the water balance is straightforward once Equation 9.125 has been solved. The volumetric flowrate of water in the concentrate is

$$Q_w^C = \frac{1 - s^C}{\rho_w s^C} \sum_{ijk} W^C p_{ijk}^C \tag{9.129}$$

where ρ_w is the density of the water. From Equation 9.119:

$$Q_w^C = \frac{1 - s^C}{\rho_w s^C} \sum_{ijk} \gamma_{ij} k_j \Phi_j(d_{pi}) S_{av} \theta^T W^T p_{p_{ijk}}^T \tag{9.130}$$

The water flow in the tailing stream is obtained by difference:

$$Q_w^T = Q_w^F - Q_w^C \tag{9.131}$$

If the solid content, s^T, of the tailing stream is specified:

$$Q_w^T = \frac{1 - s^T}{\rho_w s^T} \sum_{ijk} W^T p_{ijk}^T \tag{9.132}$$

and the water flowrate in the concentrate stream is obtained by difference:

$$Q_w^C = Q_w^F - Q_w^T \tag{9.133}$$

9.9 Symbols used in this chapter

b	Distance from front to back of flotation cell, m.
C_{ij}	Concentration of particles of type ij in the pulp or in the slurry entrained in the froth, kg/m^3.
C_D	Drag coefficient of a rising bubble.
D_{be}	Volume equivalent bubble diameter, m.
D_{bh}	Horizontal projection of bubble diameter, m.
d_{pi}	Particle size in class i, m.
$d_{p\,max}$	Largest particle that can remain attached to a bubble, m.
E_{Aij}	Attachment efficiency for particles of type ij.
E_{Cij}	Collision efficiency for particles of type ij.
E_{Dij}	Detachment efficiency for particles of type ij.
$\mathscr{F}$	Function that defines the effect of the radial coordinate on the streamline around a rising bubble.
$F_{\theta c}$	Distribution of particle collision angles.
$f_{\theta c}$	Distribution density for particle collision angles.
$F_{ij}(t_{ind})$	Distribution of induction times for particles of type ij.
$f_{ij}(t_{ind})$	Distribution density for induction times for particles of type ij.
G	Aeration rate in a flotation cell, m^3/s.
G_v	Specific aeration rate in flotation cell per unit volume of cell, s^{-1}.
g_j	Particle grade in grade class j.
K_{ij}	Specific flotation rate constant for particles of type ij, s^{-1}.
L	Fraction of bubble surface that is covered by adhering particles.
L_{max}	Load carried by completely covered bubble.
Q_s	Volumetric flowrate of solid, m^3/s.
Q_w	Volumetric flowrate of water, m^3/s.
r	Radial coordinate, m.
r'	$\dfrac{r}{R_b}$
r_p'	$\dfrac{R_p}{R_b}$
R_b	Radius of bubble, m.
R_B	Bubble bursting rate in the froth.
R_p	Radius of particle, m.
R_o	Distance from collision centerline far in front of bubble, m.
R_{oij}	R_o for particle of type ij having a grazing trajectory, m.
Re_b	Bubble Reynolds number.
$\mathfrak{R}_{ij}$	Ultimate recovery of particles of type ij.
S_{av}	Average available bubble surface area per unit volume of cell, m^{-1}.
S_F	Bubble surface area per unit volume of froth, m^{-1}.
S_v	Surface area of bubble per unit volume of cell, m^{-1}.
t_{ind}	Induction time, s.
t_s	Sliding time, s.
$\bar{t}_{ind}$	Average of the induction time distribution, s.
u	Velocity vector, m/s.

U_b	Rise velocity of bubble, m/s.
U_G	Superficial velocity of gas through the pulp–froth interface, m/s.
v	Velocity vector, m/s.
v_{Tij}	Terminal settling velocity of particles of type ij, m/s.
$\overline{V}$	Average velocity of draining slurry relative to the froth, m/s.
W	Mass flowrate of solid, kg/s.
Wh	Weir height, m.
x'	Normalized horizontal coordinate in the froth, x/b.
y'	Normalized vertical coordinate in the froth, y/b.
α	Parameter for froth stability
α_{ij}	Parameter for distribution of induction time for particles of type ij.
β_{ij}	Parameter for distribution of induction time for particles of type ij.
γ_{ij}	Froth transmission coefficient.
Γ_{ij}	Concentration of particles of type ij attached to bubble surface in the froth, kg/m^2.
δ	Thickness of lamellar film in froth, m.
η_{ij}	Packing density for particles of type ij on bubble surface, kg/m^3.
θ	Polar coordinate.
θ_c	Collision angle.
θ_L	Loading angle.
θ_p	Angle subtended by particle radius when particle touches the bubble surface.
θ_T	Residence time of pulp in flotation cell, s.
κ	Specific flotation rate constant, m/s.
μ_w	Viscosity of water, Pa s.
ρ_w	Density of water, kg/m^3.
σ	Variance of induction time distribution, s^2.
σ_{ij}	Fraction of particles of type ij that remain attached to bubble surface during a bubble burst event in the froth.
τ	Residence time in the froth measured along a froth streamline, s.
Φ	Dimensionless number $C_D Re_{bh}^2$.
$\Phi(d_p)$	Function to describe the variation of flotation rate constant with particle size.
Ψ	Normalized stream function.
ψ	Stream function.

Bibliography

The fundamental basis of the collision model for flotation kinetics is due to Sutherland (1948) and is discussed in detail in Sutherland and Wark (1955). This model has been widely discussed in the literature and many researchers have contributed to its development. Some of the key ideas are developed in Flint and Howarth (1971), Reay and Ratcliffe (1973) and Yoon and Luttrell (1989) and the work of these authors forms the basis of the model for collision efficiency that is presented here. Many of the hydrodynamic considerations of bubble motion are discussed in Clift *et al.* (1978). Schulze (1983) has presented a comprehensive analysis of many aspects of the model.

The model for the rise velocity of bubbles is based on the analysis of Karamanev and Nikolov (1992) and Karamanev (1994). Data on measured bubble rise velocities is given by Allen (1900), Fuerstenau and Wayman (1958) and Afruns and Kitchener (1976).

Tomlinson and Fleming (1965) investigated experimentally the effect of bubble loading on the rate of flotation and described its effects. Measured data on bubble loading is given by King *et al.* (1974) and Bradshaw and O'Connor (1996).

The distributed kinetic constant model has been widely discussed and used because laboratory batch experiments can almost invariably be described using the concept. The first formal statement that mineral particles showed the peculiar behavior of possessing a non-floatable component appears to be due to Morris (1952).

The effect of particle size on the flotation recovery is reviewed by Trahar and Warren (1976). The model for particle detachment was based on the presentation of Drzymala (1994).

Questions relating to the conditioning of the surfaces of the common types of minerals that are treated by flotation are discussed in King (1982). Some quantitative results are presented but the information is largely qualitative and therefore not directly applicable to model calculations and simulation.

The potential flow model for the froth phase is due to Moys (1978, 1984). Moys (1989) and Cutting (1989) have summarized measurements that have been made by taking samples from the froth of operating flotation cells. The model of the froth behavior presented here is based on Murphy *et al.* (1996), Neethling and Cilliers (1999) and Neethling *et al.* (2000) who have investigated the behavior of froths using experimental and simulation techniques. Their methods produce models of the behavior of the froth phase that are significantly more realistic than the simple model that is used here.

The discrete distributed flotation model originated with Zaidenberg *et al.* in 1964. The application of this model to the simulation of flotation plants is based on King (1973, 1975) and Sutherland (1977).

References

Afruns, J.P. and Kitchener, J.A. (1976) The absolute rate of capture of single particles by single bubbles. *Flotation*. Gaudin Memorial Volume (M.C. Fuerstenau, ed.). SME, pp. 625–637.

Allen, H.S. (1900) The motion of a sphere in a viscous fluid. *Philosophical Magazine*, Vol. 50, pp. 323–338.

Bradshaw, D.J. and O'Connor, C.T. (1996) Measurement of the sub-process of bubble loading in flotation. *Minerals Engineering*, Vol. 9, pp. 443–448.

Carslaw, H.S. and Jaeger, J.C. (1959) *Conduction of Heat in Solids*. Oxford University Press, Oxford.

Clift, R., Grace, J.R. and Weber, M.E. (1978) *Bubbles, Drops and Particles*. Academic Press.

Cutting, G.W. (1989) Effect of froth structure and mobility on plant performance. *Mineral Processing and Extractive Metallurgy Review,* Vol. 5, pp. 169–201.

de Ruiter, M.A. (1979) Particle size effects in the flotation of cassiterite. MSc. Thesis University of the Witwatersrand, Johannesburg.

Dippenaar, A. (1982) The destabilization of froths by solids. *International Journal of Mineral Processing,* Vol. 9, pp. 1–22.

Drzymala, J. (1994) Characterization of materials by Hallimond tube flotation. Part 2: Maximum size of floating particles and contact angle. *Intl Jnl Mineral Processing,* Vol. 42, pp. 153–168.

Flint, L.R. and Howarth, W.J. (1971) The collision efficiency of small particles with spherical air bubbles. *Chemical Engineering Science,* Vol. 26, pp. 115–168.

Fuerstenau, D.W. and Wayman, C.H. (1958) Effect of chemical reagents on the motion of single air bubbles in water. *Trans. AIME,* Vol. 211, pp. 694–699.

Imaizumi, T. and Inuoe, T. (1965) Kinetic considerations of froth flotation. *Proc. 6th Intl. Mineral Processing Congress* (A. Roberts, ed.), Pergamon Press, Oxford.

Karamanev, D.G. (1994) Rise of gas bubbles in quiescent liquids. *AIChE Journal,* Vol. 40, pp. 1418–1421.

Karamanev, D.G. and Nikolov, L.N. (1992) Free rising spheres do not obey Newton's Law for free settling. *AIChE Journal,* Vol. 38, pp. 1843–1846.

King, R.P. (1973) Model for the design and control of flotation plants. *Applications of Computer Methods in the Mineral Industry* (M.D.G. Salamon and F.H. Lancaster, eds). S. Afr. Inst. Min. Metall. Johannesburg, pp. 341–350.

King, R.P. (1975) Simulation of flotation plants. *Trans. AIME,* Vol. 258, pp. 286–293.

King, R.P. (1978) A pilot-plant investigation of a kinetic model for flotation. *Jnl South African Inst. Min. Metall.,* Vol. 78, pp. 325–338.

King, R.P. (ed.) (1982) *Principles of Flotation.* S. Afr. Inst. Min. Metall. Johannesburg.

King, R.P., Hatton, T.A. and Hulbert, D.G. (1974) Bubble loading during flotation. *Trans. Instn Min. Metall.,* Vol. C, pp. C112–C115.

Morris, T.M. (1952) Measurement and Evaluation of the rate of flotation as a function of particle size. *Trans. AIME,* Vol. 193, pp. 794–797.

Moys, M.H. (1978) A study of a plug flow model for flotation froth behavior. *Intl Jnl Mineral Processing,* Vol. 5, pp. 21–38.

Moys, M.H. (1984) Residence time distribution and mass transport in the froth phase of the flotation process. *Intl Jnl Mineral Processing,* Vol. 13, pp. 117–142.

Moys, M.H. (1989) Mass transport in flotation froths. *Mineral Processing and Extractive Metallurgy Review,* Vol. 5, pp. 203–228.

Murphy, D.G., Zimmerman, W. and Woodburn, E.T. (1996) Kinematic model of bubble motion in a flotation froth. *Powder Technology,* Vol. 87, pp. 3–12.

Neethling, S.J. and Cilliers, J.J. (1999) A visual kinematic model of flowing foams incorporating coalescence. *Powder Technology,* Vol. 101, pp. 249–256.

Neethling, S.J., Cilliers, J.J. and Woodburn, E.T. (2000) Prediction of the water distribution in a flowing foam. *Chem. Engng Science,* Vol. 55, pp. 4021–4028.

Neethling, S.J. and Cilliers, J.J. (2000) A model for free solids motion in froths. Private communication.

Pogorely, A.D. (1962) Limits of the use of the kinetic equation proposed by K F Beloglazov. *Izv. Vuz. Tsvetnaya Metaurgia,* No. 1, pp. 33–40. Available in English as NIM-TR-158 from Mintek. Randburg, S. Africa.

Reay, D. and Ratcliffe, G.A. (1973) Removal of fine particles from water by dispersed air flotation: effect of bubble size and particle size on collision efficiency. *Canadian Journal of Chemical Engineering,* Vol. 51, pp. 178–185.

Schulze, H.J. (1993) *Physico-chemical Elementary Processes in Flotation.* Elsevier, Amsterdam.

Sutherland, D.N. (1977) An appreciation of galena concentration using a steady-state flotation model. *Intl Jnl Mineral Processing*, Vol. 4, pp. 149–162.

Sutherland, K.L. (1948) Physical chemistry of flotation XI. Kinetics of the flotation process. *Jnl Physical Chemistry*, Vol. 52, pp. 394–425.

Sutherland, K.L. and Wark, I.W. (1955) Principles of Flotation. Australasian Institute of Mining and Metallurgy. Melbourne.

Tomlinson, H.S. and Fleming, M.G. (1965) Flotation rate studies. *Proc. 6th International Mineral Processing Congress* (A. Roberts, ed.). Pergamon Press.

Trahar W.J. (1976) The selective flotation of galena from sphalerite with special reference to the effects of particle size. *Intl Jnl Mineral Processing*, Vol. 3, pp. 151–166.

Trahar, W.J. and Warren, L.J. (1976) The floatability of very fine particles – A review. *Intl Jnl Mineral Processing*, Vol. 3, pp. 103–131.

Woodburn, E.T., King, R.P. and Colborn, R.P. (1971) The effect of particle size distribution on the performance of a phosphate flotation process. *Metallurgical Transactions*, Vol. 2, pp. 3163–3174.

Yoon, R.H. and Luttrell, G.H. (1989) The effect of bubble size on fine particle flotation. *Mineral Processing and Extractive Metallurgy Review*, Vol. 5, pp. 101–122.

Zaidenberg, I. Sh., Lisovskii, D.I. and Burovoi, I.A. (1964) One approach to the construction of a mathematical model for flotation processes. *Soviet Journal of Non-ferrous Metals*, English translation, Vol. 5, pp. 26–32.

10
Simulation of ore dressing plants

10.1 The nature of simulation

Many references appear in the literature to computer simulation as a technique for design and analysis of mineral processing operations. It has been a fashionable field of research since the 1960s and a great deal of good work has been done to make simulation into a viable and practical tool. Simulation is any procedure that can be used to model a process without actually running it. There are several ways in which a simulation can be achieved, but by far the most effective is by digital computer. The computer is programmed to mimic the behavior of the actual plant and can provide a description of what the plant will do and how it will perform under a variety of circumstances. This is a useful thing to do because a simulation can expose many aspects of plant performance without the inconvenience of operating the plant itself under experimental conditions. Although a good computer simulation can provide information about the behavior and performance of a mineral processing plant, it must be used with circumspection and care.

Simulation of a complex engineering system is only possible once a detailed understanding of each component of the system has been achieved and simulation provides the engineer with a tool for the prediction of system behavior even if the system does not exist in reality. However the simulator predictions can only be as good as the basic understanding of the component parts. The modeling techniques that are described in earlier chapters have intrinsic interest in their own right in that they attempt to build a quantitative picture of the operating behavior of the equipment in terms of the underlying physical and chemical principles that govern their operation. These models provide a focus for research efforts that are aimed at the development of a more complete and accurate understanding of the detailed mechanisms and processes that underlie the behavior of the unit operations. Improved understanding at this level contributes to the technical evolution of process equipment and to the development of new and better unit designs. The focus of this book has been directed towards an additional goal in that the models have been developed specifically to provide the building blocks for use in a process plant simulator. The models have all been developed with this particular application in mind. To be useful and effective in a simulator the models must all fit seamlessly together so that the simulator can function in the intended way. Different models for the same unit operation must also be interchangeable to facilitate their comparison under comparable operating conditions. The computer is an essential component of simulation for two reasons: in most systems of interest to the mineral processing engineer, the individual unit operations are so complex that they can be usefully described in mathematical terms only if these can be translated into computer code; in addition the systems of interest reveal complex interactions and interconnections among the individual units that must be accurately accounted for. In many cases these complex interactions cannot be described adequately by purely mathematical methods and the ability of the computer to transfer information from one program model to another is exploited effectively to simulate the transfer of actual material, information or energy in a real system. Purely mathematical formulations of complex systems are effective only when the systems are linear in the mathematical sense. Then the full power of linear and matrix algebra can be brought to bear on the problem. Regrettably most systems of real interest are rarely linear and it is necessary to use the heuristic capabilities of the computer to take the place of purely mathematical descriptions.

An ore dressing plant simulator is a set of computer programs that provides a detailed numerical description of the operation of an ore dressing plant. The simulator must be provided with an accurate description of the ore that is to be processed, a description of the flowsheet that defines the process and an accurate description of the operating behavior of each unit operation that is included in the flowsheet. The simulator uses these ingredients to provide a description of the operating plant. The detailed description of the ore will include information on its physical and mineralogical characteristics and a method to provide this information is described later. The flowsheet is the familiar graphical representation of the location of the unit operations in the plant together with the network of pipes and conveyors that transmit material between the units. The description of the operating behavior of the unit operations is based on the models that have been developed in earlier chapters for the individual equipment types. The simulator links together the modeled behavior of each of the unit operations and synthesizes the overall performance of the plant.

Four fundamental concepts underlie the construction of an ore dressing plant simulator.

- Ore dressing plants are collections of unit operations connected by process flow streams that transmit material from one unit to the next. The flow of materials is directed by the flowsheet structure.
- Each unit operation processes its own feed materials and will separate it or transform it in accordance with the specific objective of the unit.
- The behavior of the plant as a whole depends on the operating characteristics of each of the unit operations as well as on the nature of the material that is processed in the plant.
- A simulator reduces the actual plant operations, as defined by the flowsheet

structure and the behavior of the units, to a sequence of logical mathematical functions. The simulator can then mimic the real plant performance.

These basic concepts are independent of the precise nature of any particular plant that must be simulated and they lead to the development of simulation software that can be used for all possible plant configurations. The availability of such general purpose software makes computer simulation a useful practical tool in everyday engineering. It is a difficult task to write the necessary computer code to simulate a complex ore dressing plant. Most engineers have neither the time, inclination nor skill to do so and it would not be cost effective to write the code for each application. The cost in man hours to generate the code and debug it so that it can run reliably would be enormous. Computerization of any complex engineering systems is a highly specialized task and this is true also in mineral processing and such activities should be attempted only by specialists. The simulation system that is included on the compact disc that is included in this book is identified by the name MODSIM. This simulator was developed as a result of the success that was achieved with a flotation plant simulator that was developed as part of the research program of Mintek's Chemical Engineering Research Group in the Department of Chemical Engineering of the University of Natal, Durban, South Africa (King, 1972). It was the successful application of this flotation plant simulator to industrial flotation plants (King, Pugh and Langely, 1973) that demonstrated the potential for simulation techniques and that led to the development of MODSIM as a simulator for ore-dressing plants of arbitrary configuration and complexity and which contained any of the operations that are associated with ore dressing.

The simulation engine in MODSIM was developed as a Ph.D project by M.A. Ford (1979) who wrote the Fortran code for the loop finding, circuit decomposition and sequential calculation algorithms. This code still forms the core of MODSIM and it has required only minor modifications to keep it compatible with modern compilers and other requirements for code maintenance. This long run of successful computation provides MODSIM with a stable computational core that can be relied on to perform well even in the face of the extraordinary wide variety of plant configurations that can be simulated. The models for the unit operations are independent of the inner core of the simulator and these evolve continually in response to research endeavors. Consequently new models are frequently added to MODSIM and older models are improved whenever a new development warrants a change. This evolutionary process will continue well into the future.

The graphics user interface has evolved continually as new software systems and computing platforms have become available. The original version of the interface was presented at the 13th CMMI Conference in Johannesburg in 1982. Although MODSIM was developed originally for use on large mainframe computers, the preferred platform is now the ubiquitous desktop PC. The user interface is built to modern software engineering standards and it provides an intuitive environment that is designed to make the simulator easy and convenient to use.

Several general-purpose simulators for ore dressing plants are now available and of these MODSIM offers the greatest versatility to the user to modify and adapt the models of the unit operations that are used by the simulator. The standard models that are provided in the package rely on the methods that are described in this book. MODSIM is particularly strong in the modeling of mineral liberation phenomena. The underlying theme for the models that are used in MODSIM is the population balance method, which has been emphasized throughout this book. This methodology is ideally suited to the needs of a plant simulator because the method enables each type of particle to be tracked as the material that is processed passes through the plant. It is a natural framework within which the models for the unit operations are cast. The structure of MODSIM reflects its basis in the population balance method and the user should feel comfortable with this approach after reading the contents of Chapter 2.

In order for a general-purpose simulator to yield useful information on the actual process to be simulated, it must have access to three important classes of information. These are defined in general terms as follows:

- The structure of the flowsheet – what unit operations are included and how they are connected.
- The nature of the material to be processed – its mineralogical composition and structure, the size distribution and the amount that must be processed.
- The operating characteristics of each unit in the flowsheet. This requires the full description of the unit operations (the unit models) and a specification of the unit parameters that define the operating characteristics of the individual units.

MODSIM has been used primarily as an academic tool to enhance the educational experience of students of mineral processing. It appears to have been successful since students regularly comment on the improved insights that they achieve when they can try out the models, either in single units or in combinations with other units. Many of the operating characteristics of the unit operations that are discussed in the classroom can be easily demonstrated using the simulator. The modular open-ended structure of MODSIM makes it easy to add new models to the simulator, although this is not possible with the student version that is included with this book. Professors and lecturers can obtain the academic version of MODSIM, which is full-featured and does permit the introduction of new models. The method that is used to write the necessary code and add the models is fully documented in the user manual included on the compact disc.

10.2 Use of the simulator

A good simulator is a useful tool to the process plant engineer. Essentially the simulator can demonstrate, at least approximately, what a plant will do under any particular operating conditions. It can do so cheaply and without any real risk to the production rate of an operating plant or it can do so before a plant has been built and it does so in the engineer's office.

(a) Design studies:
At the design stage a good simulator can be used to:
- Help the design engineer to find the best flowsheet.
- Ensure that design specifications will be met under all required operating conditions.
- Choose the most suitable unit operations to be included in the plant.
- Size the units correctly and so eliminate wasteful over-design and avoid the catastrophe of under-design.
- Optimize the plant operation by achieving best economic combinations of grade and recovery.
- Identify potential production bottlenecks.
- Provide comparative assessment of competing manufacturers' equipment.
- Define the performance guarantees that should be met by suppliers.
- Find out what will happen if performance guarantees are not met.

(b) Operating plant performance:
A good simulator can be used to assist in the operation of a running plant in ways such as the following:
- Optimum performance of the plant can be approached.
- The plant can be tuned to suit variations in feed quality and flowrate.
- Plant bottlenecks can be found and appropriate steps taken to eliminate them.
- Performance changes can be investigated by asking 'what if' questions.
- Operations that are not properly understood can be identified and appropriate research effort undertaken to remedy such a situation.

Simulation can be seductive and the user of this technique must be skeptical and vigilant. A simulator can be effective only if it gives a reliable and valid description of plant operations. Generally this means that before simulation can be used the models that are contemplated must be calibrated against operating data that are representative of the conditions that apply to the actual plant operation.

10.3 The flowsheet structure

A plant flowsheet consists of a number of unit operations connected together by process flow streams. Material enters the plant and passes successively through the process units in accordance with the flowsheet structure, sometimes recycling on itself until finally it leaves the plant in one or other product stream. The material in a flow stream is processed by each unit in order to change the nature of the material or to separate it into its constituent parts. The number of different types of unit operation encountered in ore dressing plants is comparatively limited and these can be classified into three generic types.

10.3.1 The ore dressing unit operations
There are only three basic types of ore dressing unit operations:

- separation units that separate solid particles without changing their characteristics, for example flotation and gravity separation;
- transformation units that alter the characteristics of the solid particles, for example crushing and grinding operations; and
- units that separate solid and liquid constituents.

It is remarkable that the entire repertoire of ore dressing unit operations can be classified by such a simple scheme and this simplicity is exploited effectively in the simulator.

A further characteristic of the unit models is the flow stream connections that are possible. Separation units will always produce at least two product streams, a concentrate and a tailing, or an underflow and an overflow. Sometimes a single separation unit will produce a middling product as well. On the other hand, transformation units produce only a single product stream. Water separation units always produce two product streams, one of which is always predominantly water.

A convention is used in MODSIM that allows only a single feed stream to enter any unit. In cases where the flowsheet requires a unit to accept more than one stream, these streams are considered to be mixed before the unit and the combined stream is the unit feed. This does not cause any loss in generality and provides an important advantage for users who wish to develop and insert their own models into the simulator.

The simple structure just described defines the requirements that any mathematical model must meet if it is to be suitable to describe the operation of a unit operation within MODSIM. The mathematical model of a transformation unit must provide a description of the product stream from the details of the feed stream and the operating conditions of the unit. For example, the model of a grinding mill must calculate the particle size distribution of the product from the particle size distribution of the feed material and the breakage characteristics of the solid material that is processed. This latter will depend on the type of material as well as on the type of grinding mill (ball, rod, pebble, autogenous, etc.) and on the milling environment in the mill (ball charge, pulp filling, mill diameter, pulp viscosity, etc.).

The mathematical model of a separation unit must calculate the composition and characteristics of the various product streams that are produced. For example, the model of a cyclone should calculate the size distributions in both the underflow and overflow using the particle size distribution, flowrate and density of the feed as well as details of the cyclone geometry. Likewise the model of a flotation cell must calculate the mineralogical composition of the concentrate and tailings streams from the composition, size distribution and flowrate of the feed to the cell and from the volume, aeration rate and froth height in the cell.

The calculated product stream characteristics are then passed on by the simulator as feeds to the next units in the flowsheets. The models of these units then produce corresponding product streams that are passed on by the simulator. Eventually the material is traced completely through the plant and all the process streams are then defined. In the event that a stream is physically recycled, an interactive procedure is required because the feed streams to some units will not be known until the product streams from some subsequent units have been calculated. MODSIM will handle all such recycle loops iteratively without any interaction from the user. Efficient convergence procedures are used to ensure that computation time is minimized for the user.

The sequential calculation procedure described above leads to an obvious division of activities within the simulator – the calculation of the performance of each unit and the transmission of the details of the various flow streams from unit to unit. The latter is taken care of by the simulator executive structure and the former by individual unit models that are inserted as modules into the system and which can be called by the executive as required. This approach to simulation is known as the sequential modular approach and has proved to be superior to any other method because of its unique advantages. In particular it allows models to be inserted as modules into the simulator so that new models can be added whenever required.

The advantages of the sequential modular structure can be summarized as:

- Versatility to describe any ore dressing plant.
- Different preferences for unit operation models by different users can be easily accommodated.
- A variety of models is available for most unit operations. It is usually desirable to investigate and compare several of these during simulation studies.
- Unit models can be constructed and tested as standalone modules.
- Models must always be verified and calibrated against operating data.
- This is done most effectively by comparison with data from single-unit tests which are easy to simulate.
- It is cost effective to make use of existing models with minor modifications and extensions to describe new units or operating regimes.
- Exchange of models among users is easy.

10.3.2 The numerical analysis of simulation

The models of the various unit operations that have been described in previous chapters have one thing in common – they are all structured in such a way as to make it possible to calculate the characteristics of each of the product streams from a knowledge of the characteristics of the particulate material that makes up the feed stream to the unit.

The units that have been modeled can be classified into two categories and a general mathematical description of each category is useful.

Transformation units

These units are exemplified particularly by the comminution operations – crushing and grinding. These unit operations are designed specifically to transform the size and grade distributions of the feed stream to a product containing finer more liberated material.

The behavior of the unit as a whole can be described in matrix notation by

$$W_p = TW_f \qquad (10.1)$$

where W is a vector of mass flowrates of the particle classes in a particular stream and T is a transformation matrix that defines how the particle size and grade distributions of the feed are transformed to those in the product. Subscripts p and f refer to product and feed respectively. The transformation matrix need not be constant and elements of the matrix can be functions of the total product flowrate or even of the individual elements of W_p.

Separation units

The majority of units in an ore-dressing plant are designed to separate particles on the basis of some physical parameter. The efficiency of transfer of particles of a particular type to the concentrate, middlings, and tailings can be expressed in matrix-vector notation in terms of three diagonal recovery matrices, one for each product stream, E_C representing recovery to the concentrate, E_T the recovery to the tailings and E_M to the middling. The ith diagonal elements of these recovery matrices represent the recovery of particle of type i to the relevant product.

$$W_C = E_C W_f$$
$$W_T = E_T W_f \qquad (10.2)$$
$$W_M = E_M W_f$$

10.3.3 *The sequential method of simulation*

Any ore-dressing plant can be synthesized as a complex interconnection of the basic unit. Products from one unit become the feed to another unit or perhaps a final product. However, because of the recycle streams in the plant, it is not possible to calculate the composition of the product streams by moving sequentially from one unit to the next.

However the flowsheet can be rendered acyclic by appropriate tearing. To achieve this some streams are torn open to break open all the loops in the plant.

The structure of the simulator calculation procedure can be seen most clearly using a concrete example. Consider a flotation plant that includes a regrind mill as shown in Figure 10.1. Streams 3 and 5 can be selected as tear streams because if these streams are torn the entire flowsheet is rendered acyclic. Alternatively streams 11 and 12 can be torn instead of stream 5 but this would lead to a total of three tear streams rather than two. In general it is better to use the smallest number of tear streams that will make the plant acyclic.

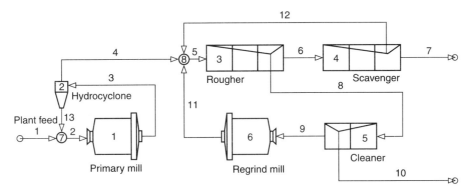

Figure 10.1 An ore-dressing plant flowsheet showing a flotation plant that includes a regrind mill

Let W_i represent the vector of mass flowrates of particle classes in stream i and let W_i' be the assumed value of the vector for tear stream i. The formal matrix-vector models of the units as given in Equations 10.1 and 10.2 can be applied in series as follows:

$$W_2 = W_{13} + W_1$$

$$W_{13} = E_{T2}\,W_3'$$

$$W_2 = E_{T2}W_3' + W_1$$

$$W_4 = E_{C2}W_3'$$

$$W_3 = T_1 W_2 = T_1 E_{T2} W_3' + T_1 W_1$$

$$W_6 = E_{T3}W_5'$$

$$W_7 = E_{T4}W_6 = E_{T4}E_{T3}W_5'$$

$$W_{12} = E_{C4}W_6 = E_{C4}E_{T3}W_5' \qquad (10.3)$$

$$W_8 = E_{C3}W_5'$$

$$W_9 = E_{T5}W_8 = E_{T5}E_{C3}\,W_5'$$

$$W_{11} = T_6 W_9 = T_6 E_{T5} E_{C3} W_5'$$

$$W_5 = W_4 + W_{12} + W_{11}$$

$$= E_{C2} W_3' + E_{C4} E_{T3} W_5' + T_6 E_{T5} E_{C3} W_5'$$

$$W_{10} = E_{C5} W_8 = E_{C5} E_{C3} W_5'$$

The feed to the plant consists of a single stream (stream 1) and is represented by the vector

$$F = W_1 \tag{10.4}$$

The plant has two product streams (streams 7 and 10) and the product is represented by the composite vector:

$$P = \begin{bmatrix} W_7 \\ W_{10} \end{bmatrix} \tag{10.5}$$

The two tear streams are combined to form a composite vector:

$$Y = \begin{bmatrix} W_3 \\ W_5 \end{bmatrix} \tag{10.6}$$

and the calculated values of the tear streams:

$$X = \begin{bmatrix} W_3' \\ W_5' \end{bmatrix} \tag{10.7}$$

The operation of the plant is concisely described by the two matrix equations:

$$Y = \Phi X + \Psi F \tag{10.8}$$

and

$$P = \Xi X + \Theta F \tag{10.9}$$

These matrices are composites formed from sums of products of the efficiency and transformation matrices of the units in the plant, as shown below:

$$\Phi = \begin{bmatrix} T_1 E_{T2} & 0 \\ E_{C2} & E_{C4} E_{T3} + T_6 E_{T5} E_{C3} \end{bmatrix} \tag{10.10}$$

Each component matrix on the diagonal of matrix Φ represents the paths from the tear stream back to itself. It is easy to see from Figure 10.1 that the path from tear stream 3 back to itself passes through the tailings of unit 2 and the tailings of unit 1. Unit 1 is a transformation unit while unit 2 is a separation unit and therefore the diagonal composite matrix is the product $T_1 E_{T2}$. On the other hand, two paths connect the tear stream 5 back to itself: one passes through the tailings of unit 3 and the concentrate of unit 4 and the other through the concentrate of unit 3, the tailings of unit 5 and the tailings of unit 6. The mixing point at unit 8 is accounted for by adding the two product terms that represent each of the distinct paths. This path structure is reflected in the lower right diagonal of the matrix Φ. Likewise the off-diagonal component matrices of Φ represent paths in the flowsheet from one tear to another. Note that, in Figure 10.1, there is no direct path from tear 5 back to tear 3 hence the zero matrix on the upper right position. There is a direct path from tear 3 to tear 5 through the concentrate of unit 2 hence the matrix E_{C2} in the lower left position of Φ.

The matrix Ξ is constructed in the same way as matrix Φ except that the elements of Ξ track direct paths from the tear streams to the products. In this example there are two such paths from tear 5; one through the tailings of units 3 and 4 and one through the concentrates of units 3 and 5. There are no direct paths from tear 3 to any product stream – all paths from tear 3 to the products pass through another tear. Hence both component matrices in the left hand column of Ξ are zero.

$$\Xi = \begin{bmatrix} 0 & E_{T4}E_{T3} \\ 0 & E_{C5}E_{C3} \end{bmatrix} \tag{10.11}$$

The component matrices of Ψ represent direct paths from the plant feeds to the tear streams while the component matrices in Θ represent direct paths from the plant feed streams to the product. In the present example, no paths connect the plant feeds to the plant products without passing through a torn stream. Hence both elements of Θ are zero in this example.

$$\Psi = \begin{bmatrix} T_1 \\ 0 \end{bmatrix} \tag{10.12}$$

$$\Theta = \begin{bmatrix} 0 \\ 0 \end{bmatrix} \tag{10.13}$$

Algorithms for the identification of all loops in a complex plant are well known using standard theorems of mathematical graph theory and all loops can be reliably identified. The choice of which streams should be torn is not uniquely defined although the set of tear streams should be chosen to ensure that the iterative calculation procedure converges as efficiently as possible. MODSIM has loop-finding and stream-tearing algorithms built in so that the user never has to be concerned about this aspect of the problem.

The simulation problem is solved as soon as a vector X is found that satisfies Equation 10.14.

$$X = Y = \Phi X + \Psi F \tag{10.14}$$

This calculation procedure is shown schematically in Figure 10.2.

The development of solution procedures is a specialized task and two

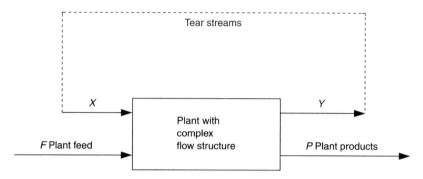

Figure 10.2 Schematic representation of Equation 10.14

methods of solution have been considered: matrix inversion and iteration.
    A formal solution by matrix inversion is easy to develop:

$$X = \mathbf{\Phi}X + \mathbf{\Psi}F \qquad (10.15)$$

$$X[I - \mathbf{\Phi}] = \mathbf{\Psi}F \qquad (10.16)$$

$$X = [I - \mathbf{\Phi}]^{-1}\mathbf{\Psi}F \qquad (10.17)$$

from which the vector of product stream can be evaluated.

$$P = \mathbf{\Xi}X + \mathbf{\Theta}F \qquad (10.18)$$

The method is attractive because only the terminal and tear streams occur in
the description and all solution calculations can be completed on these streams.
Other streams can be regenerated after solution is achieved. Thus a great
economy of description has been achieved. In the case of a linear plant $\mathbf{\Phi}$ is
a constant matrix. The apparent simplicity of this method is deceptive however.
Often $\mathbf{\Phi}$ is not constant and many of its elements depend on the solution
vector X in a non-linear and complex fashion. Furthermore algorithms to
calculate all of the elements of the component matrices are difficult to develop
for general plant configurations. Direct iterative methods have become popular
as a result. In this method, the numerical elements of the unit transformation
and efficiency matrices are not evaluated explicitly. Rather the model of each
unit operation is coded to accept the unit feed and produce a description of
the unit product or products. In other words the transformation and efficiency
matrices for the units exist only as computer code (often extremely complex)
that defines the transformation from feed to product for each unit in the
plant. The various paths from plant feeds and torn streams to the tears are
tracked sequentially – the product from one unit being passed along as feed
to the next unit in the path. The calculation must be started by making
assumptions regarding the nature of the torn streams. The composition finally
calculated at each tear is compared with the original assumption and the
initial assumptions are improved by iteration.
    Various methods for the iterative solution are available.

(i) Direct substitution

$$X^{(k+1)} = \Phi X^{(k)} + \Psi F \tag{10.19}$$

where $X^{(k)}$ is the kth iterate towards the solution and Φ and Ψ may also be functions of $X^{(k)}$.

(ii) Newton-Raphson

$$X^{(k+1)} = X^{(k)} - (I - \Phi'(X))^{-1}(X^{(k)} - \Phi X^{(k)}) \tag{10.20}$$

where $\Phi'(X)$ is the matrix of derivatives of $\Phi(X^{(k)})$ with respect to the elements of X. In practice $\Phi'(X^{(k)})$ is difficult to obtain and various approximations must be used. MODSIM includes both the direct substitution and Newton-Raphson methods and these algorithms have proved to provide rapid and reliable convergence of this iterative calculation. In most practical problems the Newton-Raphson method generally converges more quickly than direct substitution and the Newton-Raphson method is the default. Occasionally the region of convergence of the Newton-Raphson method is quite small and direct substitution will converge when Newton-Raphson will not.

There is no guarantee that a solution to Equation 10.14 does in fact exist and, if a solution does exist, that it is unique. Many model equations that implicitly generate the elements of the transformation and efficiency matrices can yield an overall matrix Φ that does not allow a solution to Equation 10.14. Unfortunately, it is not possible to predict *a priori* when a solution does not exist and this only shows up in practice when the iterative scheme fails to converge. When this happens the parameters that are set up to fix the operation of individual units must be adjusted until a valid solution exists.

10.4 Simulation of single unit operations and simple flowsheets

MODSIM includes a repertoire of unit models that can be used to simulate the performance of all of the common unit operations of mineral processing. In this section a few of these models will be demonstrated and compared to actual operating data where possible. These examples will assist the user to appreciate how the models that have been described in previous chapters can be used for practical engineering calculations. The reader is encouraged to run the exercises using MODSIM and to experiment with the wide range of possibilities that MODSIM provides to investigate the behavior of the models under many different conditions. Most of these exercises are short and take no more than a few minutes to set up and run.

Each example in this section includes only a single unit or a few units in a simple flowsheet configuration so that the reader can concentrate on the behavior of the model that is used and can easily compare the behavior of alternative models that are available for the unit in question. All the examples in this chapter are included on the compact disc that accompanies the book and the reader is encouraged to run these examples and to explore the possibilities that the simulator offers to investigate alternative models for the unit operations and to investigate how the behavior of the units may be expected to respond to variations in any of the parameters that describe the unit or the circuit.

Example 10.1: Vibrating screen

The model for vibrating screens that is recommended for routine simulation of industrial screening operations is the Karra model (see Section 4.2.1). This model has been found to produce reliable results under a wide variety of screening conditions. In this example the Karra model will be used to simulate one of the many experiments that are reported in USBM Report J0395138 (Hennings and Grant, 1982). These experiments were performed on a well controlled experimental vibrating screen. Essential details of the test are given in Table 10.1.

Table 10.1 Details of vibrating screen test

Screen size: 16 ft × 0.88 ft.
Screen type: Woven wire 1.25 inches aperture 0.5 inch wire diameter.
Angle of inclination: 20°.
Bulk density of material: Dolomite 110 lb/ft^3.
Measured size distributions:

Mesh size (inches)	Feed (% passing)	Oversize (% passing)	Undersize (% passing)
3.000	100.00	100.00	100.00
2.000	95.52	85.32	100.00
1.500	88.58	62.61	100.00
1.250	81.87	40.64	100.00
1.000	72.73	13.12	98.94
0.750	61.00	0.44	87.62
0.500	44.47	0.22	63.93
0.375	36.36	0.20	52.26
0.250	25.40	0.20	36.48

The first step is to draw a flowsheet with a single vibrating screen as shown in Figure 10.3. The basic system data is defined on the MODSIM system data input form as shown in Figure 10.4. The data that is specified for the screen and which is required for the Karra model is shown in Figure 10.5. The simulation of the operation of this screen is summarized in the size distributions that are calculated. These are shown in Figure 10.6 together with the data that was collected during the test. The agreement is satisfactory and this level of agreement with measured data is not unusual to find with this particular model for the vibrating screen. MODSIM provides additional information on the operation and in particular the area utilization factor was found to be 1.58 indicating that the screen was somewhat overloaded during the test. The reader should simulate this operation using the other models for the screen that are offered by MODSIM and note the differences in the size distributions that are calculated. The width of the screen that was used in this test work was reduced by mounting rigid vertical skirts along the length of the screen. This was done to limit the amount of material that was required for the test work. An interesting simple exercise for the reader is to investigate how the area utilization factor for this screen varies as the width of the screen varies at constant feed rate.

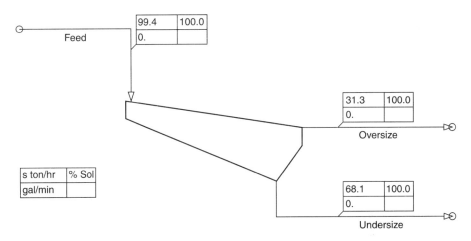

	99.4	100.0
Feed	0.	

	31.3	100.0
	0.	
		Oversize

s ton/hr	% Sol
gal/min	

	68.1	100.0
	0.	
		Undersize

Figure 10.3 Flowsheet for a single vibrating screen

Figure 10.4 Specification of the system data for the screen experiment

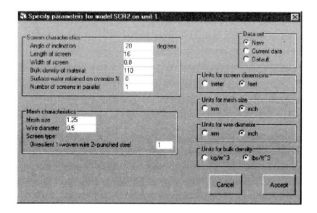

Figure 10.5 Parameter specification for the Karra screen model

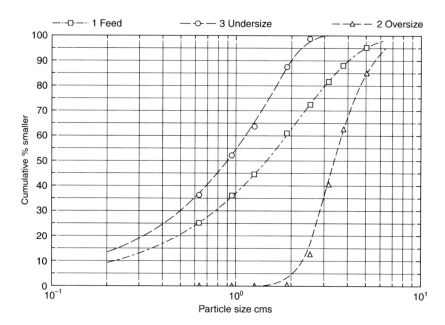

Figure 10.6 Simulated size distribution compared to the data that were measured during the screening test

The data used in the next three examples are taken from ICRA (1995). These data resulted from closely controlled sampling campaigns on the ball mill circuit of Noranda's Geco mine. In the next two examples, the behavior of the hydrocyclone classifier and the ball mill are simulated as isolated single units and then the closed ball mill circuit is simulated. Finally the changes that can be expected in the performance of the milling circuit are determined by simulation when the sizes of the vortex finder and apex of the hydrocyclone classifier are changed. The closed ball mill circuit is probably the most common subject of simulation in mineral processing and a number of variations of the ball mill model are available in MODSIM. For the purposes of this example we use a model that is based on the Herbst-Fuerstenau method for the calculation of the specific rate of breakage. This is appropriate in this case because the power drawn by the mill was measured during the sampling campaigns.

Example 10.2: Hydrocyclone classifier

This example demonstrates the use of the simulator to match the simulation to the actual operation of the cyclone in the milling circuit. A simple flowsheet consisting of a single hydrocyclone is set up in MODSIM as shown in Figure 10.7. The particle size distribution and flowrate of the hydrocyclone feed is specified as shown in Figure 10.8. The parameters for the hydrocyclone model are specified as shown in Figure 10.9. The simulated particle-size distributions are compared with the measured data in Figure 10.10. The Plitt model was used to model the behavior of the hydrocyclone (see Section 4.5.7). In order to accurately represent the measured data, the three calibration parameters for this model were set to 1.00 for d_{50c}, 0.36 for the sharpness factor and 2.0 for the flow split factor. These adjustments to the standard model indicate that

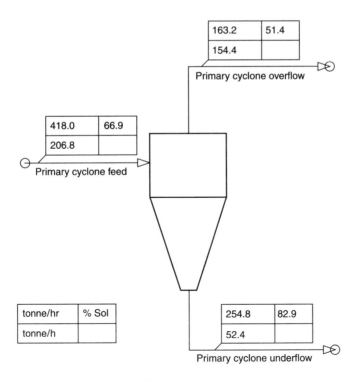

Figure 10.7 Flowsheet for single hydrocyclone

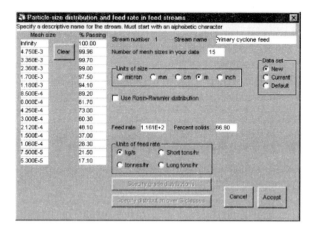

Figure 10.8 Specification of particle-size distribution in the cyclone feed

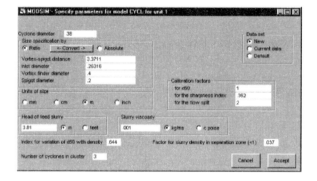

Figure 10.9 Specification of parameters for the hydrocyclone

the cyclones are operating less efficiently than the 'standard' cyclone as defined by the Plitt model. The sharpness parameter in the model must be reduced to about $^1/_3$ of its value as calculated from Equation 4.170 in order to match the data. The flow split correction factor of 2 indicates that the cyclones are sending about twice the volumetric flow to the underflow than would be calculated by the model. With these adjustments to the model the simulations represent the measured data well.

The dimensions of the cyclone are given in Table 10.2.

Example 10.3: Ball mill

The ball mill from the same circuit is simulated in isolation in this example, as shown in the simple flowsheet in Figure 10.11.

The dimensions of the mill are given in Table 10.3.

The Herbst-Fuerstenau method for calculating the specific rate of breakage

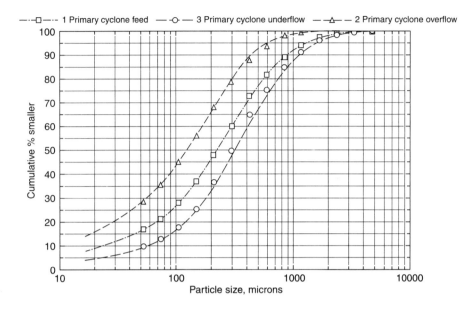

Figure 10.10 Simulated particle-size distributions compared to the measured data

Table 10.2 Dimensions of the hydrocyclone

Diameter	0.38 m
Inlet diameter	0.1 m
Vortex finder diameter	0.152 m
Apex diameter	0.076 m
Cylinder length	0.28 m
Cone angle	15°

is used because the power drawn by the mill is known. The data for this mill model is specified as shown in Figure 10.12.

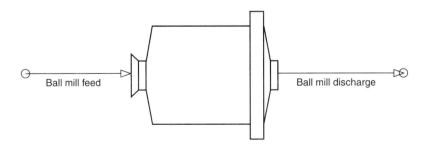

Figure 10.11 Simple flowsheet for a single ball mill

Table 10.3 Dimensions of the ball mill

Internal diameter	3.51 m
Internal length	4.11 m
Media load	35%
Media size	76.2 mm
Power demand	616 kW

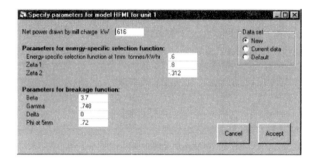

Figure 10.12 Specification of parameters for the ball mill

The simulated size distribution in the ball mill discharge is shown in Figure 10.13. The size reduction that is achieved by the ball mill is not particularly

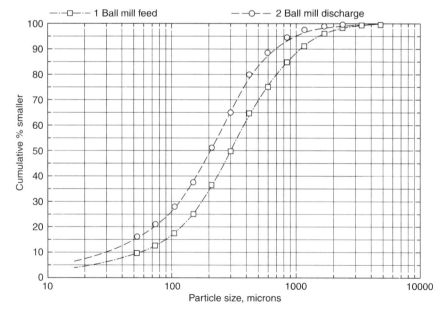

Figure 10.13 Simulated size distribution of the ball mill discharge using the measured feed size distribution as input

large in this example and the circuit relies on the recirculating load to produce a sufficiently fine product.

Example 10.4: Rod and ball mill circuit

The two units that were simulated in Examples 10.2 and 10.3 are now combined into a closed ball milling circuit and an open circuit rod mill is added to produce the feed to the ball mill circuit. This flowsheet now matches the plant from which the data were obtained. The dimensions of the hydrocyclone and the ball mill are given in Tables 10.2 and 10.3 and those of the rod mill are given in Table 10.4.

Table 10.4 Dimensions of the rod mill

Internal diameter	3.51 m
Internal length	4.11 m
Media load	35%
Media size	89 mm
Power demand	560 kW

The closed milling circuit shown in Figure 10.14 and the simulated size distributions around the circuit are shown in Figure 10.15.

The models used lead to a good simulation of the size distributions around the circuit.

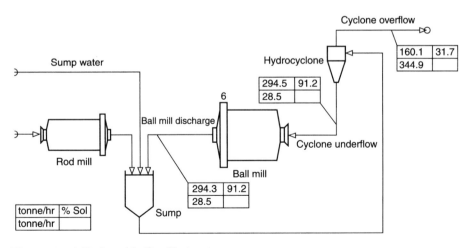

Figure 10.14 Rod and ball mill circuit

10.5 Integrated flowsheets

In this section we simulate a few integrated flowsheets and show how the models work together to simulate complex plants.

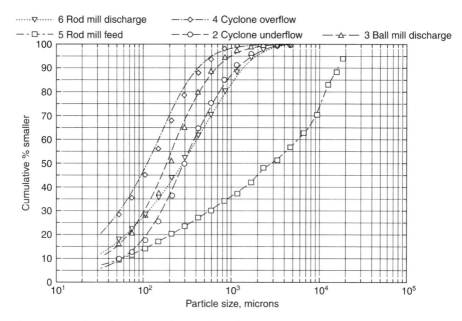

Figure 10.15 Simulated particle-size distributions in the ball mill circuit

Example 10.5: Optimization of a coal washing plant

The calculation of the performance of coal washing units that are in common use is not difficult and the principles of making such calculations have been known for many years. Because the nature of individual coal particles do not change during passage through the various washing units, it is possible to calculate the performance of integrated plants by appropriate combinations of the calculation procedures for each unit operation. When the flowsheet includes comminution operations, the computations are somewhat more complex since the size distribution and washability distributions are transformed during the size-reduction processes. The population balance method is appropriate for the simulation of both comminution and washing units and this example shows how simulation can be used to set up a conventional coal washing plant for optimal performance.

Simulators for coal washing plants have been available since the early 1970s (Gottfried, 1975), but these early programs were unwieldy and cumbersome to use, requiring data input through punched card medium and no facilities were provided for interactive specification of flowsheet structure and data and no graphical output of results was possible except graphics that were crudely produced by line printers. By the early 1980s the early coal washing plant simulators had been fully documented and these could provide effective simulations (Goodman and McCreery, 1980; Gottfried et al., 1982).

Coal washing plants generally have comparatively simple flowsheet structures and the principles for the calculation of the performance of individual units in isolation and in simple combinations has been common practice for many years. Simulation is particularly suitable for establishing practical and useful set-up points for coal washing plants that include several different sub-circuits to clean coals of various sizes. This is a common problem because different unit operations must be used to handle the different size ranges. A plant that cleans the full size spectrum of sizes in a number (usually two or three) of subsections must be set up to optimize the combined operation. This example shows how this optimization can be done using MODSIM.

The flowsheet for the plant is shown in Figure 10.16 which is fairly typical of modern coal washing plants that treat a wide size-spectrum feed. The coarse material is cleaned in a dense-medium drum, the intermediate size in a dense-medium cyclone and the fines in a water-only cyclone. In order to set up a plant such as this optimally, it is necessary to make five key decisions: the sizes at which the feed should be cut and the target specific gravities at which the dense-medium drum, the dense-medium cyclone and the water-only cyclone should be operated to get the best combination of ash rejection and coal yield. It is convenient to consider these as two separate optimization problems. Optimal settings for the target specific gravities can be obtained and then the optimal cut sizes can be computed. This example deals with the first of these problems and the optimal specific gravities are found using simulation. The simulations reveal that significant improvement in overall plant performance can be obtained by operating at the optimal settings.

The flowsheets to be simulated consist primarily of coal washing units each with a target specific gravity of separation specified for each washing

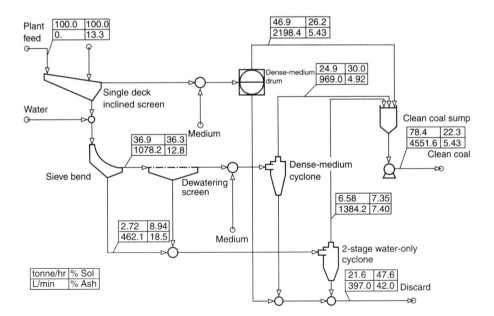

Figure 10.16 A typical coal washing plant flowsheet

unit. The feed is allocated to each washing unit on the basis of separation by size in the screening operations. The target specific gravity of separation in any unit is the separation point to be expected when the partition curve of the composite feed to clean coal is plotted without considering each narrow size fraction separately. All gravity separation units are known to show significant variation of separation performance as a function of particle size. In general the cut point increases with decreasing particle size and the efficiency of separation decreases with decreasing particle size. This is particularly noticeable at particle sizes smaller than a few millimeters – so-called fine coal separations.

The cut point can be controlled by variation of an operational control in the plant or by means of a physical adjustment to the equipment. Dense-medium separators are set by controlling the medium density and the separation density in autogenous separators such as water-only cyclones, concentrating tables and jigs are controlled by physical adjustments. At the highest level, simulation models should accept as input data a specification of the equipment type and the settings of all controllable physical variables and be capable of predicting the cut points and separation efficiencies that will be achieved in practice. Models at this level are not always available and usually require careful calibration against pilot or full-scale plant data. In the present case no suitable data are available to calibrate high-level models and more restricted models which accept a target cut point are used for the simulation. Models that use this principle are described in Section 7.4.

To assess the efficacy of the optimization procedure, the plant was simulated using the target specific gravities in the dense-medium drum, the dense-medium cyclone and the water-only cyclone that are currently set up in the plant. This is the base case against which the optimum solution can be judged. The base case specified target cut points of 1.35, 1.65 and 1.95 in the dense-medium drum, dense-medium cyclone and water-only cyclone respectively.

Before the plant can be optimized, it is necessary to investigate the performance of each of the individual washing units under conditions that it experiences during operation within the plant flowsheet. This is the first task of the simulator which can be used to investigate the range of performance of each washing unit in the plant. The performance of each washing unit in flowsheet 10.16 was simulated over a range of target cut points. The results are summarized in Figure 10.17. The figure shows the cumulative yield against the cumulative ash for each of the three washing units together with the theoretical yields calculated from the washability data of the feed material to each unit.

The arrangement of three separation units in parallel lends itself to optimization so that the plant can produce clean coal at the greatest possible yield at a given ash content or at the minimum ash content for a given yield. The optimal combination of target cut points for the three washing units is easy to establish in this case using the simulated data from Figure 10.17. The calculated yields are replotted as the product of yield × ash content versus yield for each of the three units as shown in Figure 10.18. Each calculated

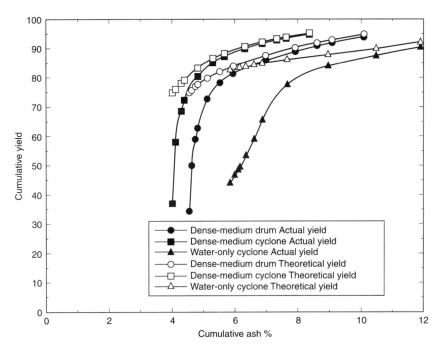

Figure 10.17 Performance curves for each of the three washing units compared to the washabilities of their respective feeds. Each point shown represents a different target cut point for each unit

point is indexed by the target cut point in the unit. Optimal combinations of the three target cut points for the individual washing units are located at points on these curves where the slopes of the curves are identical. Seven such optimal combinations are shown as points connected by broken lines in Figure 10.18. A typical set of parallel lines that touch the three curves at points where the slopes are equal are also shown. The appropriate target cut points can be established by interpolation between the target cut points that label each calculated point on the curves. It is immediately obvious that the recommended base case target specific gravities of 1.35, 1.65 and 1.95 currently set up in the plant do not constitute an optimal combination. The seven optimal target specific gravity combinations are given in Table 10.5 together with the optimal yields, ash content and calorific values of the clean coal product that are achieved in the plant. It is interesting that the optimal sets of target cut points do not depend on the relative flow rates to the three washing operations.

The performance of the plant at the seven optimal combinations of target cut points is shown in Figure 10.19 and the simulation shows that about 10% of potential yield would be lost if the plant were in fact set up to operate at the base case target specific gravities.

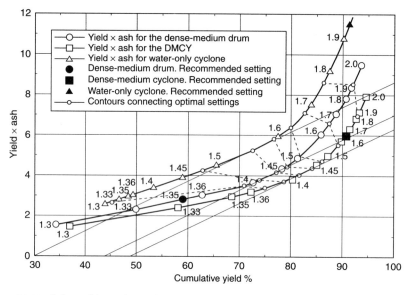

Figure 10.18 Ash-yield products for the three individual washing units. Points where these curves have equal slopes define optimal combinations of target specific gravities

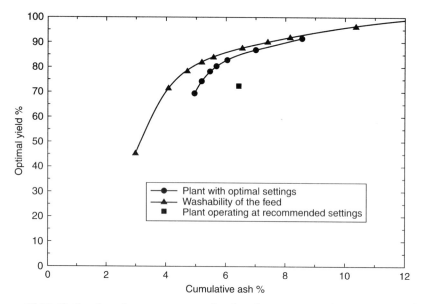

Figure 10.19 Optimal performance curve for the plant compared to the washability of the feed and to the base case which is suboptimal

Table 10.5 Optimal settings for the washing units in the plant and the resulting plant performance

Optimal combination of target specific gravities			Calculated plant performance				
Water-only cyclone	Dense-medium cyclone	Dense-medium vessel	Product	Ash content (%)	Yield (%)	Calorific value (MJ/kg)	Sulfur content (%)
1.40	1.35	1.35	Clean coal	4.92	69.5	33.60	1.10
			Discard	32.5		22.80	1.30
1.48	1.38	1.39	Clean coal	5.16	74.5	33.50	1.11
			Discard	37.2		21.00	1.31
1.56	1.42	1.44	Clean coal	5.43	78.4	33.40	1.12
			Discard	42.0		19.20	1.31
1.59	1.54	1.55	Clean coal	5.65	80.5	33.40	1.13
			Discard	45.11		18.00	1.30
1.64	1.70	1.68	Clean coal	6.01	82.9	33.20	1.13
			Discard	49.0		16.50	1.29
1.69	1.79	1.78	Clean coal	6.96	87.0	32.80	1.13
			Discard	55.9		13.80	1.36
1.78	2.00	1.90	Clean coal	8.52	91.5	32.20	1.12
			Discard	65.2		10.30	1.63

The coal washability data is entered in the simulator as shown in Figure 10.20.

Data completion

The data available for this simulation was incomplete in two important respects and required completion. The effective densities of the two outer washability fractions were not given and the washability analysis of the 200 mesh $\times$ 0 size fraction was missing. These omissions are commonly found in coal washability data sets. The effective densities of the outer washability fractions determine the behavior of this material that is predicted by the unit models and a rational assignment of effective density is necessary particularly when the units will be set up to cut at either high or low specific gravities. The washability distribution of the 200 mesh $\times$ 0 fraction was estimated by assuming that the lower nine density fractions were distributed identically to the those in the

Table 10.6 The properties of the raw coal used in the simulation

(a) Raw-coal size distribution

Size Passed	Size Retained	Weight (wt%)	Ash (wt%)
	$1\frac{1}{2}$ in.	15.8	18.12
$1\frac{1}{2}$ in.	$\frac{3}{4}$ in.	22.5	13.44
$\frac{3}{4}$ in.	$\frac{3}{8}$ in.	18.2	12.17
$\frac{3}{8}$ in.	28 mesh	33.2	11.08
28 mesh	100 mesh	4.5	10.88
100 mesh	200 mesh	1.7	12.25
200 mesh		4.1	27.07

(b) Raw-coal analysis

Specific gravity Sink	Specific gravity Float	Representative value	Ash (wt %)	Sulfur (wt %)	Heating value (Btu/lb)
	1.30	1.299	2.96	0.99	14812
1.30	1.35	1.325	5.93	1.13	14287
1.35	1.40	1.375	11.04	1.83	13350
1.40	1.45	1.425	16.07	1.68	12502
1.45	1.50	1.475	22.78	1.26	11424
1.50	1.60	1.550	28.99	0.91	10386
1.60	1.70	1.650	36.43	0.83	9122
1.70	1.80	1.750	45.71	0.58	7493
1.80	2.17	1.985	58.99	0.88	5296
2.17		2.500	81.52	2.38	1850

Table 10.6 (Cont'd.)

(c) Raw coal size-by-size washability

Specific gravity		$1\frac{1}{2}$ Weight (wt %)	$1\frac{1}{2} \times \frac{3}{4}$ Weight (wt %)	$\frac{3}{4} \times \frac{3}{8}$ Weight (wt %)	$\frac{3}{8} \times 28$ mesh Weight (wt %)	28×100 mesh Weight (wt %)	100×200 mesh Weight (wt %)	200×0 Weight (wt %)
Sink	Float							
	1.30	37.7	42.4	41.0	50.3	63.0	60.7	48.94
1.30	1.35	31.1	28.0	30.1	23.4	14.2	12.9	10.40
1.35	1.40	1.9	7.4	9.0	8.1	5.9	4.0	3.22
1.40	1.45	1.4	2.9	4.4	4.3	3.2	4.9	3.95
1.45	1.50	2.4	2.0	1.9	1.7	1.4	2.1	1.69
1.50	1.60	4.0	5.4	3.6	2.4	2.2	2.9	2.34
1.60	1.70	3.7	3.4	2.4	1.7	1.3	1.7	1.37
1.70	1.80	2.9	1.9	1.8	1.4	0.9	1.0	0.81
1.80	2.17	9.0	4.0	3.3	2.8	2.4	2.7	2.18
2.17		5.9	2.6	2.5	3.9	5.5	7.1	25.10

100 mesh × 200 mesh fraction and the percentage in the highest density fraction was adjusted to make the ash content of the 200 mesh × 0 fraction equal to the experimentally determined value which was available. The resultant distribution in the fine size fraction is shown in Table 10.6. The choice of the effective densities in the outer washability fractions was also based on the fact that accurately determined values of the ash content were available for these intervals. The relationship between ash content and particle density is simple for many coal samples since coal can often be described as a two-phase mixture of two mineral types of different apparent densities usually about 1.29 for the combustible component and between 2 and 3.3 for the ash. Then ash content will be a linear function of the reciprocal of the density or specific gravity. The data available are plotted as ash content against the reciprocal average specific gravity of the interval in Figure 10.21. The linear dependence is clearly evident and the effective densities can be evaluated from the equation of the regression line through the points at the known ash contents of the outer fractions. In this way these fractions were assigned specific gravities of 1.303 and 2.534 respectively as shown in Figure 10.21.

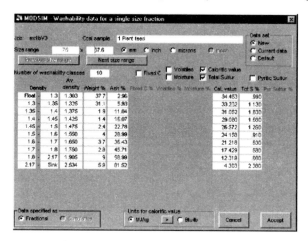

Figure 10.20 Data input form for coal washability data

Example 10.6: Ball mill circuit with liberation and concentration

The simulation of the liberation of magnetite from chert is covered in this example in which the milling circuit of an iron ore recovery plant is simulated. The plant processes a typical taconite ore and the plant includes wet magnetic drum separation in the milling circuit. The plant was comprehensively sampled so that a direct verification of the validity of the simulation and the unit models is possible. The 'liberation size' of the magnetite in the taconite is in the region of 100 μm, which is also close to the d_{80} size in the size distribution of the plant product. Thus the liberation characteristics of the ore play an important role in determining the performance of the magnetic separators and therefore of the entire process.

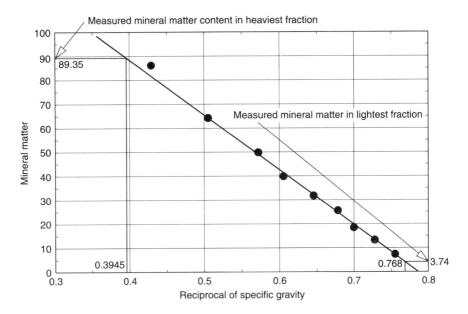

Figure 10.21 Extrapolation to find the effective density of the two outer washability fractions

A simplified flowsheet of the plant is shown in Figure 10.22. The feed to the grinding circuit consists mainly of a concentrate stream from cobber magnetic separators which concentrate the rod mill discharge from the primary grinding stage. The ball mill circuit is closed with a hydrocyclone and it includes a rougher magnetic separator in the ball mill discharge. The circuit is complicated further by feed back of water and solids from a dewatering drum magnetic separator. The magnetic rougher tails are cleaned in a scavenger magnetic separator and the concentrate from this scavenger is fed to the cyclone feed sump together with the rougher concentrate. The performance of the magnetic separator depends strongly on the magnetite content of individual particles, which in turn is dominated by the liberation processes that occur in the ball mill.

This plant cannot be characterized only by the size distributions of the various process streams. The liberation distributions are equally important. Samples were collected from the plant during a period of normal operation and the size and liberation distributions at each size were measured. The size distributions were measured using standard sieve analysis and the liberation spectra were determined by image analysis using the linear intercept method and appropriate stereological transformation. The image analysis technique used for the measurement of the liberation spectra is described in Chapter 2 and the liberation spectra measured in the four plant feed streams (cobber concentrate, dewatering drum concentrate, dewatering drum tails and scavenger concentrate) are shown in Figures 10.23 to 10.26. The composition of these streams and the distribution of liberated and unliberated particles is

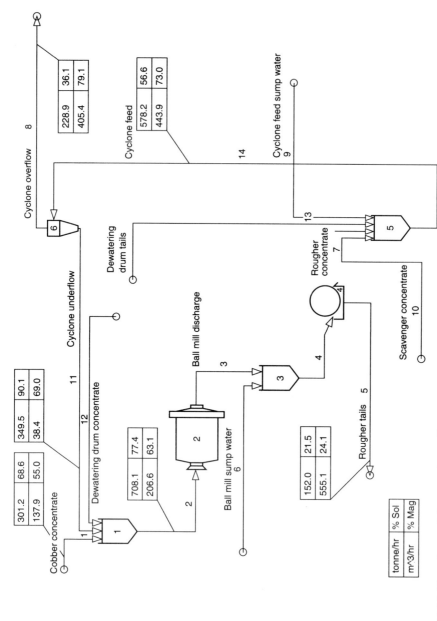

Figure 10.22 Flowsheet of the taconite milling circuit

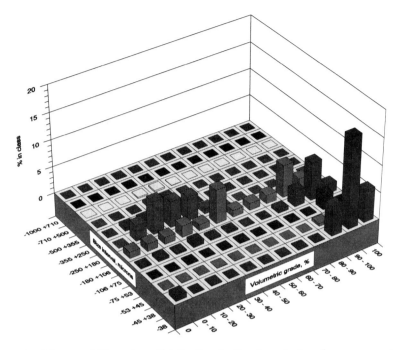

Figure 10.23 Measured liberation distribution in the dewatering drum concentrate that is returned to the circuit as a feed

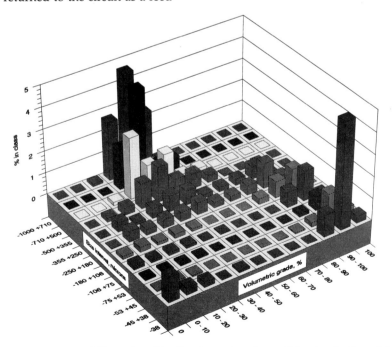

Figure 10.24 Measured liberation distributions at each particle size in the cobber concentrate which is the dominant feed to the plant

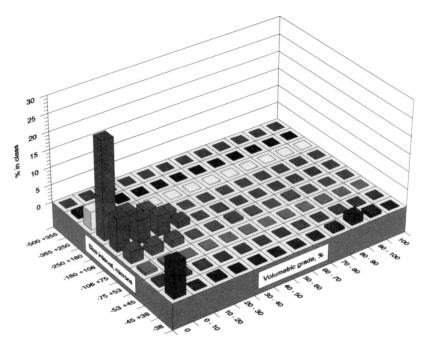

Figure 10.25 Measured liberation distribution at each particle size in the dewatering drum tails which is returned to the circuit as a feed

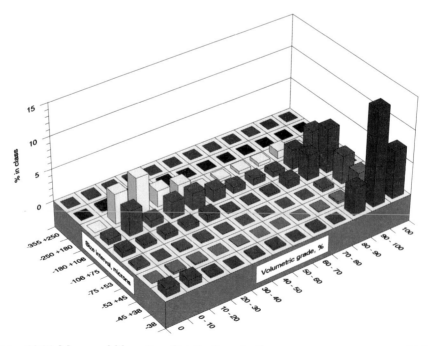

Figure 10.26 Measured liberation distributions in the scavenger concentrate which is returned to the circuit as a feed

clearly evident form these graphs. Note in particular the absence of liberated magnetite in the dewatering drum tails and the small amounts of liberated chert in the dewatering drum concentrate and the scavenger concentrate. This liberated chert is returned from these units by entrainment of the chert with the magnetite on the drum separators. This is typical of the operation of wet drum magnetic separators.

The plant was simulated with these feed streams using the Austin model for the ball mill including the Andrews-Mika diagram method to simulate the liberation that occurs during grinding. The liberation data is entered into MODSIM using the data input screen as shown in Figures 10.27 and 10.28.

The results of the simulation can be compared to the actual plant behavior by comparing the calculated particle-size distributions and the liberation distributions in each of the sampled streams. The comparison of the particle-size distributions is shown in Figure 10.29. The liberation distributions are compared in Figures 10.30 to 10.41.

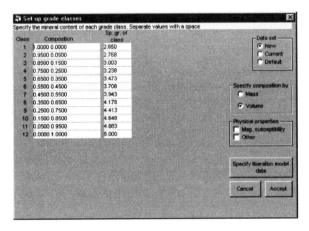

Figure 10.27 Specification of the grade classes in MODSIM

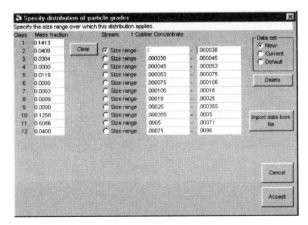

Figure 10.28 Specification of the liberation distribution in the feed stream

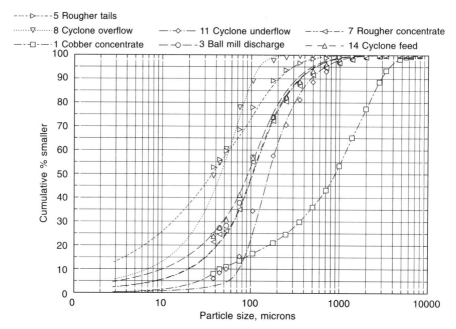

Figure 10.29 Comparison between calculated and measured size distributions in five plant streams. Plotted points are measured and the lines are calculated by the simulation

It is obvious from these graphs that the simulator provides a good representation of both the size distributions and the liberation distributions in this plant. The matches between measured and simulated data are excellent.

The measured and simulated data make it possible to diagnose several important aspects of the plant operation. Figures 10.38 and 10.39 show the character of the final product from the circuit in the cyclone overflow. This material is comparatively fine with a d_{80} size of 82 μm and both the measured and simulated data indicate that the final product is largely fine liberated magnetite, which is the desired product. However there is a small component of incompletely liberated particles in the size range 75 μm to 250 μm. This is the largest source of silica contamination in the product. Some reduction in the cut size that is achieved in the hydrocyclone could be beneficial.

The character of the particles in the hydrocyclone underflow is of particular interest in this plant. The measured and simulated liberation distributions are shown in Figures 10.36 and 10.37. It is clear that the hydrocyclones are returning coarse material to the mill as they should and that this material is mostly unliberated. This material is reground in the mill and will eventually leave the circuit as fine liberated magnetite. However the hydrocyclone is also returning liberated magnetite in significant amounts to the mill, and that is wasteful because excess energy is required in the mill to grind already liberated material. This liberated magnetite enters the hydrocyclone underflow because it has a density significantly greater than that of the chert.

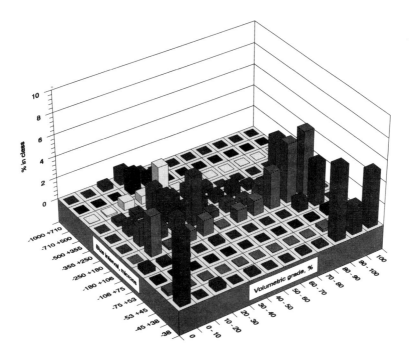

Figure 10.30 Measured liberation distributions at each particle size in the ball mill discharge

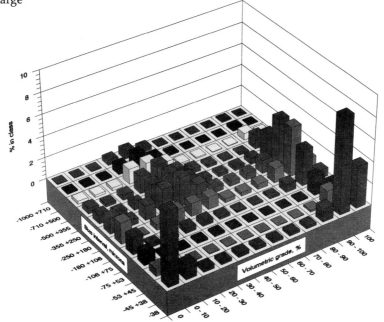

Figure 10.31 Simulated liberation distributions at each particle size in the ball mill discharge

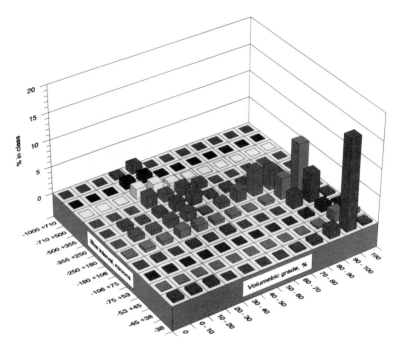

Figure 10.32 Measured liberation distributions at each particle size in the rougher magnetic separator concentrate

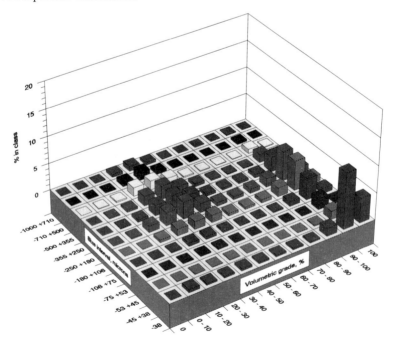

Figure 10.33 Simulated liberation distributions at each particle size in the rougher magnetic separator concentrate

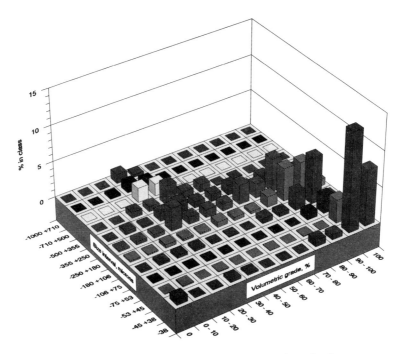

Figure 10.34 Measured liberation distributions in the cyclone feed

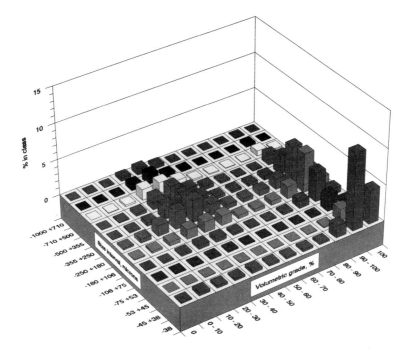

Figure 10.35 Simulated liberation distributions in the cyclone feed

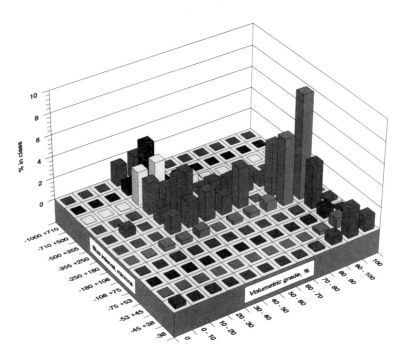

Figure 10.36 Measured liberation distributions in the cyclone underflow

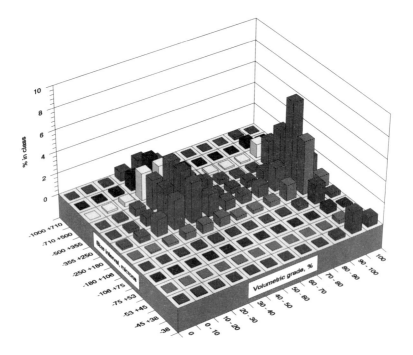

Figure 10.37 Simulated liberation distributions in the cyclone underflow

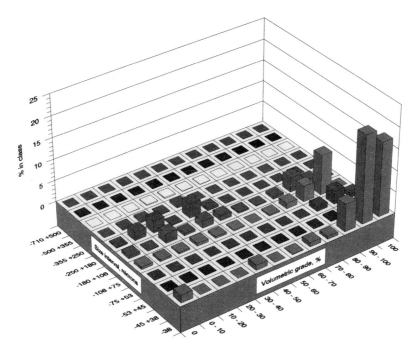

Figure 10.38 Measured liberation distributions in the cyclone overflow

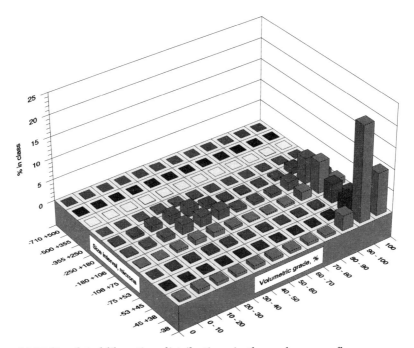

Figure 10.39 Simulated liberation distributions in the cyclone overflow

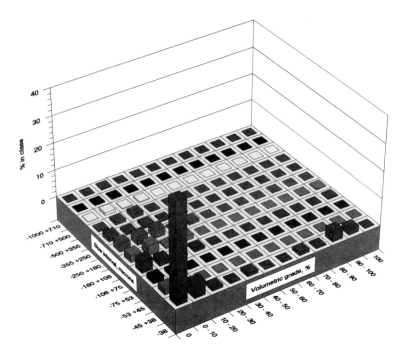

Figure 10.40 Measured liberation distributions in the rougher magnetic separator tails

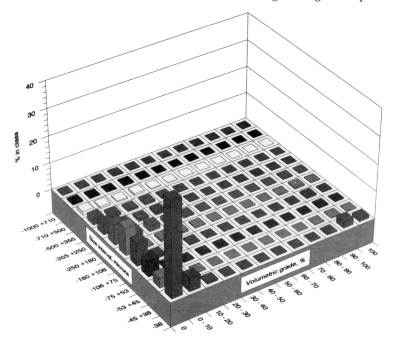

Figure 10.41 Simulated liberation distributions in the rougher magnetic separator tails

The reader is encouraged to run this simulation and investigate possible strategies to improve the performance of the circuit. The main objective should be to reduce the energy consumption without reducing the recovery or the grade of the final product. The wasteful recirculation of liberated magnetite can be addressed by using fine screens rather than hydrocyclones for classification and this can be easily investigated using the simulator. A complete account of this simulation project can be found in Schneider (1995).

10.6 Symbols used in this chapter

E   Unit recovery matrix.
F   Composite vector of mass flowrates in the plant feed streams.
I   Identity matrix.
P   Composite vector of mass flowrates in the plant product streams.
T   Unit transformation matrix.
W   Vector of mass flowrates in a stream.
X   Composite vector of assumed mass flowrates in the tear streams.
Y   Composite vector of calculated mass flowrates in the tear streams.
Θ   Composite matrix representing calculation paths from feed streams to product streams.
Ξ   Composite matrix representing calculation paths from tear streams to product streams.
Φ   Composite matrix representing calculation paths from tear streams back to tear streams.
Ψ   Composite matrix representing calculation paths from feed streams to tear streams.

References

Ford, M.A. (1979) Simulation of Ore dressing Plants. Ph.D Thesis, University of the Witwatersrand, Johannesburg.

Goodman, F.K. and McCreery, J.H. (1988) Coal Preparation Plant Computer Model Vols I and II. U.S. Environment Protection Agency EPA-600/7-80-010 a & b.

Gottfried, B.S. (1975) Computer Simulation of Coal Preparation Plants. U.S. Bureau of Mines, Grant No. GO-155030 Final Report, Dept of Industrial Engineering, Univ. of Pittsburgh, Nov 1975 and August 1977.

Gottfried, B.S. and Jacobsen, P.S. (1977) Generalized Distribution Curve for Characterizing the Performance of Coal-Cleaning Equipment USBM RI 8238.

Gottfried, B.S., Luckie, P.T. and Tierney, J.W. (1982) Computer Simulation of Coal Preparation Plants. U.S. Dept of Energy Final Report under contract AC22-80PC30144. DOE/PC/30144-T7, 284 pp.

Hennings, K. and Grant, D. (1982) A simulation model for the prediction of the performance of vibrating screens. USBM Contract J0395138, U.S. Department of the Interior. Washington D.C.

ICRA (1995) Simulation of Comminution Operations. Standard Problems for Simulators. International Comminution Research Association. 5th Workshop, Napa.

King, R.P. (1972) Data preparation and use of a computer program for the calculation of the performance of a flotation plant. National Institute for Metallurgy (now Mintek), Johannesburg, Report number 1436, 9 pp. Reissued in revised form August 1973.

King, R.P., Pugh, G. and Langley, J. (1973) Application of a flotation model to an industrial plant. National Institute for Metallurgy (now Mintek), Johannesburg. Report number 1562, 16 pp.

Schneider, C.L. (1995) The Measurement and Calculation of Liberation in Continuous Grinding Circuits. Ph.D Thesis. University of Utah.

Appendix: Composition and specific gravities of some common minerals

Mineral name	Composition	Percent metal	Specific gravity	Mass average atomic number
Albite	$NaAlSi_3O_8$	Al_2O_3 – 19.5%	2.6–2.7	10.71
Andalusite	Al_2SiO_2	Al_2O_3 – 63.2%	3.2	10.71
Angelsite	$PbSO_4$	Pb – 68.3%	6.1–6.4	59.41
Apatite	$Ca_4(CaF)(PO_4)_3$	P_2O_5 – 42.3%	3.2	14.10
Arsenopyrite	FeAsS	Fe – 34% As – 46.0%	5.9–6.3	27.25
Azurite	$2CuCO_3 \cdot Cu(OH)_2$	Cu – 55.0%	3.8–3.9	19.43
Barite	$BaSO_4$	BaO – 65.7%	4.3–4.6	37.34
Bauxite	$Al_2O_3 \cdot 3H_2O$	Al – 34.9%	2.6	9.46
Bentonite	$(CaMg)O \cdot SiO_2 (AlFe)_2O_3$	No metal source	2.1	16.55
Beryl	$Be_3Al_2(SiO_3)_6$	Be – 5% Al_2O_3 – 19%	2.6–2.8	10.18
Biotite	$(HK)_2(MgFe)_2Al_2(SiO_4)_3$	No metal source	2.7–3.1	14.70
Bismuth	Bi	Bi – 100%	9.7	83
Borax	$Na_2B_4O_7 \cdot 10H_2O$	B_2O_3 – 36.6% Na_2O – 16.2%	1.7	7.65
Bornite	Cu_5FeS_4	Cu – 63.3%	4.9–5.4	25.34
Braunite	$3Mn_2O_3 \cdot MnSiO_3$	Mn – 78.3%	4.8	18.28
Calamine	$H_2(Zn_2O) \cdot SiO_4$	ZnO – 67.5%	3.4–3.5	20.59
Calcite	$CaCO_3$	CaO – 56%	2.7	12.57
Calomel	HgCl	Hg – 85% Cl – 15%	6.5	70.54
Cassiterite	SnO_2	Sn – 78.8%	6.8–7.1	41.08
Cerussite	$PbCO_3$	Pb – 77.5%	6.5–6.6	65.29
Chalcedony	SiO_2	No metal source	2.6–2.7	10.80
Chalcocite	Cu_2S	Cu – 79.8%	5.5–5.8	26.38
Chalcopyrite	$CuFeS_2$	Cu – 34.6%	4.1–4.3	23.54
Chert	SiO_2	No metal source	2.6	10.80
Chromite	$FeO \cdot Cr_2O_3$	Cr – 46.2%	4.3–4.6	19.92
Chrysocolla	$CuOSiO_2 \cdot 2H_2O$	Cu – 36.2%	2.0–2.2	16.40
Chrysotile	$H_4Mg_3Si_2O_9$		2.2	10.17
Cinnabar	HgS	Hg – 86.2%	8.0–8.2	71.18
Cobaltite	CoAsS	Co 35.5%	6.0–6.3	27.58
Colemanite	$Ca_2B_6O_{11} \cdot 5H_2O$	No metal source	2.4	9.69
Copper	Cu	Cu – 100%	8.8	29
Corundum	Al_2O_3	Al – 52.9%	3.9–4.1	10.65
Covellite	CuS	Cu – 66.5%	4.6	24.64
Cryolite	Na_3AlF_6	Al – 13% F – 54.4%	3.0	10.17
Cuprite	Cu_2O	Cu – 88.8%	5.9–6.2	26.65

Mineral name	Composition	Percent metal	Specific gravity	Mass average atomic number
Diamond	C	C – 100%	3.5	6
Dolomite	$CaMg(CO_3)_2$	CaO – 30.4% MgO – 21.9%	2.8–2.9	10.87
Epsom salt	$MgSO_4 \cdot 7H_2O$	Mg – 9.9%	1.7	9.03
Fluorite	CaF_2	F – 48.9%	3.0–3.3	14.65
Galena	PbS	Pb – 86.6%	7.4–7.6	73.15
Garnet	Various	No metal source	3.2–4.3	
Garnierite	$H_2(NiMg)SiO_4$	Ni – 25% to 30%	2.4	16.05
Gibbsite	$Al(OH)_3$	Al – 34.6%	2.4	9.46
Gold	Au	Au – 100%	15.6–19.3	79
Graphite	C	C – 100%	2.2	6
Gypsum	$CaSO_4 \cdot 2H_2O$	CaO – 32.6%	2.3	12.12
Halite	NaCl	Na – 39.4%	2.1–2.6	14.64
Hausmannite	Mn_3O_4	Mn – 72%	4.7	20.25
Hematite	Fe_2O_3	Fe – 70%	4.9–5.3	20.59
Ilmenite	$FeTiO_3$	Ti – 31.6%	4.5–5.0	19.04
Iridium	Variable	Variable	22.7	77.00
Kaolinite	$H_4Al_2Si_2O_9$	Al_2O_3	2.6	10.24
Leucite	$KAl(SiO_3)_2$	K_2O – 21.5% Al_2O_3 – 23.5%	2.5	12.13
Limonite	$2Fe_2O_3 \cdot 2H_2O$	Fe – 59.5%	3.6–4.0	18.65
Magnesite	$MgCO_3$	Mg – 28.9%	3.1	8.87
Magnetite	$FeO \cdot Fe_2O_3$	Fe – 72.4%	5.2	21.02
Malachite	$CuCO_3 \cdot Cu(OH)_2$	Cu – 57.5%	4.0	19.90
Manganite	$Mn_2O_3 \cdot H_2O$	Mn – 62.5%	4.2–4.4	18.54
Marble	Chiefly $CaCO_3$	Ca – 40%	2.7	12.57
Marcasite	FeS_2	Fe – 46.6%	4.9	20.66
Marmatite	(ZnFe)S variable	Zn – 46.5% to 56.9%	3.9–4.2	
Mercury	Hg	Hg – 100%	13.59	80
Molybdenite	MoS_2	Mo – 60%	4.7–4.8	31.58
Monazite	$(CeLaDy)PO_4 \cdot ThSiO_4$	ThO_2 – 9%	4.9–5.3	57.56
Muscovite	$H_2KAl_3(SiO_4)_3$	Variable	2.8–3.0	11.33
Olivine	$(MgFe)_2SiO_4$	No metal source	3.3	17.40
Opal	$SiO_2 \cdot nH_2O$	No metal source	1.9–2.3	
Orpiment	As_2S_3	As – 61%	3.5	26.35
Orthoclase	$KAlSi_3O_8$	Al_2O_3 – 18.4%	2.5–2.6	11.85
Pentlandite	(FeNi)S	Fe – 42.0% Ni – 22.0%	4.6–5.0	24.61
Platinum	Pt	Pt – 100%	17.0	78
Pyrite	FeS_2	Fe – 46.7%	5.0	20.65
Pyrolusite	MnO_2	Mn – 63.2%	4.8	18.74
Pyrophyllite	$HAl(SiO_3)_2$	Al_2O_3 – 28.3%	2.8–2.9	10.58
Pyroxene	$Ca(AlMgMnFe)(SiO_3)_2$	No metal source	3.3	16.437
Pyrrhotite	Fe_5S_6 to $Fe_{16}S_{17}$	Fe – 61.5% Variable	4.6	21.92 to 22.21
Quartz	SiO_2	Si – 46.9%	2.65–2.66	10.80
Realgar	AsS	As – 70.1%	2.6	27.90
Ruby	Al_2O_3	Al – 52.9%	4.0	10.65
Rutile	TiO_2	Ti – 60%	4.2	16.39
Scheelite	$CaWO_4$	W – 63.9%	4.9–6.1	51.81
Serpentine	$H_4Mg_3Si_2O_9$	Mg – 43%	2.5–2.6	10.17
Siderite	$FeCO_3$	Fe – 48.3%	3.9	16.47

Mineral name	Composition	Percent metal	Specific gravity	Mass average atomic number
Silver	Ag	Ag – 100%	10.5	47
Sphalerite	ZnS	Zn – 67.1%	3.9–4.1	25.39
Spinel	$MgOAl_2O_3$	Al_2O_3 71.8% MgO – 28.2%	3.5–4.1	10.58
Spodumene	$LiAl(SiO_3)_2$	Al_2O_3 – 27.4% Li_2O – 8.4%	3.1–3.2	10.35
Stannite	$Cu_2S \cdot FeS \cdot SnS_2$	Sn – 27.5% Cu – 29.5%	4.5	30.53
Stibnite	Sb_2S_3	Sb – 71.8%	4.5–4.6	41.09
Sulfur	S	S – 100%	2.0	16
Sylvite	KCl	K – 52.4%	1.98	18.05
Talc	$H_2Mg_3(SiO_3)_4$	Mg – 19.2% Si – 29.6%	2.7–2.8	10.51
Tantalite	$FeTa_2O_6$	Variable Ta_2O_5 – 65.6%	5.3–7.3	55.75
Tetrahedrite	$4Cu_2S \cdot Sb_2S_3$	Cu – 52.1% Sb – 24.8%	4.4–5.1	32.18
Titanite	$CaTiSiO_5$	TiO_2 – 40.8%	3.4–3.6	14.73
Topaz	$(AlF)_2SiO_4$	No metal source	3.4–3.6	10.59
Tourmaline	$HgAl_3(BOH)_2Si_4O_{19}$	No metal source	3.0–3.2	28.49
Uraninite	$UO_3 \cdot UO_2$ Variable	Radium source	9.0–9.7	82.05
Vermiculite	$3MgO \cdot (FeAl)_2O_3 \cdot 3SiO_2$	Variable	2.7	14.64
Willemite	Zn_2SiO_4	Zn – 58.5%	3.9–4.2	21.67
Wolframite	$(FeMn)WO_4$	W – 51.3%	7.2–7.5	47.24
Zincite	ZnO	Zn – 80.3%	5.4–5.7	25.68
Zircon	$ZrSiO_4$	ZrO_2 – 67.2%	4.2–4.7	24.84

Index